MICROINDENTATION TECHNIQUES IN MATERIALS SCIENCE AND ENGINEERING

A symposium sponsored by
ASTM Committee E-4 on
Metallography and by the
International Metallographic Society
Philadelphia, PA, 15–18 July 1984

ASTM SPECIAL TECHNICAL PUBLICATION 889
Peter J. Blau and Brian R. Lawn
National Bureau of Standards
editors

ASTM Publication Code Number (PCN)
04-889000-28

1916 Race Street, Philadelphia, PA 19103

International Metallographic Society

Library of Congress Cataloging-in-Publication Data

Microindentation techniques in materials science and engineering.

(ASTM special technical publication; 889)
"ASTM publication code number (PCN) 04-889000-28."
Includes bibliographies and index.
1. Materials—Testing—Congresses. 2. Hardness—Testing—Congresses. 3. Metallography—Congresses. I. Blau, Peter J. II. Lawn, Brian R. III. American Society for Testing and Materials. Committee E-4 on Metallography. IV. International Metallographic Society. VI. Series.
TA410.M65 1986 620.1′126 85-28577
ISBN 0-8031-0441-3

Library of Congress Catalog Card Number: 85-28577

NOTE

The Society is not responsible, as a body, for the statements and opinions advanced in this publication.

Printed in Ann Arbor, MI
February 1986

Foreword

This publication, *Microindentation Techniques in Materials Science and Engineering*, contains papers presented at the Microindentation Hardness Testing Symposium and Workshop, which was held 15–18 July 1984 in Philadelphia, PA. The event was jointly sponsored by ASTM, through its Committee E-4 on Metallography, and the International Metallographic Society. Chairing the symposium were Peter J. Blau and Brian R. Lawn, both of the National Bureau of Standards, who also served as editors of this publication.

Related ASTM Publications

Practical Applications of Quantitative Metallography, STP 839 (1984), 04-839000-28

MiCon 82: Optimization of Processing, Properties, and Service Performance Through Microstructural Control, STP 792 (1983), 04-792000-28

MiCon 78: Optimization of Processing, Properties, and Service Performance Through Microstructural Control, STP 672 (1979), 04-672000-28

Damage Tolerance of Metallic Structures: Analysis Methods and Applications, STP 842 (1984), 04-842000-30

A Note of Appreciation to Reviewers

The quality of the papers that appear in this publication reflects not only the obvious efforts of the authors but also the unheralded, though essential, work of the reviewers. On behalf of ASTM we acknowledge with appreciation their dedication to high professional standards and their sacrifice of time and effort.

ASTM Committee on Publications

ASTM Editorial Staff

Contents

Introduction

Microindentation hardness testing and its associated methodology continue to be used widely in materials evaluation. The subject matter in this book, however, goes beyond the mere obtaining and interpreting of microindentation hardness numbers. It deals with the use of indentation methods in the study of intrinsic deformation properties, residual stress states, thin-film adhesion, and fracture properties in a variety of materials.

The last such collection of contributions to the general field of microindentation hardness testing in the United States was published more than a decade ago.[1] Since then, a considerable body of work has improved our understanding of indentation behavior as it relates to fundamental material properties and has extended the range of applications to engineering practice. The symposium from which the content of this book derives was organized as a joint venture between the International Metallographic Society, the American Society for Metals, and ASTM. It was held on 15 and 16 July 1984, in Philadelphia, PA, in conjunction with the 17th annual International Metallographic Society technical meeting. Contributors and attendees at the symposium represented eleven countries in addition to the United States, and their technical interaction provided a forum for discussion of microindentation research and technology.

This volume is organized into three sections dealing with fundamentals, testing techniques, and engineering uses of microindentation-based methods for metals, ceramics, and polymers. The reader will find that the classification of papers into the three sections is somewhat arbitrary. Nevertheless, as one proceeds through the book one will note something of a progression from scientific principles to practical applications.

The papers in the section on fundamentals question some of the traditional theories of indentation behavior and examine how these theories relate to intrinsic material properties. This section covers metals, ceramics, and polymers. There is an emphasis in many of these papers on a relatively new approach to quantifying microindentation behavior through the use of the load-displacement response of materials. The section on techniques addresses such topics as hardness scale interconversions, measurement methods, errors, standardization, and time and size effects. The third section, on applications, contains six papers which exemplify some of the many engineer-

[1] *The Science of Hardness Testing and Its Research Applications*, American Society for Metals, Metals Park, OH, 1973.

ing uses of microindentation techniques. Two of these papers deal with sliding wear and abrasion damage assessment, one with the mounting and microindentation testing of small particles, and three with various aspects of coatings testing.

The editors would like to express their sincere gratitude to the contributors and reviewers of this volume for their cooperation and efforts. The International Metallographic Society and the U.S. Office of Naval Research provided travel support for some of the symposium contributors from outside of the United States. Chris Bagnall and James McCall of the International Metallographic Society are also acknowledged for their help with many of the necessary details required to organize the symposium facilities and funding.

Peter J. Blau
Brian R. Lawn

National Bureau of Standards, Gaithersburg, MD, 20899; symposium cochairmen and editors.

Fundamentals of Indentation Testing

Leonard E. Samuels[1]

Microindentations in Metals

REFERENCE: Samuels, L. E., "**Microindentations in Metals,**" *Microindentation Techniques in Materials Science and Engineering, ASTM STP 889,* P. J. Blau and B. R. Lawn, Eds., American Society for Testing and Materials, Philadelphia, 1986, pp. 5–25.

ABSTRACT: The mechanisms involved when an indentation is made in the surface of a metal by a blunt indenter have received a good deal of attention, but little of this work has appeared in publications that might easily come to the attention of those who actually carry out hardness tests. Even less of the work has been analyzed for its implications in practical hardness testing. The main models of indentation which have been proposed to date are reviewed, with particular attention to the one that most closely relates to the realities of hardness testing. This model, developed by Mulhearn, proposes that indentation occurs by radial compression, the formation of the indentation being likened to the expansion of a hemispherical cavity. The implications of this model to indentation hardness testing, and specifically to microindentation testing, are considered. This involves consideration of the current status of views on the effects of indentation size on the apparent hardness number.

KEY WORDS: microindentation hardness testing, indentation, indentation hardness test, microhardness test, mechanical properties, compression, models

The distinguishing feature of the hardness indentations with which this publication is concerned is their size. A microindentation can be arbitrarily defined as one which has a diagonal length of less than 100 μm, noting that increasing interest is being taken in indentations with diagonal lengths less than 10 μm. The force that has to be applied to an indenter to produce indentations of this size is important to the design and operation of a hardness testing machine but not necessarily to the mechanism of the indentation process.

Little or no work has been carried out on the mechanism of indentation at this scale. It is, consequently, necessary to draw on the work carried out on larger macroindentations and extrapolate downwards, relying in the first instance on the principle of geometric similarity. This principle is fundamental

[1]Consultant, Melbourne, Victoria, Australia. Formerly director, Materials Research Laboratories, Department of Defence, Melbourne, Australia.

to macroindentation testing and, although it does not follow that it will be applicable down to the smallest indentations, the principle should not be abandoned lightly. It will be accepted here as applying unless there is good evidence to the contrary.

Some of the investigations referred to in the treatment which follows were carried out on Brinell macroindentations. Only pyramidal indenters, principally of the Vickers (136° included face angle) and Knoop types, are used in microindentation tests, but the general principles that emerge from the study of Brinell indentations can still be applied because of the geometric similarity of the indentations. This is a particularly reasonable assumption if the radial compression mechanism of indentation which will be discussed is accepted. Note also that some of the investigations have been carried out using two-dimensional wedge indenters. This is done for experimental simplicity and because it is possible to treat theoretically only the two-dimensional situation; it is then assumed that a pyramidal indentation has a radial symmetry. Finally, little work has been carried out directly on Knoop indentations. It has to be assumed that a Knoop indenter behaves in essentially the same way as a Vickers indenter, noting that it is the blunter of the two.

Mechanisms of Indentation

Cutting Mechanisms

Analyses of the mechanism of indentation of metals by blunt pyramids commonly has been based on a slip-line field solution developed originally by Hill, Lee, and Tupper [*1*]. This is a two-dimensional treatment of an ideal rigid-plastic material; that is, a material that is perfectly rigid up to a yield stress but then deforms plastically without work hardening. The application of this treatment, and several of its subsequent developments, have been reviewed by Shaw [*2*].

The slip-line field solutions are all based on the supposition that the indenter cuts the specimens, for example, along Plane ab in Fig. 1. It follows that this creates two new surfaces, which rotate about Point b as the indentation forms. The material originally located at Point a is thus relocated to Point c, and the material in Volume bdefc is plastically deformed with a sideways and upward motion. The rigid-plastic boundary is bdef. Relative movement is required between the surfaces of the indenter and the specimen, and consequently friction should have a significant effect, which can be taken into account by the solution (compare the left and right sides of Fig. 1). Work hardening of the specimen material is predicted to have a similar effect to an increase in the coefficient of friction.

Mulhearn [*3*] carried out a detailed quantitative analysis of indentations made by wedges, cones, and pyramids with a range of included angles, investigating a material whose mechanical characteristics closely approached

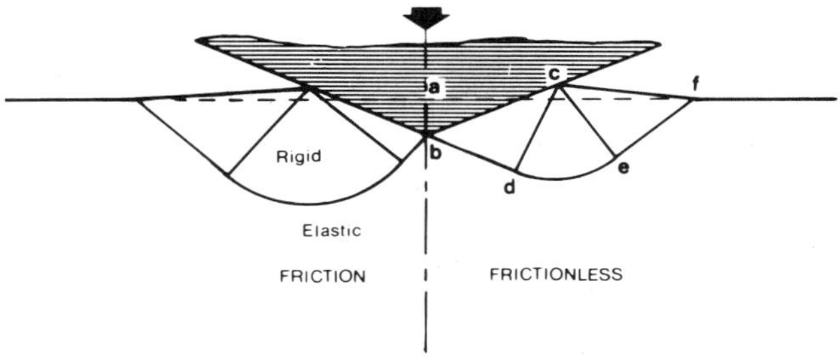

FIG. 1—*Slip-line field solution for the indentation of a rigid-plastic material by a blunt wedge, assuming a cutting mechanism of indentation. The angles of the slip-lines have been modified slightly in the lefthand half of the sketch to account for the effects of friction.*

those of an ideal elastic-plastic solid—that is, a solid which is elastic up to a yield stress and then deforms plastically with no work hardening; slip-line field theory can reasonably be assumed to apply to such a material. Mulhearn [*3*] established that the characteristics of the indentations conformed to the cutting model when the wedge angle was less than about 60°. Indentations made with wedges that had larger angles, however, did not conform. It is worth recounting the evidence that established this:

1. A cutting mechanism requires that the elastic-plastic boundary should not extend much below the tip of the indentation. In fact, as described later in more detail, it extends for a considerable depth below the tip of the indentation (Fig. 2).
2. A cutting mechanism proposes that the displacement of points on the specimen surface would have a large component parallel to the surface. In fact, the displacement is relatively small (Fig. 3 [*3,4*]).
3. If cutting occurred, the height of the lip raised adjacent to the indentation would be approximately one third the depth of the indentation. In fact, it is only a fraction of this, never more than half the required amount.

The discrepancy between predictions and observations increases as the wedge angle increases above 120° and becomes marked by a wedge angle of 140°. Mulhearn concludes that the cutting model is not valid at this point.

Mulhearn [*3*] has also established that the same discrepancies arose with a material (annealed 30% zinc) whose mechanical characteristics are considerably different from those of an ideal elastic-plastic solid. Brass has a low yield stress and work hardens to a considerable extent. Woodward [*5*] subsequently investigated in more detail the locations of various isostrain boundaries beneath indentations made in a similar brass and compared his observations with the boundaries predicted by several slip-line field solutions based on a cutting mechanism. He concluded that the discrepancies became so large when the included angle of the indenter exceeded 90° that the slip-line field solutions could no longer reasonably be applied. Hirst and Howse [*6*]

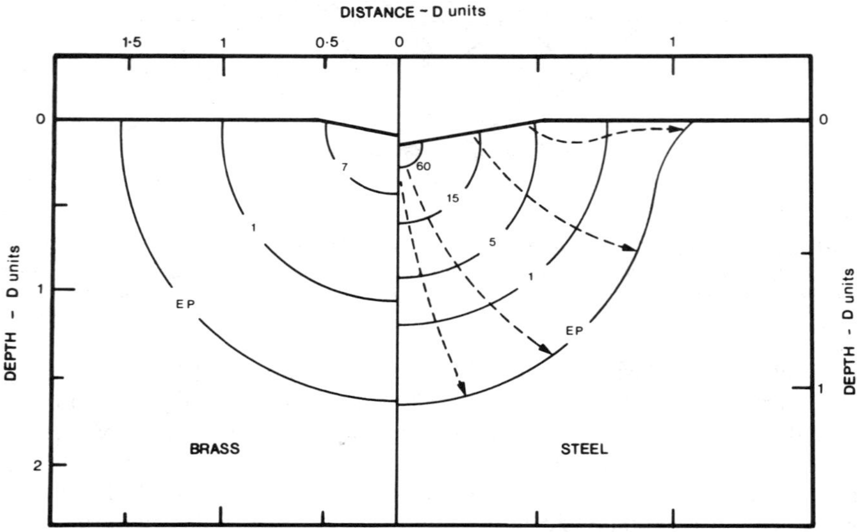

FIG. 2—(Left) *Isostrain boundaries in the deformed zones beneath a Vickers indentation in annealed 70-30 brass. The strain boundaries were determined by the metallographic investigation of a section cut through the indentation* [4]. (Right) *Isostrain boundaries in the deformed zone beneath a Vickers indentation in a cold-worked low-carbon steel. The isostrain boundaries were determined on a split specimen on the parting surface of which a grid had been ruled. The broken lines are displacement trajectories* [3]. *The figure associated with each boundary is the engineering strain as a percentage of compression.*

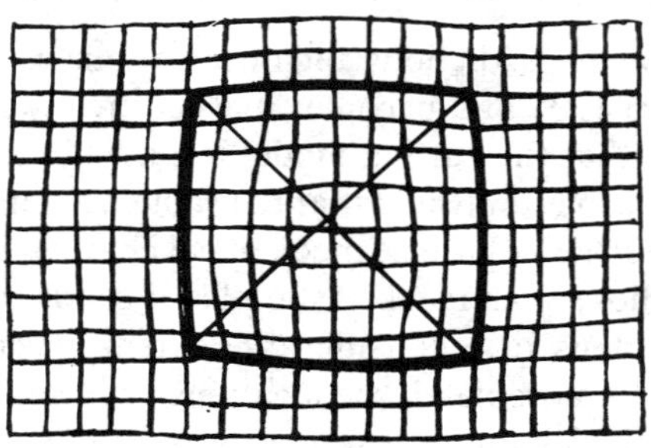

FIG. 3—*A tracing of a square grid which had been ruled on the surface of a cold-worked low-carbon steel and then distorted when a Vickers indentation was made on the surface. The outline of the indentation is indicated by the heavier lines.*

have also confirmed this point for a range of materials and have shown that the transition wedge angle is smaller as the ratio of Young's modulus to the elastic limit of the material becomes smaller.

The evidence that slip-line field analyses, based on the hypothesis that indentation occurs by a cutting mechanism, cannot usefully be applied to indentation hardness testing appears to be conclusive. It is time this type of analysis be dropped from the hardness-testing literature.

Elastic Mechanisms

Shaw and De Salvo [7] have also noted some of these disqualifying deficiencies of the slip-line field solutions and have sought an alternative mechanism of indentation. The model that they advanced was based on two presumed features of unloaded indentations: (1) that the form of the plastic zone beneath the indentation is very similar to a line of constant shear stress derived by a Hertzian analysis of elastic contact of a blunt indenter on a flat surface and (2) that the specimen surface adjacent to the indentation remains flat. Their proposal was as follows:

1. Plastic deformation beneath the indentation occurs during loading in a zone of the form sketched in Fig. 4, the whole plastic zone sinking into the specimen. There is no upward flow adjacent to the indentation.
2. The elastically strained zone surrounding the plastic zone decreases in volume (increases in density) to account for the volume of the indentation. A proviso is that the specimen should extend for at least ten indentation diagonals in all directions from the indentation.
3. A second phase of plastic deformation during unloading occurs in a volume smaller than the loading plastic zone and in the opposite direction. Biaxial residual stresses are thereby induced in planes parallel to the free surface, and these internal stresses maintain the indentation.

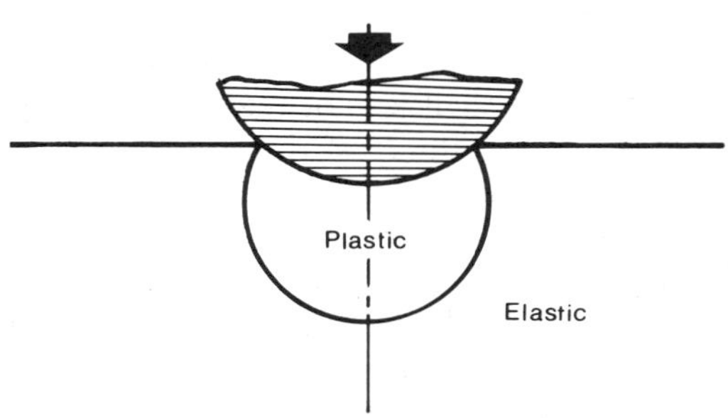

FIG. 4—*Illustration of the elastic mechanism of indentation proposed by Shaw and De Salvo* [7].

This hypothesis also encounters a number of difficulties compared with experimental observation:

1. The plastic zone beneath an indentation is not of the form sketched in Fig. 4, certainly not in metals of low-to-medium yield stress (see the discussion further on). More sensitive methods than those used by Shaw and De Salvo in their experiments are necessary to discern this [*3,4*].

2. Shaw and De Salvo did not support experimentally their proposal that the density of the elastically deformed bulk of the specimen is permanently increased sufficiently to account for the indentation volume. Precise measurements of the density of indented specimens made later by Woodward and Brown [*8*], in fact, failed to detect density changes of sufficient magnitude. No change at all in density was detected in some materials. A small change was detected in others, but this could be attributed at least partly to microstructural changes in the plastic zone. Very precise measurements in the fundamental investigations of Clareborough et al [*9*] also establish that density changes in the plastic zone could not account for the volume of the indentation either.

3. There is, in fact, significant upflow adjacent to the unloaded indentation to an extent which varies with the specimen material. Between 50 and 90% of the indentation volume can be accounted for by the upflow [*3*].

Moreover, there are fundamental difficulties with the hypothesis because it requires that both tensile and compressive stresses exist to maintain the residual stress condition necessary for indentation.

The Shaw and De Salvo elastic model of indentation thus does not adequately explain several basic physical phenomena associated with actual hardness indentations.

Compression Mechanisms

Mulhearn [*3*] proposed that indentation by blunt indenters occurs by means of a compression mechanism which has the following characteristics:

1. Deformation of the specimen occurs during loading by the radial compression of hemispherical shells centered at the point of the indenter (Fig. 5). For all practical purposes, the shells can be regarded as being centered at the point of first contact. The magnitude of the strains in the shells decreases progressively as the elastic-plastic boundary is approached and, except for regions close to the indenter, the pattern of deformation is very similar for blunt indenters of all geometrics. (The patterns for Brinell and Vickers indentations, for example, are identical.) Again, except for the regions close to the indenter tip, the strains in the plastic zone are comparatively small (Fig. 2). The formation of the indentation has been likened to the expansion of a hemispherical cavity.

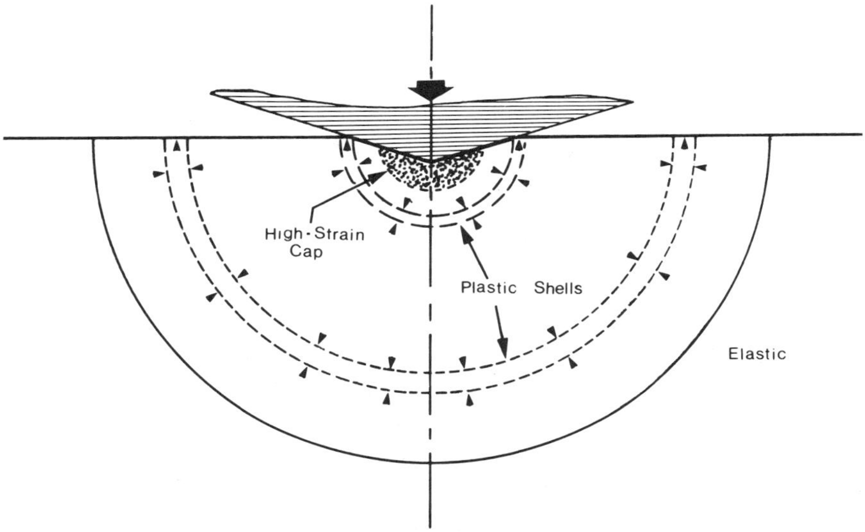

FIG. 5—*Illustration of the compression mechanism of indentation proposed by Mulhearn* [3]. *The circular continuous line is the elastic-plastic boundary. The broken lines indicate several hypothetical plastic shells, and the arrows indicate the directions of straining of the shells.*

2. Differences in the deformation pattern with indenter geometry are confined to those regions close to the indenter, approximately, to the shaded region sketched in Fig. 5. Moderately large plastic deformations occur within this region (Fig. 2). The region of higher strains can be regarded descriptively as a cap of metal which advances with the indenter, growing in the process. In this sense, it is an adjunct to the indenter.

3. The indentation surface is formed by the original test surface folding down progressively about four axes which are the edges of the indentation at the time (Fig. 6). The fold axes advance sideways as the indentation deepens.

4. Most of the surface of the indentation is therefore original surface, which has been pressed approximately vertically downwards. However, the surface area of the indentation is a few percentage points larger than the area of the original surface (Fig. 6). The additional surface area is produced either by stretching the original surface or by producing a new surface by cutting at the corners of the indentation or by a combination of these two mechanisms. Which particular mechanism operates depends on the relationship between the yield stress of the specimen material (a low yield stress encourages stretching) and the coefficient of friction between the specimen and indenter surfaces (a high coefficient of friction discourages stretching and hence requires that cutting occur). For diamond indenters, which have a low coefficient of

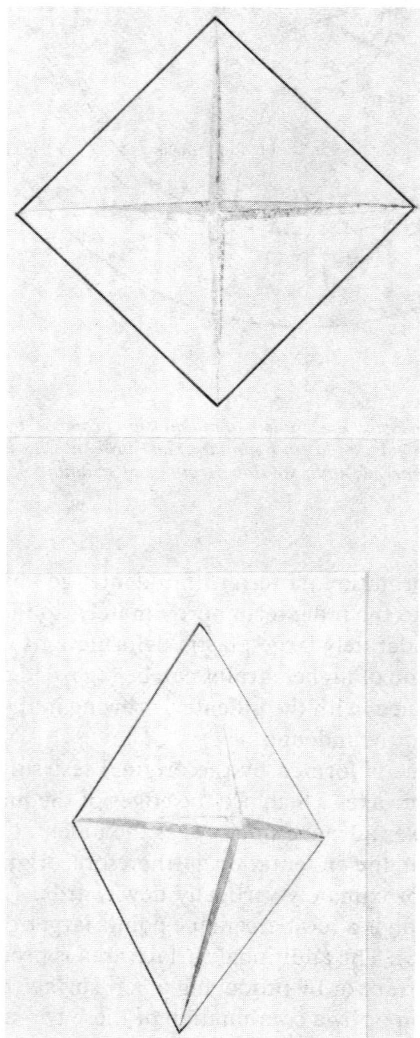

FIG. 6—*Photograph of a model illustrating the mechanism of formation of the indentation cavity, as proposed by Mulhearn* [3]*—an oblique view at the left and the plane view at the right. A cavity with the geometry of a Vickers indentation has been machined in a block. A sheet has been placed over the cavity and cut along the diagonals of the indentation. The four segments of the sheet have then been folded about the indentation edges until they touch the surfaces of the cavity. The additional indentation surface which has to be formed either by cutting a new surface or by stretching the old surface can be seen as the darker areas at the corners of the indentation.*

friction, it seems that stretching predominates with specimens softer than 100 HV and cutting with specimens harder than 200 HV.

5. Circumferential extension of the plastically deformed shell occurs at the free surface during loading (Fig. 5), but there is a corresponding downward deflection of the whole plastically deformed region. The free surface remains approximately flat under load. The downward deflection accounts for the displaced volume of the indentation and is accommodated by elastic deformation in the specimen as a whole, as was later also proposed by Shaw and De Salvo [*7*]. However, a significant amount of elastic recovery occurs during unloading, producing an extruded lip in the free surface adjacent to the impression. The volume of the lip relative to the volume of the indentation depends on the degree of elastic recovery, which is in turn related to the work-hardening characteristics of the specimen material.

Characteristics 1 to 4 are based on direct and conclusive experimental evidence. For example, experimentally determined isostrain boundaries beneath pyramidal indentations in two materials of widely different elastic-plastic characteristics are shown in Fig. 2. Annealed 70-30 brass deviates widely from ideal elastic-plastic behavior and has a low yield stress. The steel (0.15% carbon, cold worked) approaches ideal elastic-plastic behavior and has a medium yield stress. The elastic-plastic boundary is hemispherical in both cases; its diameter, in terms of impression diagonals, is smaller the higher the yield point of the material is. The isostrain boundaries beyond the immediate neighborhood of the indentations are also approximately hemispherical.

The strain distribution beneath indentations in the brass can be investigated metallographically on sections through normal macroindentations. Etching techniques are available to develop indications of prior deformation with a range of sensitivities [*10*], and the position of even the elastic-plastic boundary can be established with a high degree of certainty. Also available are metallographic etching techniques that can reveal the location of the elastic-plastic boundary around normal macroindentations in low-carbon annealed steels [*11*] and medium-carbon quench-and-tempered steels [*12*], although with less precision. The elastic-plastic boundaries in both types of steel are found to be of the general form implied by Fig. 2. The diameter of the boundary again is found to be smaller the higher the yield stresses of steel.

The quantitative data in Fig. 2 for the steel were obtained on a split specimen indented by a 136° pyramid at the parting line. A grid had been precision ruled on the parting face of the specimen, and displacements of the grid nodes were measured after indentation, again by precise methods. It is apparent from the magnitude of the strains determined in this way that precision methods are necessary to ascertain the full extent of the plastic zones. This investigation technique also allowed the displacement trajectories (the broken lines in Fig. 2) and the magnitude of the permanent displacements to be determined. Approximately, the trajectories radiate outward from the point of indentation.

The evidence supporting the fifth characteristic, however, which proposes that the specimen surface remains flat under load but uplifts variable amounts by recovery during unloading, is not so direct. This concept was advanced by Mulhearn [*3*] as the most plausible one available to explain the variability and extent of the uplift found after unloading. It is contrary to the consensus view, which is that the uplift occurs during loading. Examination of indented surfaces while under load would be necessary to distinguish positively between these two views. The apparatus developed by Müller [*13*] might be suitable for this purpose, but Müller has applied it only to indentations made in elastomers and polymers.

As has already been mentioned, the magnitude of the uplift and the extent to which it can account for the volume of the indentation have always been matters of concern and dispute. The uplift distances are small and again can be investigated sensibly only by precision methods. Such methods have not always been used. Mulhearn [*3*] did do so, using multiple beam interferometry. He found that the volume of the uplift varied from 45% of the indentation volume for a material with a long work-hardening range to 85% for a material with a small work-hardening range. This correlation with work-hardening characteristics was one of the reasons he concluded that the uplift occurred during unloading. His estimates are also minimum values. First, it is difficult to be certain of the datum plane of the surface, and small errors here would result in significant underestimates of the uplift volume. Second, recovery within the indentation itself, where the strains are highest, was not taken into account; these could account for perhaps an additional 5 to 10% of the loaded indentation volume. Third, in some materials some of the indentation volume (perhaps 5 to 10%) might be accounted for by microstructural changes beneath the indentation, such as by the closing up of cavities or inclusions [*8*].

Of those models available, the compression model is certainly the one that is best supported by the experimental evidence. It is, in fact, well supported, but it does assume that the material is mechanically homogeneous and isotropic. This assumption is usually justified with macroindentations but is not necessarily justified with microindentations. The most common example of an instance when it is not justified is when the size of the indentation is small compared to the grain size. The indentation is then effectively carried out in an anisotropic single crystal.

Microindentations have not been investigated in as much detail as macroindentations, but there is evidence [*14*] that the nature of the indentation deformation is similar to that described earlier when the indentation is made in a reasonably isotropic crystal and when it is not too small. There is also evidence [*14*], however, that the deformation mode might be different with very small indentations, a point which will be discussed later.

Implications of the Compression Model of the Indentation in Hardness Testing

A number of practical implications can be drawn from the Mulhearn indentation model that should be applicable whenever the principle of geometric similarity holds.

The Meaning of Indentation Hardness

The compression model implies that hardness is simply a measure of yield stress in compression—specifically, the averaged yield stress of the material in the work-hardened zone beneath the indentation. Tabor [*15*] had realized this earlier in a general way and suggested a value of 8% for the average compressive strain. This figure was based on an empirical best fit between hardness and the compressive yield stress of materials work hardened to varying degrees. It is a reasonable figure in terms of the data in Fig. 2.

Dugdale [*16*] subsequently attempted to develop a more soundly based correlation, using an analysis of the compression model of an expanding hemispherical cavity. He showed that, although a solution is available for an expanding spherical cavity, an exact solution is not possible at present for a hemispherical cavity. But by relating an inexact solution with empirical data, he was able to establish that hardness depends critically on the integrated compressive yield stresses of the test material up to strains of about 12% compression. It follows that the work-hardening characteristics of the materials at larger strains do not influence hardness. This, too, is a reasonable conclusion in terms of the data in Fig. 2, because only a relatively small volume of material is strained in larger amounts than 12% compression.

Relationships between Hardness Scales

The model provides a partial explanation for the good agreement between the hardness values determined by the Vickers, Knoop, and Brinell tests when using nondeformable indenters. It also suggests that these hardness values should not be sensitive to minor manufacturing errors in indenter shape. Moreover, it indicates that a logical basis exists for conversions between these scales. The conversions would be even more logical if all of the hardness scales were based on the projected areas of the indentation (as for the Knoop scale) instead of the surface area (as for the Vickers and Brinell scales). All of these conclusions follow from the observation that the stress pattern around blunt indenters is not greatly affected by the indenter shape.

Vickers, Knoop, and Brinell hardness numbers are actually based on the projected surface dimensions of the unloaded indentation, and it is assumed that recovery of these dimensions during unloading is small. The Mulhearn model suggests that this is an acceptable assumption. Plastic displacements parallel to the surface are small during indentation (Fig. 3), which predicts that elastic recovery of any dimension in this direction would be very small.

The situation is different when these scales are compared with a Rockwell scale, although the indentation deformation patterns may be similar. The Rockwell scales are sensitive to recovery of the indentation in a direction normal to the specimen surface. Recovery in this direction, particularly in the region of the tip of the indentation, can be expected to be larger because the strains in this region are large; they are also more likely to vary with indenter geometry. The relationships between these scales and Rockwell scales consequently are indirect and can be arrived at only on an empirical basis.

Shape of the Indentation

The Mulhearn model proposes that the specimen surface is flat adjacent to the indenter under load. It follows that any departures from straight edges in the unloadeded indentation are due to recovery during unloading. If so, the departures would have no influence on the hardness number as calculated, because the area over which the indenting load was supported would not be affected. The irregularities referred to include the development of convex or concave edge shapes and more local irregularities in the indentation edges.

The consensus view, however, is that these irregularities in indentation outline are due to either uplift or depression of the specimen surface during loading. This view implies that an error is introduced because the corners of the indenter pierce an uplifted ridge, for example. The diagonal length then measured is longer than it should be. This would cause an error in the hardness number but not a variation in the apparent hardness number with impression size if the uplifted ridges preserved geometric similarity. Bückle [*17*] suggests, however, that they do not. He suggests that the error in length is proportionately smaller in small indentations, and hence the calculated hardness number is proportionately larger. To this he attributes the apparent increase in hardness number with a decrease in the size of the indentation, features that will be discussed in more detail later. Bückle's concept implies that the principle of geometric similarity is transgressed. He explains this in terms of increased blocking of dislocations as the indentation deepened. This means, in effect, that the mechanism of indentation changed.

The consensus view on the stage at which surface uplift occurs is perhaps the obvious one, and this is the least well supported part of Mulhearn's model. Nevertheless, the shapes of unloaded indentations that are observed in practice can be interpreted equally well by either hypothesis. This is another reason the question of whether the uplift occurs during loading or unloading needs to be settled.

The problem is more apparent with Vickers indentations than with Knoop indentations because the two diagonals of an unloaded indentation are measured and averaged to calculate the hardness number. Usually, the differences in diagonal lengths are not too great, but they can be considerable in indentations made on single crystals of materials which are highly anisotropic. The question that then arises is which diagonal measurement or measurements should be used for the calculation of the hardness number.

The problem might not be immediately apparent in Knoop indentations if only one diagonal is measured, as is usually the case, but it would still exist if anomalous changes occurred in the length of the long diagonal. The Knoop indenter is used to explore variations in properties with direction in anisotropic crystals, this being one of its useful characteristics. Wonsiewicz and Chin [*18*] have developed an analysis based on the Mulhearn model of indentation, which explains the variation in Knoop hardness with orientation in cubic crystals, but the simplifications involved in the treatment make its applicability to highly anisotropic metals uncertain.

Friction and Surface Topography

The Mulhearn model indicates that there is little or no movement between the surfaces of the indenter and the specimen; it can therefore be concluded that friction between the two does not play a significant role in hardness testing.

Variations in the surface topography of the specimen on a small scale compared with the size of the indentation, likewise, should not have a significant effect because the specimen surface is merely forced downward. A proviso here is that topographical features do not interfere with the clear marking of the surface by the corners of the indenter. The general inclination of the surface with respect to the axis of the indenter on a scale comparable to the size of the indentation is a different matter, however, for a different reason. A tilt of more than a few degrees causes a significant error because the indenter slides down the inclined surface, thus producing an enlarged indentation [*19*]. A tilt of 4°, for example, results in an error of about −4%. The existence of the effect might be noticeable as plastic distortions on the surface on only one side of an indentation. It can be identified with certainty in Vickers indentations by measuring the intercept ratio of the indentation diagonals, which, for safety, should not exceed 1.5 [*19*]. The same type of error must also occur with Knoop indentations, but its magnitude has not been investigated, and methods of recognizing its presence quantitatively have not been established.

Spacing of Indentations

Standard specifications for macroindentations require that the center spacing of pyramidal indentations be at least three indentation diagonals (3*D*). The origin of this figure seems to be lost in the early history of the development of hardness testing, and there appears to be no modern published reports of experimental investigations to support it. However, the knowledge that is now available on the plastic zone associated with the indentations permits an assessment to be made of what should be an acceptable spacing.

As a starting point, it might be said that the elastic-plastic boundaries of adjoining indentations should not overlap. This criterion would permit a center spacing of about $3D$ for metals of moderately low yield stress, and $1D$ for metals of moderately high yield stress (Fig. 2). But these would be conservative values because hardness is not very sensitive to small plastic strains even in materials which work harden considerably. A more realistic criterion perhaps would be that the isostrain boundaries for something approaching Tabor's equivalent strain should not overlap. To be on the safe side, say that the 5% isostrain boundaries should not overlap. By this criterion, the indentations could nearly touch one another without error, except perhaps in the very softest of materials. It is difficult to confirm these conclusions experimentally, and it would be particularly difficult to do so with microindentations, but some preliminary trials in the author's laboratory supported them. As a guide, it seemed that it can be taken that a serious error has not been introduced if the shape of the preexisting indentation has not been distorted noticeably by the new indentation. This criterion is particularly likely to be acceptable with microindentations in which there are so many other experimental uncertainties.

The minimum acceptable spacing from an edge of the specimen is, in principle, a somewhat different problem. The elastic restraint to the indentation deformation can be expected to be affected as an indentation approaches an edge. Samuels and Mulhearn [*4*] showed that this is, indeed, so but that the elastic-plastic boundary in an annealed 70-30 brass is not affected unless the center of the indentation comes within $2D$ of the edge. The same argument as used previously, then, suggests that the same rule can be applied as for the spacing of impressions. Standard specification require that the center of an indentation should be no closer than $2.5D$ from a specimen edge. The standards are certainly excessively conservative for microindentation testing.

Similar arguments apply to specimen thickness. Samuels and Mulhearn [*4*] showed that the elastic-plastic boundary became distorted in the annealed brass when the specimen thickness was less than $2.5D$. The minimum thickness permitted in standard specifications is $2.5D$, which again appears to be excessively conservative. A value of $1D$ would appear to be more reasonable for microindentations.

Variation of Hardness Number with Size of Indentation

Perhaps the most important and intractable problem associated with microindentation hardness testing is that concerned with the apparent change in hardness number with change in indentation size. It has been well established that certain instrumental errors, such as those resulting from excessively fast rates of approach of the indenter or from vibrations in the testing system while the load is being applied, cause erroneously large indentations to be produced and hence erroneously low apparent hardness numbers to be obtained. But it is generally found, when all of the known instrumental errors have been controlled as well as possible, that the apparent hardness number increases as the indentation size decreases. Many of the possible explanations of this have been discussed extensively in the literature, but a few still seem to be worthy of further consideration.

Measurement of the Indentation Diagonal

The indentation diagonal normally is measured by optical microscopy, which sets a limit to the precision of measurement. Analysis of this limit usually is based on consideration of the resolution of the optical system, when it is assumed that a measuring filar will be set at a point at which the edges of the indentation are just resolved as they converge to the corner. This implies that the diagonal will be measured short by an amount which is constant for indentations of all sizes but which varies with the numerical aperture of the microscope objective. The net result would be that the apparent hardness number would increase with a decrease in the size of the indentation, the measurement system being constant.

However, it is possible that the perception limit of the optical system is of greater importance than its resolution. Perception is the ability to detect the presence of a small feature (say the corner of an indentation) even if it cannot be separated from adjoining points and so be imaged clearly, or, another way, the ability of the human eye to detect some variation in contrast in the image produced by the optical system. The perception limit is difficult to treat theoretically [*20*], particularly since the contrast-response characteristics of the human eye are involved. The perception limit is smaller than the resolution limit when the contrast between the feature and the background is strong but could be poorer when the contrast is weak, as it certainly is with many microindentations. Note here that the contrast between the indentation and the specimen surface weakens with increasing numerical aperture of the microscope objective, which militates against improvement in perception.

Some old experimental evidence [*21*] suggests that the diagonal length of Vickers indentations is measured short by about 1.5 μm under what normally would be regarded as good optical conditions. This corresponds to an error of +10% for a D = 10 μm Vickers indentation, an error which increases rapidly with decreasing indentation size. The uncertainties in diagonal measurement would be expected to be greater with Knoop indentations because of the shape of the measurement corner and the poorer contrast at that corner.

Pethica [*22*] and Newey et al [*23*] have recently devised techniques that avoid this source of error and uncertainty by measuring the depth of the loaded indentation. The apparatus concerned was, in each case, developed for use with very small indentations (diagonal lengths on the order of 1 μm), but the concept is sound in principle and seems to be worthy of further development. Modern sensing devices and electronics should make this possible and bring with them the further possibility of incorporating automation of the calculation of hardness number. The tedium of optical methods of measurement would certainly be eliminated.

Pethica's results [*22*], although not extensive, tend to show that there is a definite increase in observed hardness number with a decrease in indentation depth. However, the increase that he observed, if real, occurred only with indentations smaller that those for which the increase in apparent hardness has been reported in the past.

Surface Artifacts: Mechanical

It is commonly stated that the preparation of test surfaces by mechanical methods is a possible cause of an increase in apparent hardness with a decrease in indentation size. This is attributed to the deformation of the surface layers by the preparation process. Such statements may well be true for the particular preparation procedures concerned but are meaningless as a generalization unless details of the preparation procedures are known and have been assessed. Likewise, it is usually assumed that an electropolished surface is intrinsically free from these artifact effects and so can be taken as an unaffected datum. This, too, may or may not be true, depending on the details of the abrasion and machining processes that were used before polishing and of the polishing process itself.

It is certainly true that mechanical machining and abrasion processes produce a plastically deformed layer on the surface. The layer is deep enough, and the strains in it are large enough, to alter significantly the hardness determined in microindentation tests. However, this abrasion deformation should and can be removed by subsequent polishing stages, whether mechanical or electrolytic. This may or may not happen in practice with either mechanical or electrolytic methods of polishing, and all too often it does not happen even though the methods of doing so have been well established [*24*].

Assume, however, that the abrasion-deformed layer has been removed adequately during a polishing procedure. An electropolished surface would then indeed be free from artifact strains, although it may be chemically contaminated (see next section). Most mechanically polished surfaces, however, will not be strain free because most mechanical polishing processes introduce a plastically deformed layer of their own. The layer is not deep—typically, it is less than 1-μm deep even in a soft material—but the maximum strain in it is large [*25*]. Whether or not this layer is likely to affect a microindentation hardness test depends on a number of factors such as the depth of the indentation, the nature of the specimen material, and the nature of the polishing process. This is certainly a matter that needs to be taken into consideration, but it probably can be said as a generalization that, with good modern metallographic practices, the problem should be significant only with comparatively small indentations ($D < 10$ μm) in soft specimen materials.

Surface Artifacts: Chemical

The classic work of Pethica and Tabor [*26*] established that there are two possible sources of chemical artifacts that increase the microindentation hardness of a surface.

The first is the segregation of impurity elements at the surface, a segregation which is developed during annealing treatments of the types that commonly are used to produce strain-free surfaces. The researchers found that sputtering by ion bombardment removes the segregated layer but introduces surface damage. Several annealing and sputtering cycles are necessary to produce an unsegregated strain-free surface. Very few investigations have been carried out in the past on annealed specimens that were not surface hardened by this type of surface segregation.

The second source of chemical artifacts is the oxide and other chemical films that inevitably form on normal experimental surfaces, even if only as a result of atmospheric oxidation. Pethica and Tabor [*26*] found that the presence of an oxide film only 5-nm thick was enough to increase significantly the hardness value when the indentation was small enough. Chen and Hendrickson [*14*] also found that the apparent hardness ($D \simeq 5$ μm) of a single crystal of silver was increased significantly when a chemical film was deliberately produced on the surface. Many chemical and electrochemical polishing and etching treatments produce surface films whose thicknesses are of the order under consideration.

Elastic Recovery of the Indentation Diagonals

Several hypotheses have proposed that the variation in hardness number is due to variations in the elastic recovery of the indentation diagonals during unloading. For example, some recent careful work by Blau [*27*] showed that the ratio of the short to the long diagonal of Knoop indentations varied with the impression size. The ratio in general fell progressively below that expected from the geometry of the indenter as the size of the indentation increased. Blau attributed this to anisotropy in the elastic shape recovery of the indentation by as much as 20%.

Although the consensus view is that elastic recovery of this nature does occur, it is not in accord with the compression mechanism of indentation. As noted earlier, the experimental fact is that the permanent plastic displacements which occur in directions parallel to the specimen surface, including the directions of the indentation diagonals, are small. One measured figure is about 10% outward displacement in a cold-worked low-carbon steel [*3*]. The elastic recovery on unloading must have been much smaller than this—much smaller than would be necessary to account for Blau's results. In any event, a variation in the magnitude of the elastic recovery with impression size would remain to be explained, considering the insensitivity of the strain system to the indenter geometry.

Note again that the situation is different with the elastic recovery of the depth of the indentations, which certainly is to be expected from the compression model. Lawn and Howes [*28*] have thoroughly analyzed this matter theoretically and investigated it experimentally.

Changes in Indentation Mechanisms

It is now known, therefore, that there is a range of effects associated with instrumentation and with preparation of the test surface which could cause the apparent hardness number in a microindentation test to increase with decreasing indentation size. There is, nevertheless, still a residuum of evidence which, though not conclusive, suggests that changes in the apparent hardness number which are intrinsic to the size of the indentation also occur. If so, it must be concluded either that the indentation characteristics of shallow surface layers are different from those of deeper layers or that the mechanism of indentation changes at shallow depths of penetration. These two assumptions could be related.

It has been suggested that the apparent increase in hardness occurs when the size of the indentation becomes smaller than the spacing of dislocations in the specimen material, the lack of dislocations then inhibiting the indentation deformation process. This now seems unlikely, because it is known that dislocations can be generated easily at a surface during a process such as indentation. Moreover, it is now known that deformation at the strains involved in indentation is much more complicated than was imagined by basic dislocation theory [*29*].

Evidence has been produced by Chen and Hendrickson [*14*] that the distribution of dislocations around indentations on surfaces of single crystals of silver does change when the size of the indentation falls below a certain value ($D \simeq 20$ μm in the silver crystal investigated). The evidence was based on the distribution of dislocation etch pits developed around the indentations. The pits occupied a hemispherical zone around larger indentations in agreement with the compression model of indentation, but they were confined to rosette configurations on slip planes around smaller indentations. However, this change correlated with a decrease in hardness number instead of the usually reported increase.

Certainly, it seems possible that the plastic phase of indentation with which this paper has been concerned may be preceded by a wholly or partly elastic

DISCUSSION

P. Sargent[1] *(written discussion)*—In your discussion of the many different shapes that *post facto* indentations are observed to take, you mention that these are all caused by elastic recovery as the indenter is removed (with some plasticity in a few cases). You also give the impression that you think the surface surrounding the indentation is flat as the indenter is pressed in. Could you comment on the observation of alternating pileups and sink-ins surrounding indentations made in bismuth, which is very soft? Surely these changes in topography around the indentation must have been formed as the indenter was pressed in, not afterwards, as the degree of elastic recovery in bismuth must be very small.

L. E. Samuels (author's closure)—My proposal was only that it is *possible* that effects of the type mentioned by Dr. Sargent, or at least many of them, could develop during recovery of the indentation instead of during the loading phase. I also suggested that it is not possible to determine with reasonable certainty which will develop by examining recovered impressions, and therefore the question warrants investigation by more direct means. Personally, I have an open mind on the matter.

[1]University of Cambridge, Department of Engineering, Cambridge CB2 1PZ, England.

David B. Marshall[1] *and Brian R. Lawn*[2]

Indentation of Brittle Materials

REFERENCE: Marshall, D. B. and Lawn, B. R., **"Indentation of Brittle Materials,"** *Microindentation Techniques in Materials Science and Engineering, ASTM STP 889*, P. J. Blau and B. R. Lawn, Eds., American Society for Testing and Materials, Philadelphia, 1986, pp. 26–46.

ABSTRACT: The use of indentation testing as a method for investigating the deformation and fracture properties of intrinsically brittle materials, glasses, and ceramics is examined. It is argued that the traditional plasticity models of hardness phenomena can be deficient in some important respects, notably in the underlying assumptions of homogeneity and volume conservation. The penetrating indenter is accommodated by an intermittent "shear faulting" mode, plus (to a greater or lesser extent, depending on the material) some structural compaction or expansion. These faults provide the sources for initiation of the indentation cracks. Once generated, the cracks can grow under the action of subsequent external tensile stresses, thereby taking the specimen to failure.

In this presentation the mechanical basis for describing these phenomena will be outlined, with particular emphasis on the interrelations between hardness and other characteristic material parameters, such as elastic modulus and fracture toughness. Procedures for quantitative determination of these parameters will be discussed. Extension of the procedures to the measurement of surface residual stresses in brittle materials will be made to illustrate the power of the indentation method as an analytical tool for materials evaluation.

KEY WORDS: Microindentation hardness testing, brittleness, cracks, elastic recovery, fracture mechanics, indentation, residual stress, shear faults, toughness

Indentation methods are now widely used to study the mechanical properties of glasses and ceramics. The contact of a sharp diamond point with even the most brittle surface causes some irreversible deformation and leaves a residual impression from which a measure of the hardness can be obtained. There is also evidence (for example, from the recovery of the indentations during unloading) that the elasticity of the test material plays a far from insignificant role in the contact process. But the overwhelmingly distinctive fea-

[1]Manager, Structural Ceramics Department, Rockwell International Science Center, Thousand Oaks, CA 91360.

[2]Physicist, Center for Materials Science, National Bureau of Standards, Gaithersburg, MD 20899.

ture of the indentation patterns in this class of materials is the almost invariable appearance of so-called radial cracks emanating from the impression corners. It becomes clear that the potential exists for obtaining quantitative information on fracture as well as deformation properties from a simple hardness testing facility.

In addition to its simplicity, indentation testing has several attractions as a tool for characterizing the mechanical response of brittle materials. The geometry and size of the crack patterns can be accurately controlled and the location of the contact site predetermined. We thereby have a well-defined system for analysis in terms of "fracture mechanics" methodology [*1*,*2*]. Microscopic examination of the indentation area, both during and after the actual contact process, provides valuable information on the fundamental mechanisms of deformation and fracture. Indentation damage usefully simulates individual events in a range of cumulative surface removal processes, such as abrasive wear and machining, and accordingly serves as a base for setting up detailed models of these processes. Radial cracks can be used as strength-controlling "flaws" in the failure testing of ceramics, thus allowing the determination of fracture toughness parameters with high accuracy. Experiments of this kind have provided a unique link between the mechanical response of brittle materials at the microscopic level and the more traditional approach of macroscopic fracture testing adopted by engineering researchers.

The primary aim of this paper is to survey areas of research in which indentations have been used to determine the mechanical properties of glasses and ceramics. Various aspects of this topic have been discussed at length in other review articles [*2–7*], so our coverage here is not intended to be in any way exhaustive. To begin, some basic observations of the nature of sharp-indenter damage and how these observations fit into a general fracture mechanics formalism will be presented. Comment will be made on the underlying structural processes which accommodate the indentation deformation and the associated crack configurations. Then two major practical applications will be described: the measurement of brittle fracture parameters and the evaluation of surface stresses. Our emphasis here is on physical principles rather than mathematical details, although we shall include some of the more important equations to demonstrate the power of the fracture mechanics approach.

Characteristics of Indentation Damage in Brittle Materials

General Observations

As with metals, the contact of a glass or ceramic surface with a sharp, fixed-profile indenter leaves a residual impression, indicating some form of irreversible deformation. Casual observation of the contact site shows nothing unusual about the appearance of the depressed material; the imprint of the indenter is well defined, and the surface regions within the imprint appear characteristically smooth. At first sight, therefore, one might feel justified in adopting the same, traditional plasticity models used to describe the deformation response in ductile materials [*8,9*]. And, in fact, this approach has met with a certain degree of success in the description of the hardness properties of brittle surfaces [*10*]. The picture most commonly conjured up is that of an "expanding cavity," in which the volume of the impression is accommodated by a net radial flow of material, resulting in an approximately hemispherical plastic zone surrounded by a confining elastic matrix. A distinctive feature of such radial flow models is a predicted absence of "pileup" around the indentation, a prediction borne out (at least in the harder ceramics and glasses) by experimental observation [*10*]. It is implicit in all continuum-based plasticity models of this kind that the deformation processes are volume conserving (as characterized by a well-defined yield stress), a consequence of which is that a state of residual stress must exist around the indentation site [*11*].

Closer inspection of the deformation regions beneath the contact area reveals some important departures from the idealized picture just presented. First, the deformation processes are by no means uniformly distributed within the plastic zone but are manifested (at least in part) as a cumulation of discrete shear events. These events are akin to the dislocation slip processes which occur on preferential glide planes in softer materials, but can differ in two important respects: (1) they occur at stress levels close to the theoretical shear strength of the structure in the more covalent materials, and (2) the shear surfaces are not necessarily crystallographic (similar shear events are observed in glassy and in crystalline materials) but are determined more by stress trajectory patterns [*12*]. One therefore has to be extremely cautious before using classical dislocation concepts to describe the flow properties of solids with intrinsically rigid bonding.

A second departure from ideal behavior is apparent in certain materials which show a greater tendency toward deformation in hydrostatic stress than in shear. Fused silica, for instance, undergoes structural densification when subjected to confining pressures [*13*]. (The term *anomalous* has been used to describe glasses of this kind.) Many crystalline solids undergo pressure-induced phase transformations, which may be either expansive (for example, in zirconia composites) or compactive. Compaction modes can accommodate the volume of the impression with relatively little stress mismatch at the deformation zone boundary [*14*].

A clear illustration of such effects is given in Fig. 1. The micrographs are of Vickers indentations in soda lime and fused silica glasses [*15*]. The section views are obtained by indenting across a preexisting hairline fissure and then running this fissure through the specimen. Well-defined shear faults are distinguishable in the cross-sectional view in the soda lime glass. It is envisaged

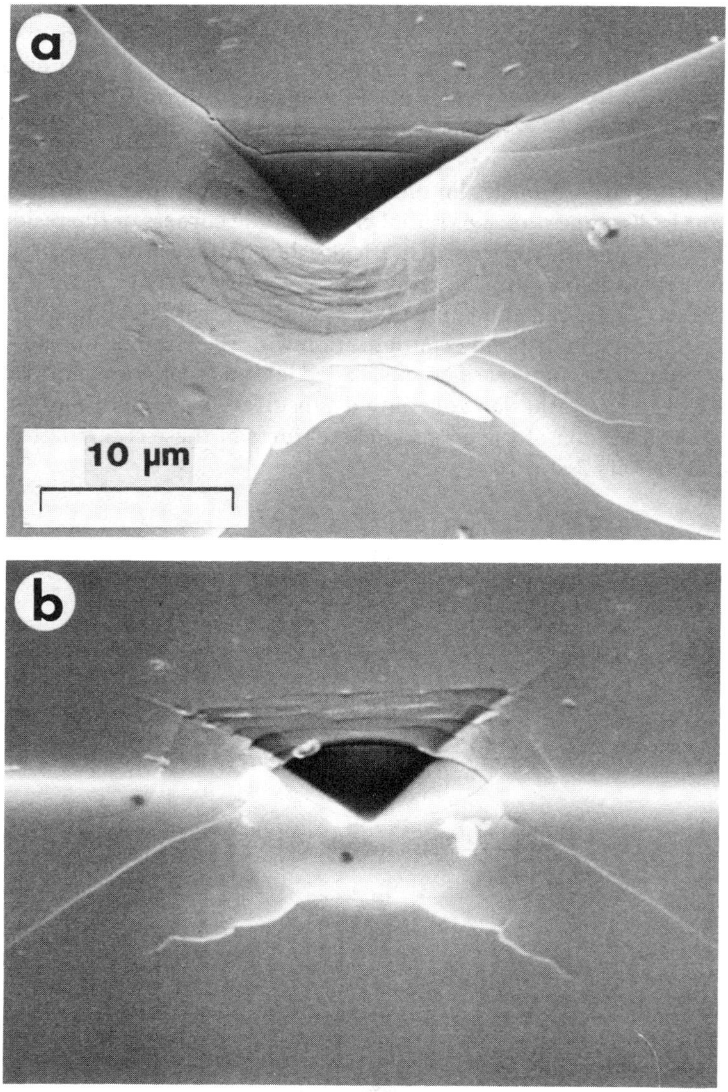

FIG. 1—*Scanning electron micrographs of Vickers indentations in* (a) *soda lime and* (b) *fused silica glasses, showing half-surface and section views.*

that these faults are produced as catastrophic slip failures as a result of the punching action of the penetrating indenter. Once such a failure occurs, the local stress intensity will be somewhat relieved, so that further penetration is needed to produce the next fault; this explains the periodic nature of the pattern. In the fused silica, the surface pattern of the slip traces is not dissimilar, but the faults do not penetrate deeply into the subsurface regions. In this latter glass, structural compaction absorbs a greater proportion of the indentation energy.

As mentioned previously, an indicator of brittleness in an indentation experiment is the appearance of radial cracks at the corners of the residual impression. Figure 2 illustrates schematically the characteristic fracture pattern for a Vickers indentation. The radial cracks are oriented normal to the specimen surface, on so-called median planes coincident with the impression diagonals, and have a half-penny configuration with their centers at the original contact point. A second set of cracks, called lateral cracks, extends from near the base of the deformation zone into a subsurface saucerlike configuration. A general discussion of the geometrical features of these crack patterns is to be found in Refs *2* and *16*.

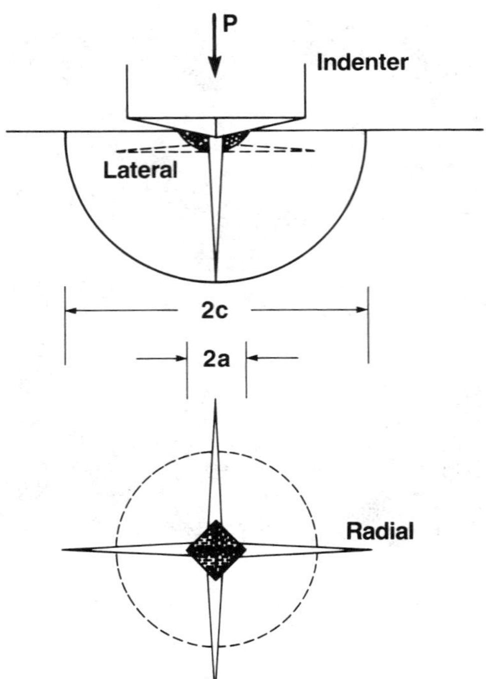

FIG. 2—*Indentation fracture pattern, depicted here for Vickers geometry.*

Let us now consider the nature of the processes responsible for *initiating* and *propagating* the cracks. From close observation of the indentation sites, for example, as in Fig. 1, it becomes clear that the shear events referred to earlier are critically important in acting as embryonic nuclei for crack formation [*12,15–17*]. (Cracks can be thus initiated in the most perfect of surfaces, such as freshly drawn optical fibers, which are free of preexisting defects down to molecular dimensions [*18*].) There is a threshold indent size, typically ≈ 10 μm, below which crack initiation does not occur, although this threshold is subject to considerable variation, depending on such things as the duration of contact and the test environment [*17*]. Once initiated, the cracks "pop in" abruptly, to a characteristic length ≈ 100 μm, at which stage they are considered to be fully propagating. It is observed that most of the crack development occurs not on loading, but on *un*loading, the indenter, indicating that it is the *ir*reversible component of the contact stress field which provides the dominant driving force for fracture. The sequence of micrographs of crack evolution during Vickers indentation in soda lime glass in Fig. 3 illustrates the point [*19*]. Note also the appearance of the stress birefringence in the final frame of this sequence; glass is not optically active, so the persistence of the "Maltese cross" in this frame confirms the existence of a substantial residual stress intensity.

An even more dramatic indication of residual stress effects in the fracture evolution is manifest in postindentation observations. At loads above threshold, the popped-in cracks are often seen to continue propagating well after completion of the contact. Below threshold, delayed pop-in can occur. These phenomena are attributable to the time-dependent enhancement of crack development in the presence of moisture, as alluded to in the previous paragraph.

Mechanics

Evaluation of the indentation cracking properties of brittle materials requires a knowledge of the underlying fracture mechanics. The starting point for the requisite formulation is the characterization of the elastic-plastic stress field, with particular focus on the residual component of this field. Unfortunately, this formulation can be a formidable task, even for the idealization of a perfectly homogeneous, continuous solid. The difficulty is especially pronounced in the modeling of crack initiation, for there one is concerned with the complex details of the contact near field [*2*]. Things are not so bad once the crack is in its fully propagating stage, in which the far field may be regarded in terms of simple "point" force solutions. Within these limitations, working equations defining the scale of the indentation damage as a function of contact load can be developed from first principles.

Let us begin with a consideration of the manner with which the indenter load, P, varies with the penetration, z, during a contact cycle. Experimen-

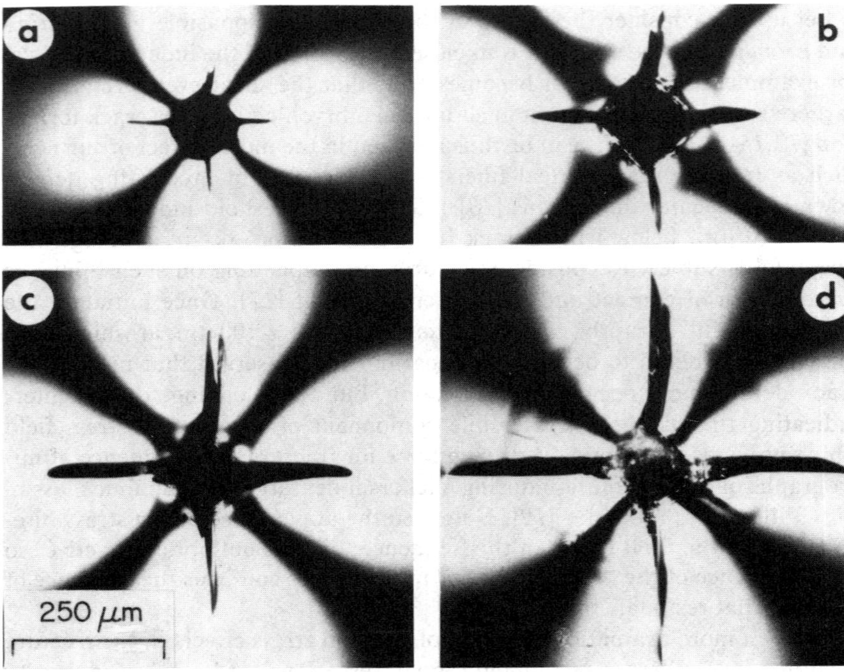

FIG. 3—In situ *observations of Vickers indentation in soda lime glass, viewed from below during the loading cycle. The sequence shows the* (a) *half-loaded,* (b) *fully loaded,* (c) *half-unloaded, and* (d) *fully unloaded stages. Note the development of radial cracks to completion during the unloading half-cycle. Polarized light reveals strong stress birefringence.*

tally, the function $P(z)$ is found to have the form shown in Fig. 4. During the loading half-cycle, the contact pressure ($\propto P/a^2$, where a is a characteristic impression dimension) remains constant (at least in the region where geometrical similarity prevails). This pressure, by definition, determines the hardness. The deformation has both elastic and inelastic components at this stage. On unloading, the deformation is entirely elastic (*re*loading simply reproduces the unloading curve). Hence, the contact pressure in the unload region is determined by Young's modulus, E. We may write the functional relations for the two half-cycles in the form [20]

$$P \propto Hz^2 \qquad \text{(load)} \qquad (1a)$$

$$P \propto E(z^2 - z_r^2) \qquad \text{(unload)} \qquad (1b)$$

where due allowance is made in Eq 1*b* for the existence of a residual impression depth, z_r. The requirement for compatibility of these two equations at the maximum penetration, z_m, yields the relation

$$\left(\frac{z_r}{z_m}\right)^2 = 1 - \frac{\eta H}{E} \tag{2}$$

where η is a numerical constant (≈6 for Vickers indenters [*20*]). For ceramics, characterized by relatively high values of H/E, the value of z_r/z_m is typically ≈0.5, indicating substantial elastic recovery. It emerges from this kind of formulation that the ratio H/E has an important place in the specification of the elastic-plastic field. The same conclusion is drawn from other treatments of the elastic recovery problem [*21,22*].

Now consider how the elastic-plastic parameters enter the fracture problem. We shall concentrate on the simplest case, that of well-developed radial cracks, as depicted earlier in Fig. 2. Mention has already been made, in the previous section, of the tendency for the radial cracks to develop into their immediate postindentation configuration during the unloading half-cycle, in which case it is the residual component of the field which provides the principal crack driving force. This component may be evaluated in terms of a concentrated force, P_r, centered at the contact origin. It can be shown that $P_r = \chi_r P$, that is, the residual crack-opening force scales directly with the contact load, where χ_r is a dimensionless parameter which determines the intensity of

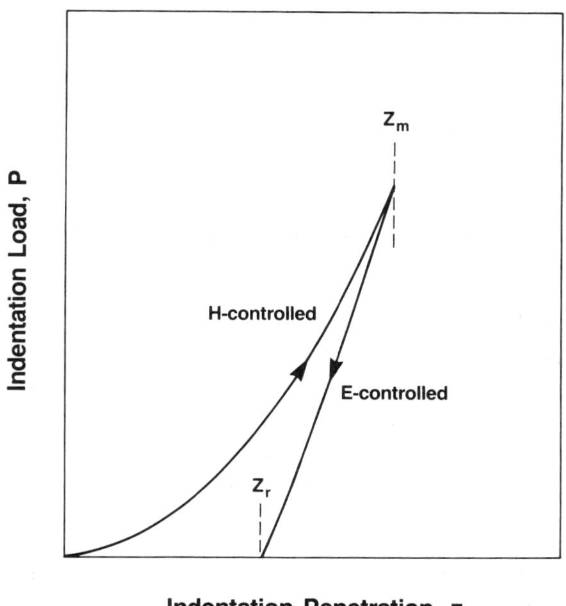

FIG. 4—*Load-displacement response for sharp indenters.*

the field [*19*]. For an ideal elastic-plastic material in which the irreversible deformation is volume conserving, analysis gives [*23*]

$$\chi_r \propto \left(\frac{E}{H}\right)^{1/2} (\cot \phi)^{2/3} \tag{3}$$

where ϕ is the indenter half-angle.

The next step is to incorporate the center-load force into an appropriate fracture mechanics formulation. This is most conveniently done using the *stress intensity factor* notation, K, which expresses the intensity of the field concentrated at the tip of the extending crack [*1*]. For pennylike cracks, the stress intensity factor takes on the form [*19*]

$$K_r = \frac{\chi P}{c^{3/2}} \tag{4}$$

where c is the characteristic crack size (Fig. 2) and $\chi \propto \chi_r$. In the absence of time-dependent crack growth effects, that is, for tests in inert environments, the condition for equilibrium extension is expressible as $K = K_c$, where K_c defines the material toughness. This toughness parameter quantifies the intrinsic resistance to fracture, just as hardness quantifies the resistance to deformation. Insertion of this requirement into Eq 4 then gives the equilibrium crack length

$$c_0 = \left(\frac{\chi P}{K_c}\right)^{2/3} \tag{5}$$

at completion of indentation.

An analogous expression to that in Eq 5 can be obtained for lateral cracks, although the analysis is somewhat more complicated because of the relatively high compliance of the thin layer of material which overlays the crack plane [*24*].

A configuration of special interest is that of an indented specimen subsequently subjected to an externally applied tensile stress, for example, as in a bend test. The radial cracks can then be emplaced so as to be normal to the tensile direction, and thereby act as strength-controlling flaws. A unique feature of such a "controlled-flaw" test is the facility to predetermine the failure site, so that the crack response can be observed directly throughout its evolution to failure [*25*]. Such observations reveal a new kind of flaw behavior. Whereas the earlier, classic study by Griffith (see Ref *1*) suggested that the size of flaws should remain stationary up to the critical load for failure, at which point the equilibrium is unstable, it has been observed in the indentation experiments that a significant amount of precursor extension precedes the instability configuration. This stabilization of the crack system is attributable to the residual driving force, K_r, in Eq 4, which augments the applied loading. It should be noted that whereas the external loading term, K_a, is always an increasing function of crack size [*1*], K_r is a decreasing function. Formally, superposition of K_a and K_r gives, at equilibrium [*25*]

$$K = \psi \sigma_a c^{1/2} + \frac{\chi P}{c^{3/2}} = K_c \tag{6}$$

where σ_a is the applied tensile stress, and ψ is a crack geometry constant (of the order unity). The condition for *unstable* equilibrium is that K should have a minimum value in Eq 6. Putting $dK/dc = 0$, accordingly determines the critical configuration

$$\sigma_m = \frac{3K_c}{4\psi c_m^{1/2}} \tag{7a}$$

$$c_m = \left(\frac{4\chi P}{K_c}\right)^{2/3} \tag{7b}$$

at failure. Note that c_m in Eq 7*b* is $4^{2/3} = 2.52$ times c_0 in Eq 5, indicating that the stage of precursor crack growth is by no means insignificant.

The fracture mechanics formulations to this point are contingent on the threshold load being exceeded. If this threshold is *not* exceeded, we are faced with the considerably more complex problem of crack initiation, in which the corresponding relations for crack instability involve detailed analysis of the near-field conditions. Phenomenological treatments of the instability condition for spontaneous radial crack pop-in from an incipient nucleus, based on the hypothesis that the *intensity* of the indentation stress field should remain invariant (since it scales directly with the hardness) while the *spatial extent* should scale directly with the indent size, a, predict a critical threshold condition [*26,27*]

$$a_c = \mu \left(\frac{K_c}{H}\right)^2 \tag{8}$$

where $\mu \propto E/H$. Hence, the initiation mechanics reflect an inherent size effect. This size effect relates to the fact that, while the resistance to deformation, H, has the units of stress, the fracture resistance, K_c, has the units of stress multiplied by dimension to the one half power (as is evident from Eq 6).

Measurement of Material Properties

Elastic Recovery

Measurements of the elastic recovery at hardness indentations are of interest for the light they shed on the partition of input deformation energy into reversible and irreversible components and for their characterization of the elastic-plastic field in the ensuing fracture mechanics formalism. (They can also be used to evaluate Young's modulus at a microscopic level.) The parameter of interest here is the ratio of hardness to modulus, H/E. One way of measuring this parameter is to monitor the load-penetration function, $P(z)$, and to make use of Eqs 1 and 2 [*20*] or some equivalent analytical expressions [*22*]. (The theme of load-penetration characteristics to determine indentation response is, incidentally, a recurring one in this volume.) Unfortunately, the apparatus needed to measure indentation displacements is not yet a standard accessory on most hardness testing facilities, so quantitative studies of elastic recovery have not occupied a prominent place in the deformation evaluation literature.

However, one way in which a measure of the recovery can be obtained without recourse to special equipment is to observe the relative contraction of the surface diagonals of Knoop impressions [*21*]. Whereas the long diagonal turns out to be relatively insensitive to "springback" effects, the short diagonal does not. (The analogy with the age-old problem of trying to break an egg by pressing along the longitudinal axis is instructive here.) Analysis gives

$$\frac{b'}{a'} = \frac{b}{a} - \frac{\alpha H}{E} \tag{9}$$

where $b/a = 1/7.11$ is the *nominal* ratio of short to long half-diagonal, b'/a' is the corresponding value after recovery, and α is a dimensionless constant (≈ 0.45 [*21*]). Hence, materials with relatively rigid structures, that is, high H/E, are likely to show greater lateral recovery. This is evident in Fig. 5, which contrasts Knoop indentations of similar length in two extreme materials, zinc sulfide (ZnS) ($H/E = 0.02$) and soda lime glass ($H/E = 0.09$) [*21*]. Figure 6 shows the validity of Eq 9 for a wider range of materials [*21*].

Fracture Toughness

Perhaps the most widespread use of indentation testing in the context of brittle materials is the evaluation of material toughness. In this application, one seeks to relate the fracture resistance to the scale of the crack pattern. It is perhaps ironical that the earlier hardness testing fraternity tried to avoid cracking like the plague, whereas now the indentation method has become

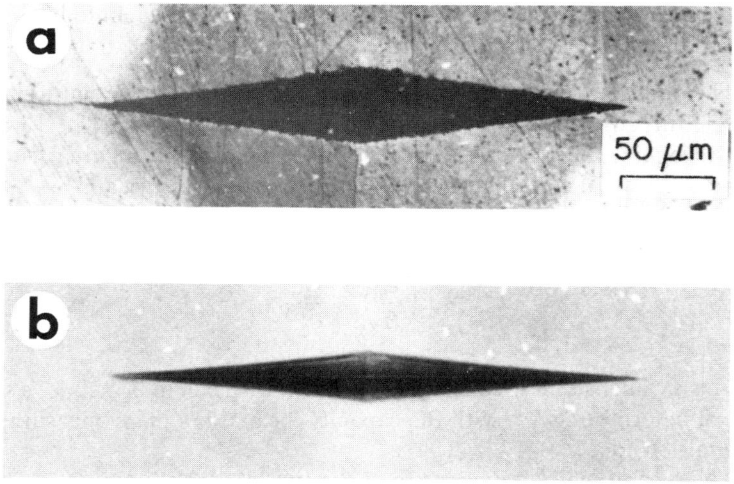

FIG. 5—*Knoop indentations in* (a) *polycrystalline zinc sulfide and* (b) *soda lime glass. Note the relatively strong elastic recovery of the short diagonal in the latter material.*

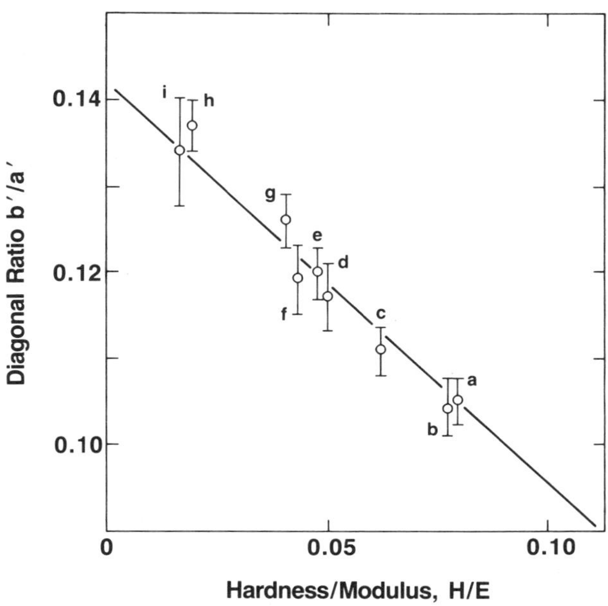

FIG. 6—*Plot of the short-to-long-diagonal ratio,* b′/a′, *from Knoop indentation measurements against* H/E *for selected materials:* (a) *soda lime glass,* (b) *glass ceramic,* (c) *silicon nitride,* (d) *alumina,* (e) *zirconia,* (f) *magnesium fluoride,* (g) *steel,* (h) *zinc sulfide, and* (i) *zinc oxide.*

the most commonly used of all toughness measurement techniques in glass and ceramics. Here we describe three variants of the indentation procedure.

1. The first and most straightforward approach involves relating toughness, K_c, directly to postindentation crack size, c_0. Palmqvist [28] was the first to recognize the potential for such a relationship, but his treatment was purely empirical. We now have, through Eqs 3 and 5, the means for deriving an appropriate expression from first principles

$$K_c = \xi\left(\frac{E}{H}\right)^{1/2} \frac{P}{c_0^{3/2}} \tag{10}$$

where ξ is a dimensionless constant [29]. Thus, we have, in principle, a simple means of determining K_c, with the capacity for making many measurements on a single surface.

2. The second method makes use of the special crack response implicit in the derivation of the instability relations in Eq 7. The test procedure involves emplacing a controlled indentation flaw in the prospective tensile face of a bend bar and then measuring the strength of the bar in an inert environment. Eliminating c_m from Eqs 7*a* and 7*b*, and combining with Eq 3, we obtain

$$K_c = \eta\left(\frac{E}{H}\right)^{1/8} (\sigma_m P^{1/3})^{3/4} \tag{11}$$

where η is another dimensionless constant [30]. Note that strength replaces crack size as a test variable in this formulation, a distinct advantage in materials in which direct crack observations are difficult (as is true with many ceramics). However, only one test result is obtained per specimen.

3. The third method is a hybrid of the first two in that it involves measurement of *both* crack size and strength. The appropriate toughness relation derives directly from Eq 7*a* [31]

$$K_c = \left(\frac{4\psi}{3}\right)\sigma_m c_m^{1/2} \tag{12}$$

Although the measurement requirements are clearly more stringent, the method has the advantage of circumventing knowledge of the E/H parameter. It will be recalled that this parameter is determined from Eq 3, the basis of which is volume conservation in the deformation process. The specific advantage of Eq 12 is that it contains no implicit assumption at all concerning the mode of deformation. Experimentally, c_m can be conveniently determined by introducing *several* nominally identical indentations within the tensile test span of the bend specimen; failure occurs from just one of the indentations, leaving "survivors" for measurement of the *near*-critical crack lengths [*31,32*].

The results of measurements on selected glasses and ceramics are shown in Fig. 7 for each of these three variant methods [*29–31*]. Each plot is reasonably well represented by a linear fit, although the lines do not pass through the

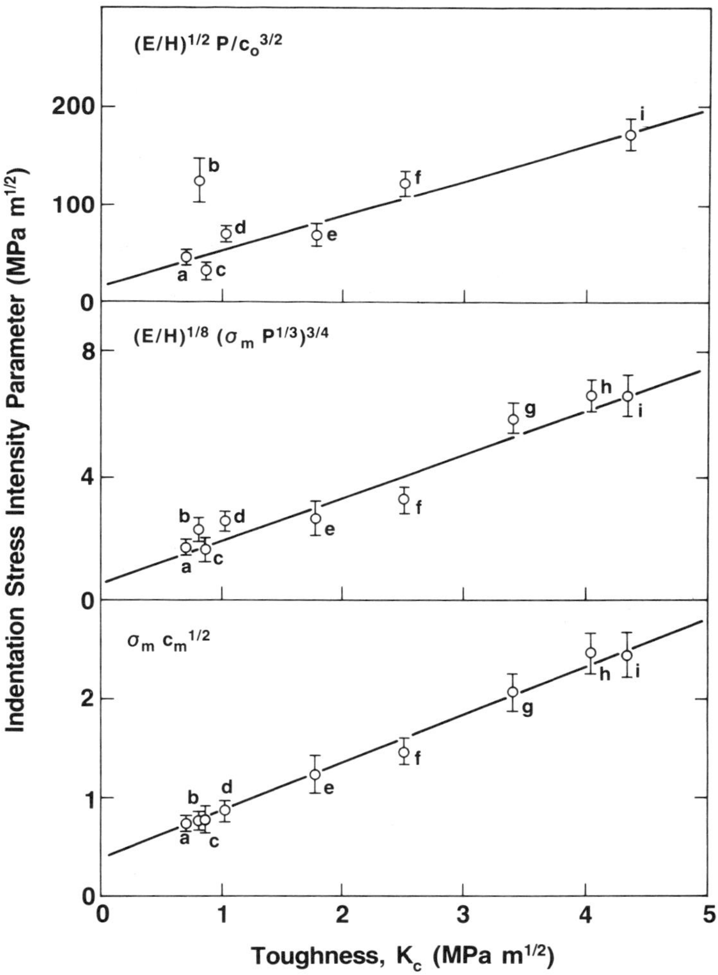

FIG. 7—*Plots of indentation test results illustrating the three toughness formulas, Eqs 10, 11, 12, for selected glasses and ceramics:* (a) *soda lime glass,* (b) *fused silica,* (c) *lead zirconium titanate,* (d) *barium titanate,* (e) *Synroc,* (f) *glass ceramic,* (g) *alumina I,* (h) *silicon carbide, and* (i) *alumina II. The vertical axis represents the measured indentation test variables; the horizontal axis represents the independently determined toughness values. The curves are linear fits to data.*

origins as required by Eqs 10 through 12. There is the implication here that some material dependence exists in the "proportionality constants" ξ, η, and ψ. The plots, nevertheless, remain useful representations for toughness calibration. It may be noted that the data points for some of the materials, especially fused silica (a material which, it will be recalled, deforms by densification), show significant departures from the line fits to the overall data in all but the third plot, reinforcing our previous allusion to Eq 12 as the most universal of the three toughness formulas.

Brittleness

The question often arises as to how one may quantify the "brittleness" of materials. Intuitively, brittleness should be a measure of the competition between deformation and fracture processes, as manifested, for instance, in the ductile-brittle transitions observed in structural metals. This competition has proved extremely difficult to formulate as a function of easily determinable material parameters. Here we shall propose that the ratio of hardness to toughness, H/K_c, be used as an appropriate index of brittleness (analogous to the adoption of H/E as an index of rigidity) [*33*]. Then the way is open, through Eq 8, to evaluate the relative susceptibilities to flow and fracture in terms of indentation threshold conditions.

To illustrate, let us consider the plot of the experimentally measured threshold indent size, a_c, as a function of $(E/H)(K_c/H)^2$ in Fig. 8 [*34,35*].

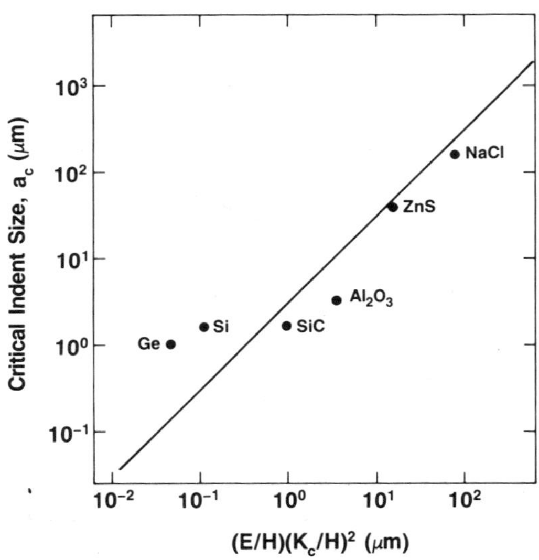

FIG. 8—*Plot of the measured threshold indentation size against* $(E/H)(K_c/H)^2$ *for selected ceramic materials. The line is best fit to Eq 8.*

Points to the lower left of this plot represent materials for which cracking is most easily produced at local stress concentrations, that is, which are relatively brittle. Conversely, the upper right of the plot represents the domain of more ductile materials, notably metals. In this interpretation, brittleness is quantified as a size effect. Physically, a_c represents the scale of damage above which the mechanical response is essentially fracture dominated and below which it is essentially deformation dominated.

The utility of this interpretation is most evident in the description of surface removal processes. If a surface is contacted with coarse particles, material removal occurs by a chipping mode (grinding), and if with fine particles, by a flow mode (polishing). The critical particle size defining the transition between these two modes is then dependent on the index H/K_c.

Surface Stress Evaluation

The surfaces of materials can undergo treatments which differ from those experienced by the underlying bulk and can thereby be left in a state of residual stress. Such stresses, although often confined to very shallow layers, can be of high intensities and can therefore exert a strong influence on mechanical properties. Indentation crack patterns can be used in brittle materials to evaluate these stresses. Basically, the surface traces of the radial crack system expand or contract, depending on whether the stresses are tensile or compressive. Figure 9, which shows Vickers indentations at the same load in annealed (stress-free) and thermally tempered (residual-compression) soda lime glass, illustrates the effect [*19*]. The fracture mechanics formulation then allows us, through appropriate modifications of the previous stress intensity relations, to establish a quantitative base for analysis.

Calculation of the surface stress is relatively straightforward when the stress is uniformly distributed over the prospective indentation crack area. We can then define a stress intensity factor conjugate to the surface stress, σ_s, $K_s = \psi\sigma_s c^{1/2}$, in the manner that K_a was defined earlier for substitution into Eq 6. Thus, in combination with Eq 4, one can solve to obtain a simple working expression for evaluating σ_s in terms of direct crack-size measurements, c. Alternatively, in a controlled-flaw test we may retain Eq 6 as our starting expression for calculating strengths, provided we replace σ_a by $\sigma_a + \sigma_s$. Accordingly, using the same logical development as in the derivation of Eq 7, we can readily show that, for a fixed indentation load, the critical applied stress at failure instability is given simply by

$$\sigma_m = \sigma_m^0 - \sigma_s \tag{13}$$

where σ_m^0 now refers to surfaces in the stress-free state. Two sets of strength tests, one on the actual surface-stressed materials and the other on stress-free controls, should then suffice to determine σ_s. Figure 10 shows the strength

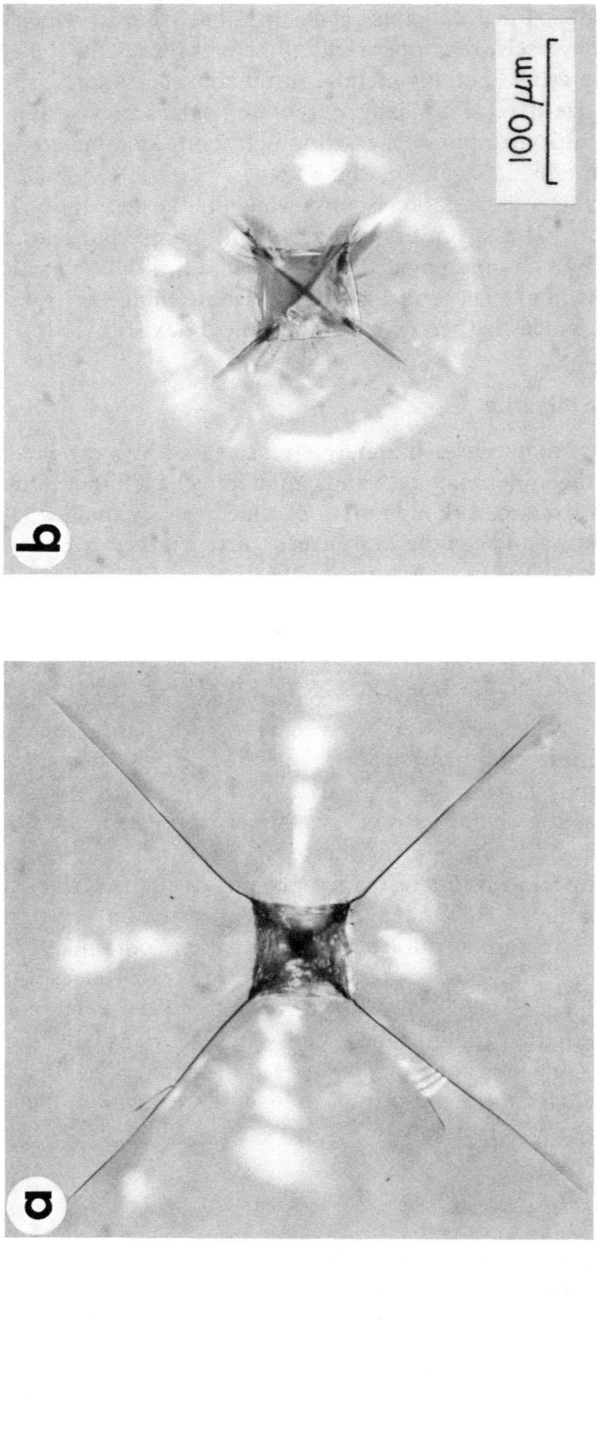

FIG. 9—*Vickers crack patterns in* (a) *preannealed and* (b) *thermally tempered soda lime glass. The indentation load was the same in both cases. Surface stresses in the tempered specimen have clearly inhibited the crack expansion.*

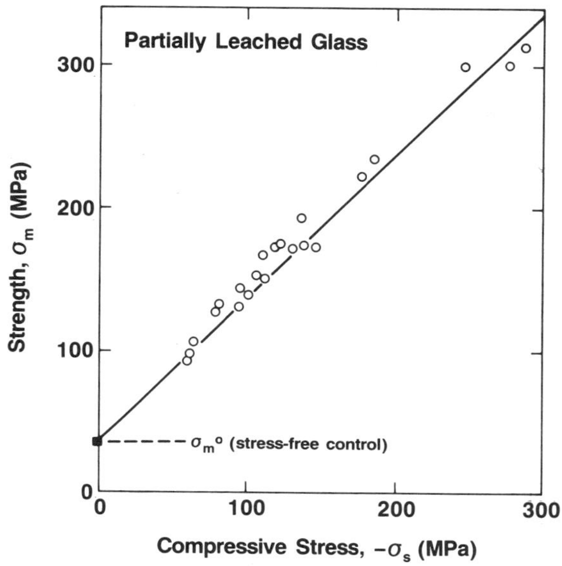

FIG. 10—*Plot of the measured strength as a function of the independently determined compression stress for surface-treated glass rods containing controlled indentation flaws (indentation load = 100* N*).*

data for glass rods subjected to various degrees of surface compression by a "partial leaching" treatment [*36*].

A more complicated, although still tractable, case is that of very thin stress layers, of depth $d \ll$ prospective crack size, c. The surface stress intensity factor in this limit is $K_s = 2\psi\sigma_s d^{1/2}$ [*37*]. In the immediate postindentation state, K_s augments the residual contact stress intensity factor, K_r, in Eq 4, so that, at the equilibrium condition $K = K_c$, we may solve to obtain [*37*]

$$\frac{c}{c^0} = \frac{1}{(1 - 2\psi\sigma_s d^{1/2}/K_c)^{2/3}} \tag{14}$$

where c^0 refers to the radial crack size in the stress-free state. Hence, the surface stresses can be evaluated directly from radial crack length measurements. Tensile stresses in proton-irradiated glass have been measured using this formulation. It will be noted that for sufficiently large positive σ_s, c is predicted to expand without limit, corresponding to the spontaneous fracture of the surface stress layer. Such surface breakup is, indeed, observed in the irradiated glass specimens [*37*].

Conclusions

We have had a brief look at the ways in which indentation testing may be applied to evaluate the deformation and fracture characteristics of glasses and ceramics. Insofar as *deformation* is concerned, we have indicated that materials with large components of covalent bonding may not satisfy the classic plasticity models. Further work needs to be done on such materials, whose hardness numbers approach theoretical strength limits. The *fracture* properties are better understood, at least in the fully propagating region, where conventional fracture mechanics principles may be applied.

Selective applications of the indentation mechanics formalism have been considered. We have seen how indentation measurements may be used to quantify such properties as elastic recovery, toughness, and brittleness. We have also shown how one may evaluate surface stress levels. Many other applications are described in the references.

Acknowledgments

The authors thank Robert F. Cook for assistance in preparation of data for some of the figures. This work was funded, in part, by the U.S. Air Force Office of Scientific Research and the U.S. Office of Naval Research.

References

[1] Lawn, B. R. and Wilshaw, T. R., *Fracture of Brittle Solids*, Cambridge University Press, London, 1975.
[2] Lawn, B. R. and Wilshaw, T. R., *Journal of Materials Science*, Vol. 10, 1975, pp. 1049–1081.
[3] Evans, A. G. and Wilshaw, T. R., *Acta Metallurgica*, Vol. 24, 1976, pp. 939–956.
[4] Lawn, B. R. in *Fracture Mechanics of Ceramics*, Vol. 5, R. C. Bradt, A. G. Evans, D. P. H. Hasselman, and F. F. Lange, Eds., Plenum, New York, 1983, pp. 1–25.
[5] Lawn, B. R. and Wiederhorn, S. M. in *Contact Mechanics and Wear of Rail/Wheel Systems*, J. Kalousek, R. V. Dukkipati, and G. M. Gladwell, Eds., University of Waterloo Press, Vancouver, 1983, pp. 133–148.
[6] Lawn, B. R., Hockey, B. J., and Richter, H., *Journal of Microscopy*, Vol. 130, 1983, pp. 295–308.
[7] Lawn, B. R. in *Strength of Glass*, C. R. Kurkjian, Ed., Plenum, New York, in press.
[8] Tabor, D., *The Hardness of Metals*, Clarendon, Oxford, 1951.
[9] Samuels, L. E. and Mulhearn, T. O., *Journal of Mechanics and Physics of Solids*, Vol. 5, 1957, pp. 125–134.
[10] Marsh, D. M., *Proceedings of the Royal Society of London*, Vol. A279, 1964, pp. 420–435.
[11] Hill, R., *The Mathematical Theory of Plasticity*, Oxford University Press, London, 1950, Chapter 5, pp. 97–105.
[12] Hagan, J. T. and Swain, M. V., *Journal of Physics D: Applied Physics*, Vol. 11, 1978, pp. 2091–2102.
[13] Ernsberger, F. M., *Journal of the American Ceramic Society*, Vol. 51, 1968, pp. 545–547.
[14] Arora, A., Marshall, D. B., Lawn, B. R., and Swain, M. V., *Journal of Non-Crystalline Solids*, Vol. 31, 1979, pp. 415–428.
[15] Multhopp, H., Lawn, B. R., and Dabbs, T. P. in *Plastic Deformation of Plastic Materials*, R. E. Tressler and R. C. Bradt, Eds., Plenum, New York, 1984, p. 681.

[16] Lawn, B. R. and Marshall, D. B. in *Fractography of Glass*, R. C. Bradt and R. E. Tressler, Eds., Plenum, New York, in press.
[17] Lawn, B. R., Dabbs, T. P., and Fairbanks, C. J., *Journal of Materials Science*, Vol. 18, 1983, pp. 2785-2797.
[18] Dabbs, T. P., Marshall, D. B., and Lawn, B. R., *Journal of the American Ceramic Society*, Vol. 63, 1980, pp. 224-225.
[19] Marshall, D. B. and Lawn, B. R., *Journal of Materials Science*, Vol. 14, 1979, pp. 2001-2012.
[20] Lawn, B. R. and Howes, T. R., *Journal of Materials Science*, Vol. 16, 1981, pp. 2745-2752.
[21] Marshall, D. B., Noma, T., and Evans, A. G., *Journal of the American Ceramic Society*, Vol. 65, 1982, pp. C175-C176.
[22] Loubet, J. L., Georges, J. M., and Meille, G., in this publication, pp. 72-89.
[23] Lawn, B. R., Evans, A. G., and Marshall, D. B., *Journal of the American Ceramic Society*, Vol. 63, 1980, pp. 574-581.
[24] Marshall, D. B., Lawn, B. R., and Evans, A. G., *Journal of the American Ceramic Society*, Vol. 65, 1982, pp. 561-566.
[25] Marshall, D. B., Lawn, B. R., and Chantikul, P., *Journal of Materials Science*, Vol. 14, 1979, pp. 2225-2235.
[26] Lawn, B. R. and Evans, A. G., *Journal of Materials Science*, Vol. 12, 1977, pp. 2195-2199.
[27] Hagan, J. T., *Journal of Materials Science*, Vol. 14, 1979, pp. 2975-2980.
[28] Palmqvist, S., *Archiv für das Eisenhuettenwesen*, Vol. 33, 1962, pp. 629-633.
[29] Anstis, G. R., Chantikul, P., Lawn, B. R., and Marshall, D. B., *Journal of the American Ceramic Society*, Vol. 64, 1981, pp. 533-538.
[30] Chantikul, P., Anstis, G. R., Lawn, B. R., and Marshall, D. B., *Journal of the American Ceramic Society*, Vol. 64, 1981, pp. 539-543.
[31] Cook, R. F. and Lawn, B. R., *Journal of the American Ceramic Society*, Vol. 66, 1983, pp. C200-C201.
[32] Cook, R. F., Lawn, B. R., and Anstis, G. R., *Journal of Materials Science*, Vol. 17, 1982, pp. 1108-1116.
[33] Lawn, B. R. and Marshall, D. B., *Journal of the American Ceramic Society*, Vol. 62, 1979, pp. 347-350.
[34] Lankford, J. and Davidson, D. L., *Journal of Materials Science*, Vol. 14, 1979, pp. 1662-1668.
[35] Chiang, S. S., Marshall, D. B., and Evans, A. G., *Journal of Applied Physics*, Vol. 53, 1982, pp. 312-317.
[36] Chantikul, P., Marshall, D. B., Lawn, B. R., and Drexhage, M. G., *Journal of the American Ceramic Society*, Vol. 62, 1979, pp. 551-555.
[37] Lawn, B. R. and Fuller, E. R., *Journal of Materials Science*, Vol. 19, 1984, p. 4061.

DISCUSSION

P. Sargent[1] *(written discussion)*—The analysis presented in this paper is based on the principle that as far as the growing cracks are concerned, there is no plasticity at the crack tip. However, the hardness value (which depends on the plasticity of the material) does enter the formulas because it is related to the magnitude of the stresses around the indentation. The effects of indentation size on hardness values are quite marked, especially for hard and brit-

[1]Department of Engineering, University of Cambridge, Cambridge CB2 1PZ, England.

tle materials. Have you thought of including some kind of indentation size effect correction in your data analysis or theory?

D. B. Marshall and B. R. Lawn (authors' closure)—The indentation hardness becomes size dependent only at small indentation diameters, typically $\lesssim 10$ μm for brittle materials. This size range generally falls within the subthreshold region, in which well-developed cracks do not form. Thus, the hardness/indentation-size effect may affect the critical threshold condition (for example, Eq 8). However, at larger indentation sizes, corresponding to well-developed fracture, the hardness is essentially independent of size.

Hubert M. Pollock,[1] *Daniel Maugis,*[2] *and Michel Barquins*[2]

Characterization of Submicrometre Surface Layers by Indentation

REFERENCE: Pollock, H. M., Maugis, D., and Barquins, M., **"Characterization of Submicrometre Surface Layers by Indentation,"** *Microindentation Techniques in Materials Science and Engineering, ASTM STP 889*, P. J. Blau and B. R. Lawn, Eds., American Society for Testing and Materials, Philadelphia, 1986, pp. 47–71.

ABSTRACT: With our instrument, the penetration depth of a three-faced indenter is continuously measured during both indentation and unloading. Useful loads are about 10 μN to 30 mN (1 to 3000 mg). The specimen can be transferred between two locations (test and microscopic observation); a particular area can be relocated by means of a computer-controlled three-dimensional movement. Scratch hardness testing with measurement of frictional force can be performed at speeds as low as 30 nm/s. The general shape of the depth/load graphs, together with measurement of hysteresis and slope, gives an initial characterization of the specimen. Further analysis involves various theoretical topics, which we review in the light of recent literature. These include the way in which a variation of yield stress with depth will affect the measurements and the surface effects, such as the contribution of surface energy to the effective applied load. A controversial topic is the extent to which the depth of indentation in metals, relative to the mean distance between preexisting dislocations or grain boundaries, can lead to a true size effect, or even a critical load below which no plastic deformation occurs. Indentation creep may be significant even for metals such as steel and nickel at ambient temperature.

Accordingly, the concept of microindentation hardness as a material property can be of doubtful value for depths in the 10 to 100-nm range; the apparent mean pressure can be characteristic of a depth much greater than that of the plastic indentation. However, it is still possible to obtain measurements within this depth range, to deduce the type of regimen (for example, elastoplastic or fully plastic), and to derive alternative semiempirical indexes of the extent of plastic deformation, elastic recovery, and time-dependent behavior.

KEY WORDS: microindentation hardness testing, microindentation, submicrometre depths, indentation creep, film hardness, surface effects, elastic recovery, ion implantation

It has been known since the early 1970s that it is possible to carry out indentation experiments in which plastic deformation is limited to the top hundred nanometres or less of the specimen. This review is concerned with three ques-

[1] Physicist, Department of Physics, University of Lancaster, Lancaster LA1 4YB, England.

[2] Researchers, Mechanics of Surfaces Research Group of the Centre National de la Recherche Scientifique, 75732 Paris, France.

tions: As the indentation depth is reduced, what are the physical and mechanical effects that tend to alter the measured values of hardness? At what stage does the concept of hardness cease to be of use? Finally, when this occurs, what alternative "indexes" may be derived from the test in order to characterize the mechanical behavior of the specimen? We shall first discuss the basic theory of three mechanical phenomena, which in practice have proved to be especially important at the submicrometre level, and after referring to some recent literature on surface and scale effects in hardness, we give examples of plasticity measurements involving indentation depths in the 10 to 300-nm range, with emphasis on the method of continuous depth recording.

The Effect of Variation of Yield Stress with Depth

According to the U.K. National Physical Laboratory's recommendations [*1*], the indentation depth should not exceed a tenth of the thickness of the region of the specimen that is being tested. If this region is a very thin surface coating, the one-tenth rule may be impossible to satisfy. Alternatively, the presence of a contaminant layer may complicate the analysis of hardness measurements on the underlying material. In either case, useful conclusions may be drawn from certain theoretical analyses of the mechanical effects of a variation of yield stress with depth. Bückle [*2*] used an empirical model to show that if, for example, a surface coating has a thickness of k = five times the indentation depth, then the measured hardness is the sum of just 70% of the hardness of the coating added to 30% of the hardness of the substrate; only when $k = 10$ is the contribution of the substrate negligible ($<2\%$). Recent theoretical work by Lebouvier et al [*3,4*] has thrown some light on the mechanical process underlying this model. They have analyzed the fully plastic indentation of a bilayer in plane strain, using a two-dimensional kinematic method [*5*] in cases in which the slip-line field method would have been extremely complex. They find that Bückle's one-tenth rule for the critical ratio of depth to thickness, at which the contribution of the substrate is negligible, is a good approximation for $Y_0/Y \gtrsim 10$, where Y_0 and Y are the yield strengths of the film and the substrate. However, the one-tenth rule is too severe in the case of relatively softer films; for $Y_0/Y < 1$, the critical ratio is closer to 4. At much greater indentation depth/thickness ratios (δ/h), the change in mean pressure as the indenter approaches the film/substrate interface is predicted to occur in two distinct ways, according to whether the film is softer or harder than the substrate. In the first case, the expected increase in mean pressure begins typically at $\delta/h \approx 0.5$ to 1.5 and is quite abrupt (this critical value of δ/h is predicted to vary slightly with Y_0/Y). In the second case, the expected decrease in mean pressure begins at a shallower depth (for example, $\delta/h \approx 0.3$ for $Y_0/Y = 1.75$) and is much more gradual. In Fig. 1 we have replotted an example of this result in the form of a theoretical depth/(load)$^{1/2}$ curve for a soft film, together with an experimental curve of Tazaki

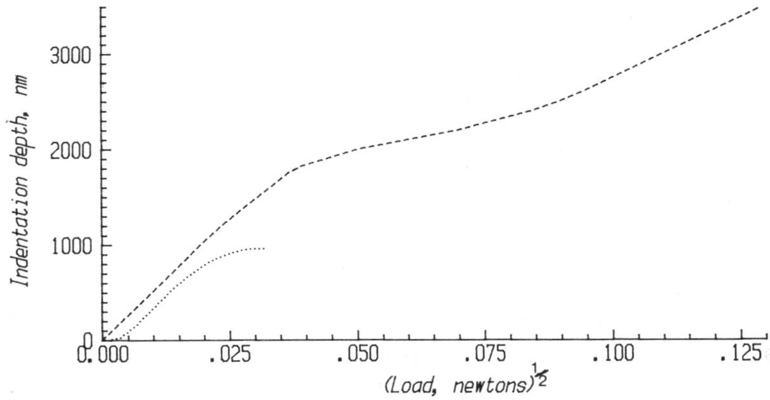

FIG. 1—*Indentation depth, δ, as a function of the square root of applied load, for a soft film (gold) on a hard substrate (glass). The film thickness is* h = *1070 nm; trigonal pyramid indenter (angle between edges 100°,* k_I = *4.67). The dotted curve shows the replotted experimental results of Tazaki et al* [6]. *The dashes show the theory of Lebouvier (Fig. 26 of Ref* 3 *converted to a prediction of mean pressure as a function of* h/δ, *case of zero friction) with shear strengths* Y_0 *(gold)* = *0.0325 GPa and* Y *(glass)* = *0.13 GPa.*

et al [*6*]. A specimen of constant hardness would give a straight line, the hardness being inversely proportional to the square of the gradient. The theoretical curve shows a particularly striking feature. At a certain depth the gradient becomes quite small: the soft film has "spread" the load so that further increases in load produce very little increase in indentation depth! When the indenter finally penetrates further into the substrate, the gradient increases. Although the experimental curve appears to show the effect of the substrate at a smaller δ/h ratio than predicted, the discrepancy is removed with the help of a more refined $\delta/\sqrt{P}$ plot [*7*] taking into account the effects of pileup, and using a corrected value of the effective cone angle for the trigonal indenter, to allow for the fact that the analysis is two-dimensional.

Thus, in principle, by matching this type of curve with the predictions of the model, it should be possible to derive reliable values of film hardness, although other factors will have an influence, such as strain hardening [*8*].

Time-Dependent Effects at the Submicrometre Level

Regarding experimental techniques, this review will emphasize those in which indentation depth, δ, is measured continuously as a function of load, P. Often δ and P are simply related, for example, $\delta \propto P^{1/2}$ for fully plastic deformation with constant hardness, but when δ varies significantly with time as well as load, then even if the rate of loading is kept constant, such simple relations will no longer apply. Accordingly, we shall first discuss time-dependent plasticity at constant load. Measurement of "indentation creep" parameters forms an important part of the characterization of the subsurface region of a specimen. It is also useful [*9,10*] to study the relationship between indentation creep and scratch hardness. The fundamentals of indentation creep have recently been summarized in Maugis's review [*11*] of creep, hot hardness, and sintering in the adhesion of metals at high temperature. Recently, Hooper and Brookes ([*12*], also in this publication) have detected an "incubation period" in lead at high homologous temperatures (≈ 0.5), at which thermal recovery offsets work hardening [*10*]; Naylor and Page [*13*] have listed relevant deformation mechanisms in ceramics. Here we shall be concerned primarily with indentation creep in metals below about $0.2T_m$, where T_m is the melting point on the absolute scale. Several examples [*14,15*] are known of cases in which indentation creep is more significant at submicrometre depths than in macroscopic tests at the same temperature. The continuous depth recording techniques used for loads below 10 mN prove to be well suited to the investigation of indentation creep.

In a constant-load indentation creep test, the stress, of course, decreases with time (unless a flat-ended punch [*16*] is used). Gerk [*15*] was able to calculate the time-dependent hardness in a case in which the relationship between dislocation velocity and stress was known. In general, however, it is important to find the appropriate relationship between the strain rate and the measured rate of increase of the diameter or depth of the plastic indentation: in a hardness test, what is the effective strain? Tabor [*17*] showed empirically that indentation by a Vickers pyramid of metals that are not fully work hardened appears to produce a strain of around 8%: this is in addition to any preexisting strain and applies to the situation at large t, that is, after all the plastic deformation and associated work hardening has taken place. If the load and the indentation depth are then increased, the strain, ϵ, remains constant at around 8%: yet the plastic flow rate may well be significant, in which case $\dot{\epsilon} > 0$! An answer to this paradox was implied by Atkins et al [*18*], who argued on the basis of a model involving an expanding hemispherical cavity that at a distance, r, from the center of the indentation, the strain rate is given by

$$\dot{\epsilon}_r = k_2 a^2 \frac{\dot{a}}{r^3} \tag{1}$$

where a is the radius of the indentation and k_2 is a geometrical constant of value $3/2$ [*17*]. Using a plasticity equilibrium condition that gave the maximum stress, σ_m, as a function of $(\partial\sigma/\partial r)_{t,a'}$, and a theoretical creep expression for $\dot{\epsilon}$ as a function of σ_m and t, they thus obtained a differential equation for the stress, σ, which they were able to integrate over the whole volume (at constant t and $\dot{a}$) in order to obtain the hardness, equal to σ_m, as a function of t. We may note in particular three aspects of the AST analysis:

1. It is not necessary to derive an expression for ϵ itself, for example, as a function of the size of the indentation.

2. $\dot{\epsilon}$ is associated with the rate at which the plastic/elastic boundary moves on into the material.

3. The procedure becomes very complex if no analytical solution to the differential equation for σ exists (this is the case for many creep expressions). However, if we can measure $\dot{\epsilon}_r$ directly with the help of Eq 1, then the rest of the AST procedure is not necessary; we may test any theoretical plastic flow expression directly, without having to calculate the time-dependence of the hardness itself.

Following this simple idea, we made the approximation that the strain rate $\dot{\epsilon}$, as predicted by any *physical* model for creep, may be identified with the value of $\dot{\epsilon}_r$, in Eq 1, at $r = a$, where it has its maximum value. For a fully plastic conical indentation, the depth, δ, is proportional to a, so that we have

$$\dot{\epsilon} \simeq k_2 \frac{\dot{\delta}}{\delta} \tag{2}$$

We describe later the technique of using continuous-depth recording to measure $\dot{\delta}$ as a function of t at constant load (decreasing stress, σ) in a single test, without the need for a series of indentations with different loading times. We thereby observe directly how $\dot{\epsilon}$ varies with decreasing σ. (It should be noted that several approximations are involved: detailed effects of indenter shape, the nonhomogeneity of stresses and strains, and the elastic properties of the hinterland are neglected, and this approach will certainly not be valid in all situations. Matthews [*19*] has shown that for time-dependent indentation by a sphere, involving progressive work hardening, the AST spherical cavity model is inappropriate and a more complex argument should be used, involving differential and integral operators acting on the stress and strain according to the properties of the appropriate rheological model. We do not claim that the model described here is sufficient to explain indentation creep curves from start to finish. However, it does appear to be the simplest model with which the measurements can usefully be compared.)

Regarding plastic flow mechanisms for predicting $\dot{\epsilon}(\sigma, t, T \ldots)$, the literature makes use of the alternative "constant structure" and "steady-state structure" formulations, which refer to microstructural state variables such as dislocation density [*20*]. These variables may, for example, be fixed, or may be uniquely determined by σ and T; in either case we refer to "steady creep."[3] On the other hand, for various reasons (such as work hardening), it

[3] Since *when σ is constant*, $\dot{\epsilon}$ also will be constant. Of course, with indentation creep at constant load, the deformation volume and σ, and hence $\dot{\epsilon}$, are varying with t; but this does not necessarily imply transient creep as defined here and as found by Atkins et al. The available microindentation data are insufficient to justify the extra complication of using a transient creep model.

may be necessary to include t, as an explicit variable, within the expressions describing the state variables and $\dot{\epsilon}$: here we have "transient creep" [$\dot{\epsilon} = f(\sigma, T, t)$]. Indentation creep below $0.25T_m$ has generally been difficult to detect. Our measurements, described later, were performed on nickel at room temperature, where $T < 0.2T_m$; here, at the stress values used, deformation-mechanism maps [*20*] predict that we should observe low-temperature plasticity rather than creep (note that, strictly speaking, the term *creep* is used only when $\dot{\epsilon}$ varies relatively slowly with σ, the dependence being typically power-law; in low-temperature plasticity, the dependence is exponential. Figure 2 illustrates the distinction. However, for simplicity we use the term *indentation creep* throughout).

We next examine the consequences of the simplest low-temperature creep model, based on a dislocation glide mechanism in which the dislocation velocity is limited by weak obstacles (activation energy barrier $\lesssim 0.2 \times \mu b^3$, where μ is the shear modulus and b the Burgers vector). Following Frost and Ashby [*20*], the strain rate is then given by

$$\dot{\epsilon} = \nu_0 \exp\left\{-\frac{\Delta F}{kT}\left(1 - \frac{\sigma}{p_0}\right)\right\} \tag{3}$$

where ν_0 is a frequency ($\approx 10^6\ \mathrm{s}^{-1}$), ΔF is the activation energy characterizing the strength of a single obstacle, k is Boltzmann's constant, and p_0 is the athermal yield pressure. Replacing $\dot{\epsilon}$ by $k_2\dot{\delta}/\delta$ (Eq 2), and defining a *characteristic time*, t_c, and a *characteristic distance*, δ_c by

$$t_c \equiv \frac{k_2}{\nu_0} \exp\frac{\Delta F}{kT}; \qquad \delta_c \equiv \left(\frac{2\Delta FP}{kTk_1p_0}\right)^{1/2} \tag{4}$$

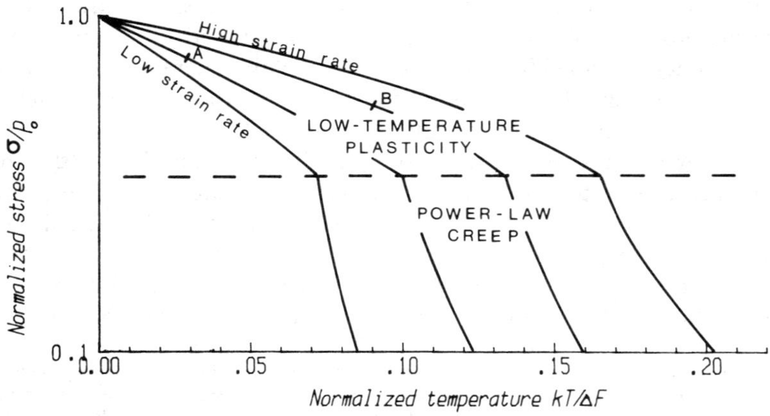

FIG. 2—*Schematic temperature/stress deformation map, following Ref* 20 *but with different normalizations.*

we obtain

$$\ell n \frac{\dot{\delta}}{\delta} = -\ell n\ t_c + \frac{\delta_c^2}{2\delta^2} \tag{5}$$

where k_1 is the geometrical constant relating δ^2 with the projected area of indent if we assume fully plastic deformation by a pyramidal indenter [$\sigma = P/(k_1\delta^2)$]. Thus, using the technique of continuous depth recording, from direct observation of δ and $\dot{\delta}$ at a known constant load, we may determine t_c and δ_c, and hence the basic creep parameters ΔF and p_0($k_1 p_0$ may be regarded as an "athermal yield index").

To illustrate the significance of the empirical parameters t_c and δ_c, we refer to Fig. 3, which shows how δ varies with t according to numerical solutions of the differential Eq 5. After an initial steep rise, δ increases more slowly, the graph being approximately linear during the period $0.1t_c < t < t_c$. During

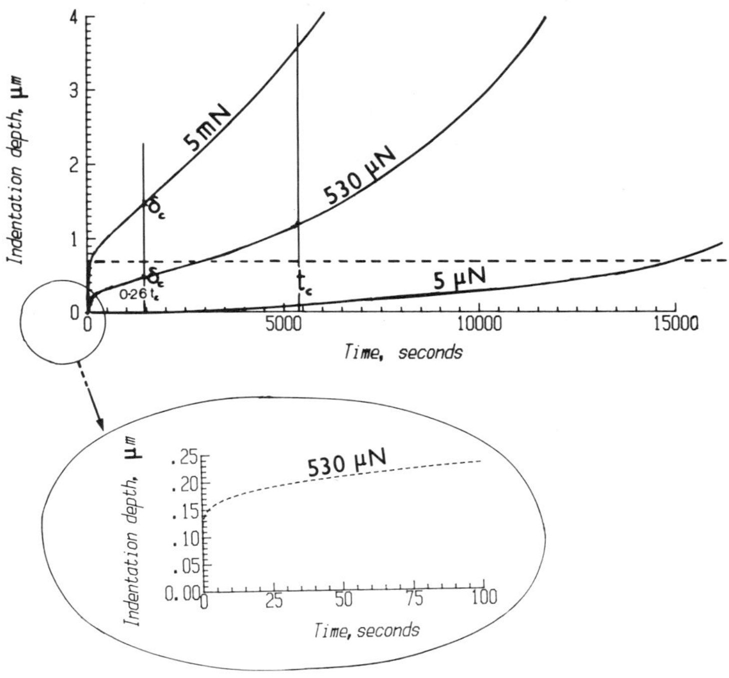

FIG. 3—*Theoretical* δ/t *curves for low-temperature indentation creep, calculated according to the following values of material parameters: athermal yield index* $k_1 p_0$ = *95 GPa; normalized activation energy* $\Delta F/kT$ = *21.32; preexponential term* $k_2/\nu_0 = 3 \times 10^{-6}$ *s. The value of the load is marked against each curve (the lower diagram is an expanded plot of the initial part of the 530-µN curve, which matches the experimental data of Fig. 9 at small* t*). The vertical lines show* t_c *and* δ_c.

this period, δ passes through the value δ_c, which is in fact the point at which $\dot{\delta}$ is at its minimum. In this elementary form, the model thus leads to two predictions: first, whatever the load, a time $t \gg t_c$ and a depth $\delta \gg \delta_c$ will, at length, be reached at which the stress term in Eq 3 becomes negligible, $\dot{\epsilon}$ becomes constant, and δ increases exponentially with t. This surprising result is somewhat unlikely to be seen in practice, however, owing to work hardening at great depths and the inaccuracy of Eq 2. A more important prediction is that if we wish to define the hardness at a given load or depth, it is best to restrict the conditions of the corresponding hardness test so that $0.1t_c \lesssim t \lesssim t_c$. As shown in Fig. 3, within this region the hardness will show the least amount of variation with loading time [at $\delta = \delta_c$ its value, $P/(k_1\delta_c^2)$, will be $\frac{1}{2}kTp_0/\Delta F$]. Further complications arise if we wish to investigate how the hardness varies with load or to interpret the results of microindentation tests in which δ is measured as a continuous function of P. Even if the material properties are independent of depth, the expected $\delta \propto \sqrt{P}$ relation will not be obtained if creep is significant. Furthermore, unless the indentation depth can be expressed in terms of separable functions of stress and time, the hardness, even if defined for a particular (constant) value of loading time, will not be independent of load. In such cases, we suggest that a full characterization of the material should include quantitative measurements of the indentation creep behavior and, where possible, determination of the values of the creep parameters, as in the example given later. These conclusions will be valid also in cases where, for example, a constant-viscosity or other creep model applies.

Elastic Recovery

The last of the three mechanical effects that particularly concern us at small indentation depths is the partial recovery in the depth of the indentation as the load is removed, as a result of the release of stored elastic strain energy. Other papers in this publication analyze how such energy release provokes fracture in brittle materials [*21*] and how it may be quantitatively determined from the measurement of elastic recovery [*22*]. Accordingly, here we restrict ourselves to a summary of how we derive the relevant elastic recovery parameter from depth/load curves. Since there is no complete theory of elastoplastic indentation, we make the following simplifying assumptions: (1) that a rigid pyramid (geometrically equivalent to a cone of semiangle ψ) is indenting an isotropic solid under conditions in which, on loading, any elastic displacement of the hinterland does not significantly affect the value of mean pressure, p_m; (2) that the specimen is fully work hardened, with a high enough ratio of Young's modulus, E, to flow stress, Y, that p_m (which equals H, the hardness) approaches a value of $3Y$ (full plasticity in a polycrystalline material). Under these conditions, we may still detect elastoplastic rather than ideal rigid-plastic behavior. A direct means of assessing the relative contributions of the two components of deformation during indentation—irreversible and reversible—is through the measurement of elastic recovery in the depth of the indentation as the load is slowly reduced to zero (the recovery in the diameter is very much less but with Knoop indentations may still be measured [*23*]).

We use the notation of Newey et al [*24*] to describe a typical depth/load cycle (Fig. 4). In principle, it is possible to define at least four types of "hardness" whose values may be derived from such curves, knowing the indenter geometry and the maximum load applied:

(*a*) an on-load or unrecovered hardness, involving δ_T,

(*b*) an off-load or recovered hardness, involving $\delta_T - \delta_e$,

(*c*) a refined value of off-load hardness (theoretically closer to p_m), involving $\delta_T - \delta_e'$, which allows for the fact that a small but significant recovery in the *diameter* begins at Point A, and

(*d*) a hardness close to the value obtained by the classic optical measurement of the diagonal (this derivation takes account of pileup).

The derivations are complex [*22*], and here we are concerned simply with the most useful choice of recovery parameter whose value may be obtained by direct measurement. We consider two alternatives, one from Lawn and Howes [*25*] and one based on the work of Loubet et al [*22*]. The first assumes that the elastic unloading is described throughout by the contact of a rigid cone with an elastic half-space. With a minor approximation involving possi-

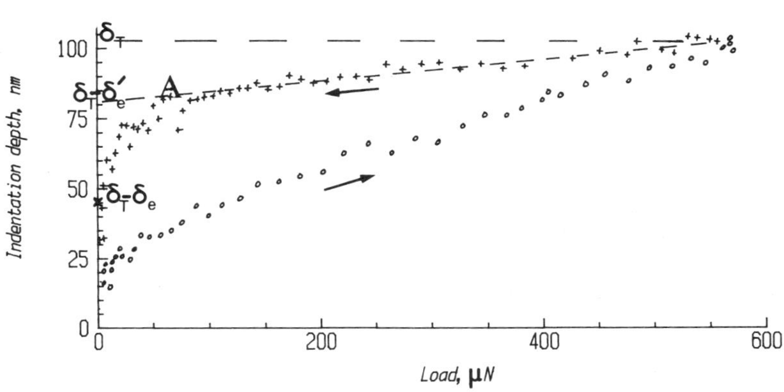

FIG. 4—*Depth/load test (150-nm-thick nickel film on silicon substrate). Loading (○) and unloading (+) curves are shown by arrows.* δ_T *is the total indentation depth due to both plastic and elastic deformation;* δ_e *is the recovery in the depth when the load has been removed;* δ_e' *is a modified value of depth recovery, obtained by extrapolation of the linear portion of the unloading curve* [22].

ble changes in the level of the contact perimeter, we may write the Lawn and Howes recovery parameter as

$$1 - \frac{(\delta_T - \delta_e)^2}{\delta_T^2} = \frac{H}{E} \times 2(1 - \nu^2) \tan \psi \tag{6}$$

(ν is the Poisson ratio of the material). However, an alternative assumption is that the elastic unloading corresponds to that of a *blunt punch*, whose constant area, πa^2, was determined by the plastic flow during loading; this implies that the elastic recovery occurs primarily in the hinterland beyond the plastically deformed zone. In practice, as has been shown in certain cases [*22*], observed unloading curves tend to lie between these two theoretical extremes. We may escape from this unsatisfactory situation by defining our recovery parameter in terms of δ_e' rather δ_e (Fig. 4), in cases in which we obtain an initially linear portion of the unloading curve, indicating punchlike behavior, with recovery only in the hinterland. (To the left of Point A in Fig. 4, the indent diameter begins to decrease [*22*]). For a maximum load, P_1, we have

$$\begin{aligned} P_1 &= \pi a^2 H = \alpha \tan^2 \psi \delta_T^2 H \\ &= \frac{E}{1 - \nu^2} 2a\delta_e' \end{aligned} \tag{7}$$

where α is the appropriate geometrical factor. Thus, the simple ratio of δ_e' to δ_T defines a recovery parameter, R, that again is independent of load: since $\delta_e' = \pi a H(1 - \nu^2)/(2E)$, we have

$$R \equiv \frac{\delta_e'}{\delta_T} = \frac{H}{E}\sqrt{\frac{\alpha\pi}{4}}(1 - \nu^2) \tan \psi \tag{8}$$

Only if the elastic recovery is very small does Eq 6 take the same form as Eq 8, which is probably the more useful of the two equations. In either case, the basic parameter derived is $H \tan \psi/E$. The model could usefully be refined with the help of X-ray topographic observations of the strain field surrounding the indentation [*26*].

Surface and Scale Effects

We have seen how the measured properties of submicrometre surface layers depend on the mechanical consequences of a variation of indentation stress with depth or with time. Why should this stress differ from the macroscopic bulk value? The physicochemical mechanisms involved may be divided (somewhat arbitrarily) into two categories: the effects arising from the proximity of the surface and the effects of scale. Regarding surface effects, space does not allow us to discuss the well-established chemomechanical effects [*27,28*]. When both the specimen and the indenter surfaces are sufficiently free from contamination, a less well known type of surface effect becomes important, namely, the increase in effective load arising from intermolecular attraction between the two solids. The case of a spherical indenter has been analyzed and the experimental evidence reviewed [*29*]: a size effect is involved, in that the smaller the value of the radius of curvature, *r*, the larger the effective force of attraction compared with the load required to give plastic deformation. Even at zero externally applied load, if the Dupré energy of adhesion for the interface is large enough or *r* is small enough, plastic deformation will be initiated, or even full plasticity will be obtained, so that the measured hardness will tend to zero at very small loads (Fig. 5) [*30,31*]. In addition to such surface effects, in many materials the bulk mechanical properties may vary significantly with the size of the indentation—in other words, below a certain size we no longer have a constant load/area ratio (hardness) even for pyramidal indentations. In a recent review, Atkins [*32*] has discussed such size effects. Various mechanisms have been described, including the ductile/brittle transition at small depths [*32–34*], the effects of microstructure in bulk specimens [*35*] and thin films [*36–38*], and the depth-dependent behavior of dislocations in single crystals [*14,39,40*].

It is not yet clear which of these mechanisms will be dominant in a particu-

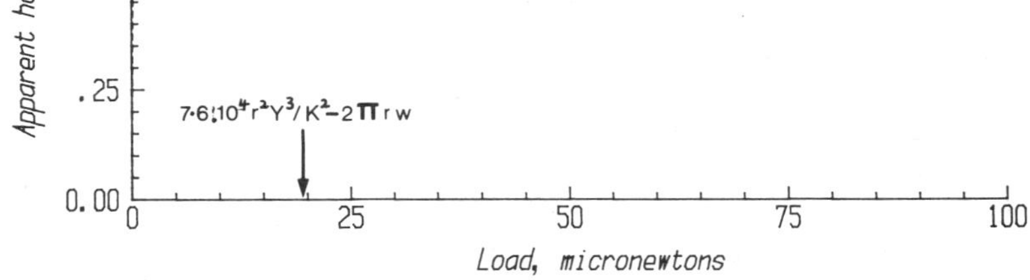

FIG. 5—*Theoretical variation of apparent mean pressure* $P/(\pi a^2)$ *with load* P, *for a specimen of yield strength* $Y = 0.33$ *GPa and with the energy of adhesion* $w = 3$ *J* m^{-2} *(*a *is the radius of the contact circle):* dotted curve*: from the equation* $P + 2\pi rw = 3\pi a^2 Y$ [30] *(the hemispherical indenter has a radius* $r = 0.3$ *μm; the initial load for full plasticity was calculated for a value of the relevant elastic modulus* K *of 100 GPa);* dashes*: from the equation* $P + 2wk_l\delta = 3k_lY\delta^2 = 3\pi a^2 Y$ [31] *(conical indenter with* $k_l = 2.6$*). The measured hardness will decrease at small loads, as a result of the force of attraction between the specimen and the indenter surfaces.*

lar situation. Very few measurements have been carried out at loads of less than around 1 mN (100 mg). Within this range, Kinosita and colleagues [*6*] in 1978 described an interesting phenomenon known as the critical load effect. Below a certain load, P_c, which varied from around 350 μN for magnesium fluoride (MgF) to 40 μN for gold, no permanent indent was detected. However, for various metals, when in the form of films of about 600 to 1100-nm thickness, the critical load effect disappeared, with $P_c \leq 10$ μN. Newey et al [*41*] found a significant critical load effect in electropolished iron foils and lapped iron disks, with $P_c \approx 200$ and 650 μN, respectively; if the disks were implanted with 190-keV titanium ions to a fluence of 4.6×10^{20} m^{-2}, the effect disappeared, as with the thin films of Tazaki et al [*6*], who ascribed the critical load effect to the finite radius of curvature ($r \approx 120$ nm) of the tip of their pyramidal indenter. They claimed that the critical load simply represents the load at which the Hertzian elastic stress reaches the elastic limit of the material. This explanation is surprising, since for polycrystalline copper, gold, and silver, taking into account an error of a factor of two in their Eq 5, their corresponding derived values of yield stress, Y, range from 30 to 60 times the hardness determined at higher loads (if Y were independent of load, we would expect a ratio of about 3). At the elastic limit, equating $P_c = K\,a^3/r$ to $P_c = 1.1\ \pi a^2\ Y$, we have

$$Y = \frac{1}{1.1\pi}\left[\frac{P_c K^2}{r^2}\right]^{1/3} \qquad (9)$$

(all the symbols as previously defined). If we postulate a larger effective value of r, a more reasonable value of Y is obtained; this could arise, for example, from an oxide or contaminant layer that "spreads" the load, but the critical load effect was seen with materials such as calcium fluoride (CaF_2), for which this explanation is inapplicable. For the specimens tested in the present study, critical loads were very small or absent (<10 μN). However, differences between specimens such as those reported in the earlier literature could perhaps require a mechanical explanation, of the type discussed in the earlier section on the effects of variation of yield strength with depth: in this connection we recall that the sudden yielding of gold indented by tungsten, as observed by Gane with the scanning electron microscope (SEM) [*39*], took place only in the presence of a thin surface layer formed by polymerization of pump oil vapors present in the microscope. Full answers to the question of when a clean surface is harder or softer than a contaminated one, and when the near-surface layer is harder than the bulk, are unlikely to be found until more microindentation experiments are carried out under conditions in which the specimen may be cleaned and characterized by means of surface physics techniques and in conjunction with direct observation of dislocation distribution and movement.

Some Examples of Measurements at the 10 to 100-nm Level

Penetration Depth as a Function of Load

The principal techniques for mechanical characterization at submicrometre depths include the following: scratch hardness testing, the measurement of the distribution and movement of dislocations as a function of applied load, *in situ* indentation in a scanning electron microscope, which is used to measure the indent diameter, and microindentation with continuous recording of penetration depth as a function of load or of time. This last technique [*24,42,43*] is a development of earlier work involving continuous recording at greater depths [*44–46*] or single-valued measurements [*47*]. Here we present some examples of how this method may be used to derive numerical values of plasticity indexes, elastic recovery parameter, creep parameters, and critical load. The equipment is shown schematically in Figs. 6 and 7: the test procedure follows that of Newey et al [*24*] except that in addition, the specimen can be transferred between two locations (test and microscopic observation). A particular area of interest may thereby be identified in the microscope and then transferred to the test position, or relocated, by means of a computer-controlled three-axis (*X*-*Y*-*Z*) movement. Scratch hardness testing with measurement of frictional force may also be performed at speeds as low as 30 nm/s. The indentation depth is measured with a capacitor bridge arrange-

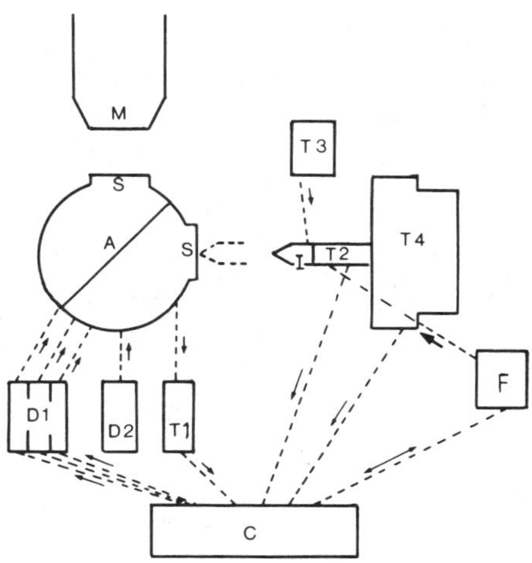

FIG. 6—*Microindentation hardness tester (schematic diagram): S, two specimen positions; A, rotation axis of specimen turret; M, microscope objective; D1, drive for three-axis movement (stepping motors); D2, additional fine-scale transverse movement for scratch hardness and friction tests; I, indenter; F, electrostatic system for application of force; T, transducers for control or measurement (T1, transverse displacement; T2, frictional force; T3, piezoelectric control of indenter position; T4, indentation depth); C, computer.*

FIG. 7—*Specimen turret and indenter assembly: I, indenter holder; F, electrostatic system for application of force; C, capacitative displacement transducer; P, piezoelectric control; S, specimens.*

ment, which is quickly recalibrated whenever the initial indenter position is altered, with the help of piezoelectrically controlled movement. The load of between 10 μN and 30 mN is applied to the three-faced pyramidal diamond indenter, for which a 90° angle between edges is chosen. For such a sharp indenter, friction between it and the specimen is likely to be significant, but this disadvantage is often outweighed by an earlier finding [*41*] that, for rougher surfaces, sharp indenters give more consistent results. The resolution of the depth measurement is estimated to be better than 1 nm, but both specimen roughness and vibration tend to limit, particularly, the accuracy with which the moment of contact (that is, the zero of the depth scale) can be identified. In practice, the zero of *plastic* indentation depth can be determined with reasonable precision, as described in the following.

A typical experimental procedure is as follows. First, we obtain a preliminary curve of indentation depth, δ, against time, t, the applied load, P, being held constant at a value approximately equal to the maximum load of interest (Appendix). The δ value always increases rapidly during the first few seconds: if, thereafter, its rate of change, $\dot{\delta}$, remains higher than, say, 0.3 nm s^{-1}, a quite different "indentation creep" procedure is essential, as described later. Otherwise, we next carry out several plots of δ as a function of increasing P on different points on the specimen surface, keeping the loading rate the same for each plot. A different maximum load is used for each plot, so that we obtain a series of loading-unloading curves whose hysteresis area, W, corresponds to the plastic work performed on the specimen. If we are interested in the off-load hardness, then, as discussed earlier, we can obtain from the unloading curves values of δ_p defined as $\delta_T - \delta_e'$ (Fig. 4). The loading curves are all superimposed and averaged by computer. Because of the uncertainty, mentioned earlier, in the zero of δ, where possible it is best to plot graphs with δ or δ_p itself (rather than some function of δ) as the abscissa: in the simplest case of a specimen showing fully plastic behavior with hardness, H, independent of δ, we would expect $P \propto \delta_p^2$ and $W \propto \delta_p^3$ [*22,48,49*]. We normally plot δ_p against $P^{1/2}$ and, where necessary, δ_p against $W^{1/3}$. By extrapolating the $W^{1/3}$ versus the δ curve to zero load, we thereby obtain the δ-zero at which plastic deformation begins, and also any critical load, an example is shown in Fig. 8. In the simplest case, both graphs are linear for at least part of the range of load and extrapolate to give the same δ-intercept at zero load. To enable absolute values of hardness to be obtained, the indent area may be measured by electron microscopy [*50*], but if time does not permit the use of this delicate procedure, the gradients of the two types of curve allow one to derive values of three "plasticity indexes" (relative hardnesses)

$$I_P' \equiv \left(\frac{d\sqrt{P}}{d\delta_p}\right)^2; \qquad I_W' \equiv \left(\frac{dW^{1/3}}{d\delta_p}\right)^3 \tag{10}$$

or, in nondifferential form, $I_P \equiv P/\delta_p^2$; $I_W \equiv W/\delta_p^3$; $I_H \equiv P^3/W^2$.

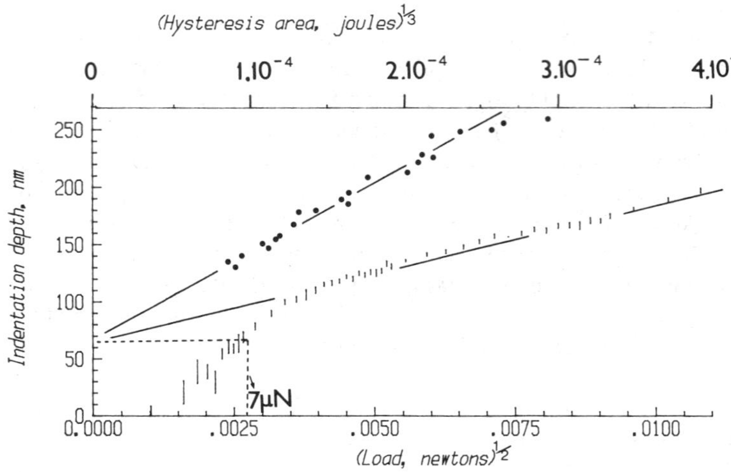

FIG. 8—*Microindentation results at very low load (platinum foil, thickness 7.5 μm, surface cleaned with* n-*propanol only):* ●, *$\delta/W^{1/3}$ points; vertical bars show $\delta/P^{1/2}$ points (each bar represents the limits of error for an average of typically 5 to 30 data points from loading curves); loading rate, 2.5 μN/s; creep rate, δ, at maximum load after 5 s, <0.2 nm/s; derived values of plasticity indexes,* I_p *= 7.7 GPa and* I_w *= 2.5 GPa; critical load, 7 μN.*

In the simplest case of full plasticity, $I_P = 3I_W$, and in addition, for a 90° pyramidal indenter, we have the hardness $H = I_P/2.6 = I_W/0.87$. Newey et al [*24*] first showed that hardness values thus obtained could agree with the values obtained by optical measurement of indent diameter at large depths to within 10 to 20%. Figure 8 shows the results of measurements performed on platinum foil. Although the two curves do not show perfect linearity, they give values of the δ_p-zero and of H which are in good agreement, with $I_p/I_w = 3.1$ compared with the theoretical value of 3. From the unloading curves, we can also calculate the recovery parameter, defined earlier, but in the case illustrated it has a very small value ($\lesssim 0.05$), the unloading curves being nearly horizontal. Later we show some results for nickel films grown on a silicon substrate, where elastic recovery is much more significant: for example, for a film thickness of 150 nm, the results shown in Fig. 4 give a value of $\delta_e'/\delta_T \approx 0.22$. From Eq 8, taking $k_1 = \alpha \tan^2 \psi = 2.6$ for the 90° indenter, we have $H(1 - \nu^2)/E = 0.15$. Thus, the unloading curves may be used to calculate the values of the elastic modulus if H has been determined from the loading curves. A similar calculation applied to measurements made with a material of known Young's modulus (Lucite/Perspex) gave an E value of 3.0 GPa [*31*], in agreement with the known value of 2.4 to 3.5 GPa.

Penetration Depth as a Function of Time

We have pointed out that when time-dependent effects, as assessed in the preliminary constant-load test, are significant, in general the material will not have a constant "hardness" even if this is defined for a particular value of loading rate or loading time. As a result it turns out that the simple analysis we have just described is inadequate even for metals such as nickel and Type 304 steel at room temperature, when indented to depths below about 1 μm. Figure 9 shows the results of a typical creep test at constant load. (The Appendix gives some details of this experimental procedure.) The specimen was a thick deposit of nickel (thickness 600 nm) on silicon. Such curves show that even at homologous temperatures below 0.2, we must at least take into account the possibility of creep when interpreting microindentation hardness measurements. The first 50 s of the experimental curve shown resembles in shape the theoretical curve for the same load in Fig. 3. The same is true for a wide range of loads and of specimens, including bulk polycrystalline nickel of a grain size 100 times larger than that of the thin film specimens and nickel implanted with ions of boron or phosphorus, and accordingly we have attempted some preliminary theoretical interpretation of such data. For example, Fig. 10 shows the results of Fig. 9 replotted in a form which would give a straight line if Eq 5 were obeyed. (We note, incidentally, the advantage of being able to measure $\dot{\delta}$ directly as well as δ in the continuous depth recording method, especially as there is no analytical solution to Eq 5.)

As δ increases, the logarithm of the strain rate varies linearly with the stress until a depth of around 170 nm is reached (approximately 40% of the final depth). In this region, the value of the characteristic time, t_c, is 5400 s, as determined from the intercept, and the characteristic distance, δ_c (the square root of twice the slope), is 0.5 μm. At $\delta \approx$ 170 nm, the rate of indentation slows down greatly, but long before t_c is reached, the results diverge from the predicted curve (Fig. 3); as the stress decreases further, with δ increasing beyond 170 nm, the curve in Fig. 9 rises more steeply, and it is tempting to try

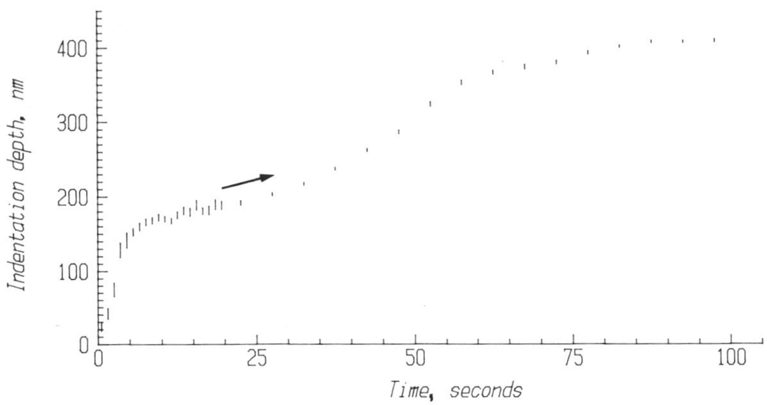

FIG. 9—*Low-temperature indentation creep: δ/*t *results for thick nickel deposit (*h = *600 nm) on silicon, for a constant load of 530 μN.*

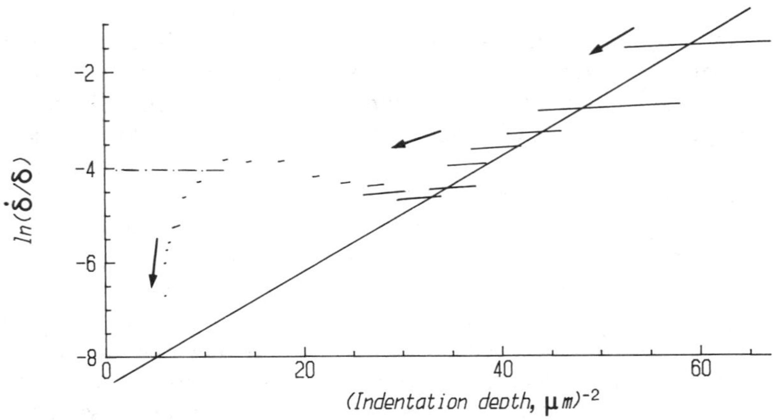

FIG. 10—*Low-temperature indentation creep. The data of Fig. 9 are replotted, giving* X-*values proportional to the applied stress and* Y-*values proportional to the logarithm of the strain rate.*

to identify this rise with the predicted transition to exponential behavior at $\delta \gg \delta_c$.

Figure 10 shows, however, that at this stage we have a striking departure from the simple theory in that the strain rate, $\dot{\epsilon}$, or its logarithm, no longer varies linearly with the stress; instead, it levels off and even increases with time, so that the slope becomes negative. Such behavior may be termed viscosity-decrease or "accelerated creep," and it is characterized by the value of the y-intercept, $(-\ell n\, t_c)$, in Fig. 10. In this case $(-\ell n\, t_c)$ has decreased from 8.5 to 4 at $\delta \approx 200$ nm. Finally, and possibly as a result of work hardening, $\dot{\delta}$ (and $\dot{\epsilon}$) fall toward zero, at a depth of about 400 nm in the example shown. (We have not yet investigated how this "final" depth is related to the deposit thickness. These effects are seen also with bulk nickel specimens.) Often the times required are so long ($\gtrsim$ a few hundred seconds) that this range of depth and stress would not be accessible in practice, in the first type of microindentation hardness test (depth as a function of load). One would have to wait so long between each increment of load that errors due to drift in the apparatus could be significant, and the total time would be unacceptably long.

As yet we have no confirmatory evidence that Eq 5, with the simple dislocation glide model, does apply, and instead a simple constant-viscosity model gives a better fit to data for certain types of specimens. It is nevertheless interesting to use the values of intercept and slope to obtain effective values of activation energy and athermal yield index and to consider the transition to accelerated creep in terms of the corresponding deformation map. From Eq 4, and assuming $\nu_0/k_2 = 0.3 \times 10^6$ [*20*], and then from the preceding values of t_c and δ_c, the transition to accelerated creep involves a decrease in $\Delta F/kT$ from 21 to 17, and the initial value of $(k_1 p_0)$ is 95 GPa. This transition is shown schematically on the deformation map as a shift from A to B in Fig. 2. Although this stress decreases, the strain rate increases, as if there had been a large rise in temperature. We find that this accelerated creep occurs with tests on a range of different specimens. [In most cases, the athermal yield index $k_1 p_0$ appears to increase greatly (small positive slope of $\ell n\,(\dot{\delta}/\delta)$ versus δ^{-2}), while in others, as in Fig. 10, the slope becomes negative, so that the concept of athermal yield index breaks down.] In conclusion, it appears that with specimens for which indentation creep is significant, constant-load δ/t tests should be performed, and characterization should include the value of H (or P/δ^2) calculated from the limiting work-hardened value of δ at very large t, together with the values of t_c and δ_c derived from creep plots such as Fig. 10. It is also worth attempting to perform $\delta/\sqrt{P}$ tests with a rapid loading rate: if in practice (as is often the case) I_p does not vary excessively with the loading rate, this parameter may be regarded (with some suspicion) as a measure of the material's plasticity before accelerated creep or work hardening takes place.

Some Measurements of Thin-Film Specimens

We will return later to the subject of indentation creep, as seen in nickel and steel, both as received and implanted with ions of various species. Before attempting to analyze measurements made on a material such as an implanted specimen whose hardness was expected to vary rapidly with depth, we wished to know whether the macroscopic theory of the indentation of a bilayer (as summarized earlier) was valid at this scale. We therefore decided to carry out both depth/load and creep tests on nickel films of varying thickness on a hard substrate (silicon), as shown in Fig. 11. The films were prepared by Takadoum and Pivin.[4] Indentation creep at $\delta_p \gtrsim 100$ nm prevented us from obtaining reliable values of the plastic deformation index, I_p, at larger depths, and accordingly we present the values for two depths only (25 and 50 nm) as a function of thickness, h. Significant elastic recovery was also observed, and details will be published elsewhere [*51*]. The minimum I_p values are two to three times greater than for bulk polycrystalline nickel, as described later. It appears that for experiments on this scale, that is, with film thickness, h, considerably smaller than that of Fig. 1, the marked rise in hardness with decreasing h predicted by Lebouvier still occurs, but at a value of $\delta/h \approx 0.3$, less than the expected value. As we have seen, he predicts that this rise would begin at even smaller values of δ/h when the film is harder than the substrate. Thus, in the case of a hard surface layer, say 200 nm in thickness, produced by ion implantation, the hardness values derived from the measurements obtained at depths of more than a very few tens of nanometres will not be truly characteristic of the implanted region.

[4]Dr. J-C. Pivin and J. Takadoum, Centre for Nuclear Spectrometry and Mass Spectrometry, 91405 Orsay, France.

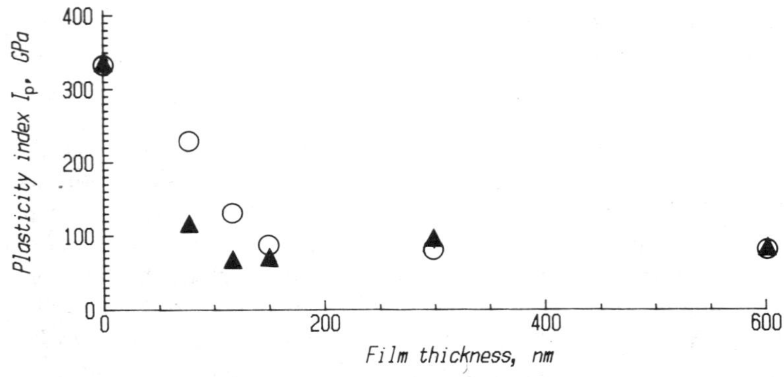

FIG. 11—*Variation of plasticity index with film thickness (nickel deposited from the vapor onto silicon substrates: substrate temperature, 320 ± 20°C; vacuum, (2 to 10) × 10^{-6} torr; deposition rate, 0.1 to 0.2 nm/s);* ▲, *values at δ_p = 25 nm;* ○, *values at 50 nm (nondifferential definition of* I_p*).*

Effects of Ion Implantation

Nevertheless, such measurements have yielded a useful indication of the mechanical properties of the combination of implanted layer plus underlying solid. As reviewed by Hartley [52], standard microhardness techniques have been used successfully to measure changes due to implantation in metals. (For more recent work see also Refs *53* and *54*.) Roberts and Page [*55*] have used an empirical "Meyer index approach," involving the measurement of the variation of hardness with depth, as an indirect way of investigating plasticity of implanted layers in silicon and silicon carbide. The continuous depth recording technique has been used by Newey et al [*41*] and later by Pethica et al [*56,57*] to detect increases in hardness in metals. Our method has recently been used to measure the various plasticity and creep parameters discussed earlier, in nickel [*51*] and in Type 304 steel [*58*], together with the effects of implantation by boron and phosphorus (into nickel) and nitrogen (into steel). The nickel specimens (Ni200), prepared by Pivin and Takadoum, were polished with 1-μm abrasive and then electropolished. The Type 304 steel specimens (austenitic $Fe_{18}Cr_8Ni$) were prepared by Singer and Bolster.[5] After the specimens had been polished with 0.2-μm abrasive, producing a known composition of strain-hardened austenite or martensite as a function of depth, various thicknesses were then removed by ion erosion. Finally, the specimens were implanted, as detailed in Figs. 12 and 13, which illustrate the principal features of the indentation creep and depth/$(\text{load})^{1/2}$ curves obtained. These were analyzed according to the methods detailed earlier. In view of the simplifications involved in the creep model, as discussed earlier, numerical values of

[5] Drs. I. L. Singer and R. N. Bolster, Chemistry Division, Naval Research Laboratory, Washington, DC.

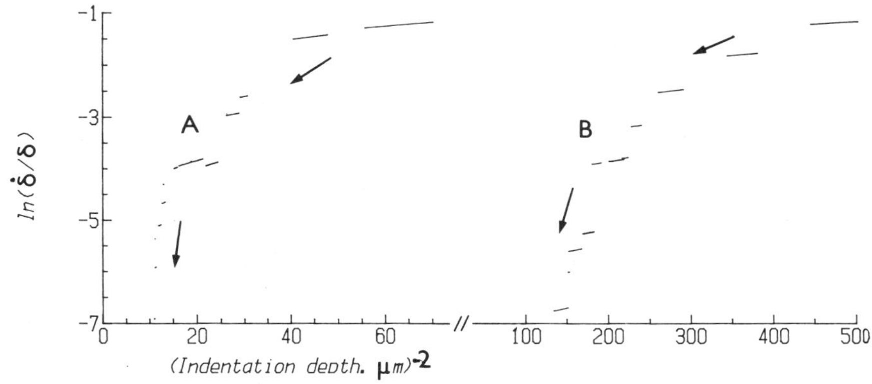

FIG. 12—*The effect of boron–ion implantation on indentation creep in bulk polycrystalline nickel (axes as in Fig. 10) at a constant load of 325 μN: Curve A, nonimplanted; Curve B, implanted with 70-keV boron ions; fluence, 1 × 10^{21} m^{-2}; peak range, 61 nm. The final values of indentation depth, at large* t, *differ by a factor of 3.5.*

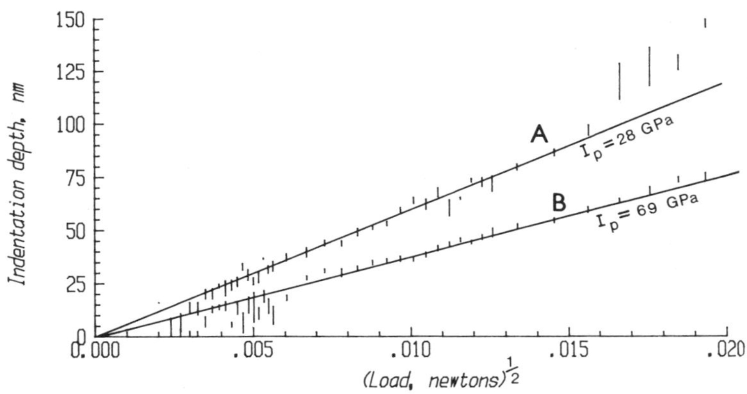

FIG. 13—*The effect of nitrogen–ion implantation on* I_p' *for Type 304 steel, unstrained γ-austenite structure, with a thickness of 380 nm removed by ion milling after polishing: Specimen A, nonimplanted; Specimen B, implanted with 40-keV nitrogen ions: fluence, 2 × 10^{21} m^{-2}; peak range, 43 nm. With the nonimplanted specimen, severe creep (giving departure from linearity) is seen at higher loads.*

creep parameters must be regarded as provisional. We may nevertheless summarize the direct conclusions as follows:

1. For unimplanted (and non-strain-hardened) specimens, a viscosity decrease or "accelerated creep" region of the curve, as described earlier, is seen when a certain depth has been exceeded.
2. When the suggested creep analysis is carried out, the values of the derived parameters indicate the following:

(*a*) the characteristic time ranges from 50 to more than 10 000 s, implying a corresponding variation in activation energy;

(*b*) in the "accelerated creep" range, its value is reduced by between one and two powers of ten but, provided that this range is avoided (that is, at small depths), it is still possible to obtain reproducible $\delta/\sqrt{P}$ curves and thus to derive values of plasticity index;

(*c*) with the nonimplanted (and unstrained) specimens, we have detected no significant variation in characteristic time, t_c, athermal yield stress, p_0, or plasticity index, I_p (and thus, of hardness), as a function of load; and

(*d*) under certain conditions, a better linear fit results from plotting the strain rate itself, rather than its logarithm, against stress (constant-viscosity effect).

3. At a final limit of depth, δ_f, the creep rate tends toward zero or toward a value too small to detect. The time involved is often inconveniently long, but it is possible to derive a single value of hardness or of plasticity index P/δ_f^2 (denoted by I_f).

4. The effects of ion implantation (above a certain value of fluence) may be summarized as follows:

(*a*) it tends to reduce greatly the range of depth at which accelerated creep is observed. Often accelerated creep is eliminated completely;

(*b*) generally we do not otherwise detect any significant change in t_c; and

(*c*) there is a threefold hardening effect. Both I_p and p_0 tend to increase; the final limit of indentation depth at a given load falls, so that I_f also increases. These effects are particularly marked in the case of boron-implanted nickel and nitrogen-implanted steel (Figs. 12 and 13). Details of the minimum fluence required, and the effect (if any) of increasing fluences, are described elsewhere [*51*,*58*].

As discussed in Singer's review of the tribological properties of ion-implanted metals [*59*], while surface hardness is important in wear resistance, it is not necessarily a controlling factor. The work-hardening rate and creep properties may play equally important parts.

Conclusions

1. Materials may be characterized at depths in the 10 to 300-nm range, through measurements of plasticity indexes, elastic recovery parameter, elastic modulus, and indentation creep parameters. For certain types of specimens, there appears also a critical load for initiation of plastic deformation.

2. Significant low-temperature indentation creep at small depths can occur in metals even at ambient temperature. When this is so, attempting to derive values of the hardness of the material is of little use. Instead, it is more profitable to measure empirical creep parameters such as characteristic time and depth.

3. The method of continuous depth recording is simple and rapid compared with methods involving measurements of indent diameter. This applies to both experimental and theoretical interpretation, especially when elastic recovery or indentation creep is observed.

4. The effect of ion implantation on creep behavior can be as important as its effect on the hardness itself. For example, in nickel and Type 304 steel, it can eliminate the "accelerated creep" which otherwise appears when the indentation depth reaches 100 to 300 nm.

5. The plasticity indexes may be related to macroscopic hardness values and the creep parameters to activation energy and athermal yield stress, if we make certain simplifying assumptions regarding full plasticity and low-temperature creep mechanisms. Both experimental work and the required theory are at an early stage, especially when surface or scale effects are involved or when the yield stress varies with the depth.[6]

Acknowledgments

We are grateful to Dr. J-C. Pivin and J. Takadoum for prompting us to undertake a large part of this work, to them and to Dr. I. L. Singer for discussions and the provision of specimens, to Dr. D. Lebouvier for valuable depth-load calculations and comments, and to Dr. R. S. Irani of the U.K. National Physical Laboratory for suggesting improvements to our apparatus. The research reported herein has been sponsored in part by the U.S. Army through its European Research Office. Pollock is also grateful to the French Centre National de la Recherche Scientifique for enabling him to work as visiting associate researcher in their Mechanics of Surfaces Research Group.

APPENDIX

Constant-Load Test Procedure

For kinetic and impact effects to be minimized, the moment of inertia of the indenter assembly should be as small as possible (1.7×10^{-5} kg · m^2 for our instrument). The full constant load is applied to the indenter, which is held away from the specimen by means of the piezoelectric control. This is then used to allow the indenter to approach the specimen at a typical final speed of 50 to 200 nm s^{-1}: within this range, varying the approach speed does not appear to affect the results. After contact, as soon as $\dot{\delta}$ falls below this value, the piezoelectric control continues to move forward and therefore plays no further part in the test. Thermal drift in the apparatus can be

[6] Jönsson and Hogmark [*60*] have recently measured the difference in load needed to produce a given value of indent diagonal on a substrate (steels, copper) with and without a chromium film, where $\delta \gtrsim h$. They derived values of film hardness for thicknesses down to 200 nm, by means of a simple model consistent with their finding that this load difference is proportional to the diagonal.

especially important in creep tests, which generally require that the specimen first be left for several hours to stabilize. If after, say, 500 s of a test, $\dot{\delta}$ fails to fall below 0.3 nm s^{-1}, the results are discarded. As a check, following each series of creep tests a control test is performed on a reference specimen attached to the same mount as the original specimen. Irregularities in the indenter geometry, due to contamination or to the grinding process, are a constant possibility. One way to check that these are not having a significant effect is to repeat the control test using a second indenter. Although this is not done in every case, each of the principal effects described has been seen with more than one indenter.

References

[*1*] Peggs, G. N. and Leigh, I. C., "Recommended Procedure for Micro-Indentation Vickers Hardness Test," Report MOM 62, U.K. National Physical Laboratory, England, 1983.
[*2*] Bückle, H., *Revue de l'Industrie Minérale*, 1970, pp. 188-216.
[*3*] Lebouvier, D., "Modelisation de l'Indentation d'un bicouche en déformation plane," Microthese de D.E.A., École Nationale des Mines de Paris, Valbonne, France, 1983.
[*4*] Lebouvier, D., Gilormini, P., and Felder, E., *Journal of Physics D, 1984*, Vol. 18, 1985, pp. 199-210.
[*5*] Collins, I. F. in *Mechanics of Solids (the Rodney Hill 60th Anniversary)*, H. G. Hopkins and M. J. Sewell, Eds., Pergamon, Oxford, 1982, pp. 135-184.
[*6*] Tazaki, M., Nishibori, M., and Kinosita, K., *Thin Solid Films*, Vol. 51, 1978, pp. 13-21.
[*7*] Lebouvier, D., Ecole Nationale Superieure des Mines de Paris, private communication, 25 June 1984.
[*8*] Shaw, M. C. and De Salvo, G. J., *Transactions, American Society of Mechanical Engineers*, Vol. 92, 1970, pp. 480-494.
[*9*] Gerk, A. P., *Journal of Physics D: Applied Physics*, Vol. 9, 1976, pp. L179-L181.
[*10*] Brookes, C. A., *Philosophical Magazine*, Vol. A43, 1981, pp. 529-543.
[*11*] Maugis, D., *Wear*, Vol. 62, 1980, pp. 349-386.
[*12*] Hooper, R. M. and Brookes, C. A., *Journal of Materials Science*, Vol. 19, 1984, pp. 4057-4060.
[*13*] Naylor, M. G. S. and Page, T. F., *Journal of Microscopy*, Vol. 130, 1983, pp. 345-360.
[*14*] Chen, C. C. and Hendrickson, A. A. in *The Science of Hardness Testing and Its Research Applications*, J. H. Westwood and H. Conrad, Eds., American Society for Metals, Metals Park, OH, 1973, pp. 274-290.
[*15*] Gerk, A. P., *Philosophical Magazine*, Vol. 32, 1975, pp. 355-365.
[*16*] Chu, S. N. G. and Li, J. C. M., *Journal of Materials Science*, Vol. 12, 1977, pp. 2200-2208.
[*17*] Tabor, D., *The Hardness of Metals*, Clarendon Press, Oxford, 1951.
[*18*] Atkins, A. G., Silvério, A., and Tabor, D., *Journal, Institute of Metals*, Vol. 94, 1966, pp. 369-378.
[*19*] Matthews, J. R., *Acta Metallurgica*, Vol. 28, 1980, pp. 311-318.
[*20*] Frost, H. J. and Ashby, M. F., *Deformation-Mechanism Maps*, Pergamon, Oxford, 1982.
[*21*] Lawn, B. R., Marshall, D. B., and Chanticul, P., *Journal of Materials Science*, Vol. 16, 1981, pp. 1769-1775.
[*22*] Loubet, J. L., Georges, J-M., Marchesini, O., and Meille, G., *Transactions, American Society of Mechanical Engineers Journal of Tribology*, Vol. 106, 1984, pp. 43-48.
[*23*] Blau, P. J., *Metallography*, Vol. 16, 1983, pp. 1-18.
[*24*] Newey, D., Wilkins, M. A., and Pollock, H. M., *Journal of Physics E: Scientific Instruments*, Vol. 15, 1982, pp. 119-122.
[*25*] Lawn, B. R. and Howes, V. R., *Journal of Materials Science*, Vol. 16, 1981, pp. 2745-2752.
[*26*] Yoo, K. C., Roessler, B., Armstrong, R. W., and Kuriyama, M., *Scripta Metallurgica*, Vol. 15, 1981, pp. 1245-1250.
[*27*] Ciftan, M. and Saibel, M., *Wear*, Vol. 56, 1979, pp. 69-80.
[*28*] Shchukin, E. D., Savenko, V. I., and Kochanova, L. A. in *Physico-Chemistry of the Solid State*, P. Lacombe, Ed., Elsevier, Amsterdam, 1984.

[29] Maugis, D. and Pollock, H. M., *Acta Metallurgica*, Vol. 32, 1984, pp. 1323-1334.
[30] Roy Chowdhury, S. K. and Pollock, H. M., *Wear*, Vol. 66, 1981, pp. 307-321.
[31] Newey, D., Ph.D. dissertation, University of Lancaster, Lancaster, England, 1981.
[32] Atkins, A. G., *Metal Science*, Vol. 16, 1982, pp. 127-137.
[33] Puttick, K. E., *Journal of Physics D: Applied Physics*, Vol. 13, 1980, pp. 2249-2262.
[34] Lawn, B. R., Evans, A. G., and Marshall, D. B., *Journal of the American Ceramic Society*, Vol. 63, 1980, p. 574.
[35] Sargent, P. M. and Page, T. F., *Proceedings*, British Ceramic Society, Vol. 26, 1978, pp. 209-229.
[36] Hoffman, R. W., *Thin Solid Films*, Vol. 89, 1982, pp. 155-164.
[37] Wierenga, P. E., Dirks, A. G., and Van den Broek, J. J., *Thin Solid Films*, Vol. 119, 1985, pp. 375-382.
[38] Dirks, A. G., Van den Broek, J. J., and Wierenga, P. E., *Journal of Applied Physics*, Vol. 55, 1984, pp. 4248-4256.
[39] Gane, N., *Proceedings of the Royal Society of London*, Vol. A317, 1970, pp. 367-391.
[40] Upit, G. P. and Varchenya, S. A. in *The Science of Hardness Testing and Its Research Applications*, J. H. Westwood and H. Conrad, Eds., American Society for Metals, Metals Park, OH, 1973, pp. 135-146.
[41] Newey, D., Pollock, H. M., and Wilkins, M. A. in *Ion Implantation into Metals*, V. Ashworth, W. A. Grant, and R. P. Procter, Eds., Pergamon, Oxford, 1982, pp. 157-166.
[42] Oliver, W. C., Hutchings, R., and Pethica, J. B., this publication, pp. 90-108.
[43] Wierenga, P. E. and Franken, A. J. J., *Journal of Applied Physics*, Vol. 55, 1984, pp. 4244-4247.
[44] Armstrong, R. W. and Robinson, W. H., *New Zealand Journal of Science*, Vol. 17, 1974, pp. 429-433.
[45] Fröhlich, F., Grau, P., and Grellmann, W., *Physica Status Solidi*, Vol. A42, 1977, pp. 79-89.
[46] Gosilin, V. V. and Kunchenko, V. V., *Zavodskaya Laboratoriya*, Vol. 46, 1980, pp. 353-356.
[47] Nishibori, M. and Kinosita, K., *Thin Solid Films*, Vol. 48, 1978, pp. 325-331.
[48] Chiang, S. S., Marshall, D. B., and Evans, A. G., *Journal of Applied Physics*, Vol. 53, 1982, pp. 298-311.
[49] Shorshorov, M. Kh., Bulychev, S. I., and Alekhin, V. P., *Soviet Physics Doklady*, Vol. 26, 1981, pp. 769-771.
[50] Pethica, J. B., Hutchings, R., and Oliver, W. C., *Philosophical Magazine*, Vol. A48, 1983, pp. 593-606.
[51] Takadoum, J., Pivin, J.-C., and Pollock, H. M., *Thin Solid Films*, in press.
[52] Hartley, N. E. W. in *Ion Implantation in Materials Science and Technology*, J. K. Hirvonen, Ed., Academic Press, New York, 1981.
[53] Li, H. T., Liu, P. S., Chang, S. C., Lü, H. C., Wang, H. H., and Tao, K., *Nuclear Instruments and Methods*, Vols. 182-183, 1981, pp. 915-920.
[54] Saritas, S., Procter, R. P. M., and Ashworth, V., *Wear*, Vol. 82, 1982, pp. 233-255; Madakson, P. B., *Journal of Applied Physics*, Vol. 55, 1984, pp. 3308-3314.
[55] Roberts, S. G. and Page, T. F. in *Ion Implantation into Metals*, V. Ashworth, W. A. Grant, and R. P. Procter, Eds., Pergamon, Oxford, 1982.
[56] Pethica, J. B. and Oliver, W. C. in *Metastable Materials Formation by Ion Implantation*, Vol. 7, S. T. Picraux and W. J. Choyke, Eds., Elsevier, New York, 1982, pp. 373-379.
[57] Pethica, J. B., Hutchings, R., and Oliver, W. C., *Nuclear Instruments and Methods*, Vols. 209-210, 1983, pp. 995-1000.
[58] Singer, I. L., Bolster, R. N., and Pollock, H. M., in preparation.
[59] Singer, I. L., *Materials Research Society Symposium Proceedings*, Vol. 27, 1984, pp. 585-595.
[60] Jönsson, B. and Hogmark, S., *Thin Solid Films*, Vol. 114, 1984, pp. 257-269.

Jean L. Loubet,[1] *Jean M. Georges,*[1] *and Gerard Meille*[1]

Vickers Indentation Curves of Elastoplastic Materials

REFERENCE: Loubet, J. L., Georges, J. M., and Meille, G., **"Vickers Indentation Curves of Elastoplastic Materials,"** *Microindentation Techniques in Materials Science and Engineering, ASTM STP 889,* P. J. Blau and B. R. Lawn, Eds., American Society for Testing and Materials, Philadelphia, 1986, pp. 72–89.

ABSTRACT: Studies of Vickers indentation curves of elastoplastic materials give information about the relationship between plasticity and elasticity. We have built equipment which gives a record of the applied depression depth, h, as a function of the carried load, P (range, 0.5 to 20 N).

Investigation of the Vickers indentation curves, $P(h)$, during loading and unloading and microscopic observations shows that the phenomena occurring during indentation are complex. We first discuss the loading curve. A comparison with a theoretical analysis is made. Young's modulus, E, and the Vickers indentation pressure during loading can be calculated from the unloading and loading curves. The meaning of the different characteristic parameters of the depth-load graphs is debated. Their correlation with optical measurements of the remanent Vickers indentation is proposed.

KEY WORDS: microindentation, elastoplastic behavior, indentation curves, elastic recovery, metals testing, microindentation hardness testing

An indentation curve is the relationship between load, P, and penetration, h, continuously measured and recorded during a hardness indentation experiment. The aim of this paper is to explain the Vickers indentation curves of homogeneous elastoplastic materials.

The contact between a flat specimen and a rigid conical indenter, normally loaded, is well known for two types of materials: elastic materials [*1*] and rigid plastic materials [*2*].

The theoretical determination of the relationship between P and h for elastoplastic materials is a difficult problem. A complete theoretical analysis will need to use elastoplasticity theory with the help of the study of the solids in large strain. As a theory does not exist, several approximations are proposed [*3–5*]. All are founded on experimental studies [*6*].

[1] Research worker, professor, and research worker, respectively, Ecole Centrale de Lyon, Laboratoire de Technologie des Surfaces, Ecully 69131, France.

The main results in the literature about the indentation curves of elastic or plastic materials are summarized here. Comparison of these results will help to explain the Vickers indentation curves of homogeneous elastoplastic materials.

Elastic Material

Sneddon [*1,7*] gives the relationship between the load, P, and the penetration depth, h, in the case of a nonadhesive rigid conical punch, normally loaded, on the plane surface of a smooth elastic body as

$$P = \frac{2Eh^2}{(1 - \nu^2)\pi} \tan\theta \tag{1}$$

where

E = the Young's modulus of the specimen,
ν = its Poisson's ratio, and
θ = the half apical angle of the cone.

Notice that the normal load, P, is proportional to the square of the penetration depth, h, in the course of loading and unloading (Fig. 1*a*). The proportionality factor is dependent on both the elastic parameters, (E, ν), and the geometrical properties of the contact, θ.

The mean pressure, p_m, is constant, independent of the normal load, P, and the radius of contact, a.

$$p_m = \frac{P}{\pi a^2} = \frac{E}{2(1 - \nu^2)} \cot\theta \tag{2}$$

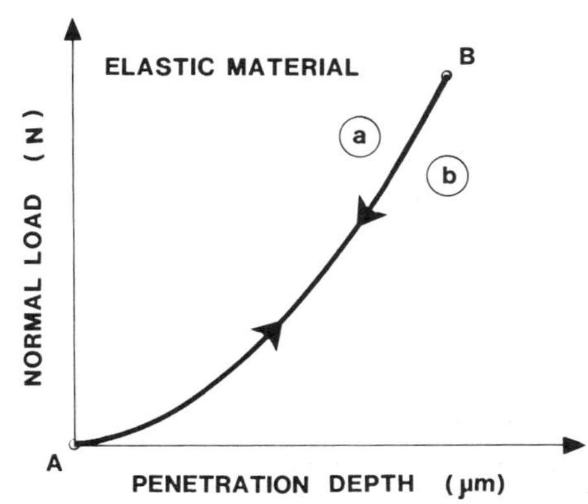

FIG. 1a—*Indentation curve of elastic materials:* ⓐ *loading curve;* ⓑ *unloading curve.*

Rigid-Plastic Material

Tabor [8], using the principle of geometric similarity, shows that the mean pressure acting on a cone indenter (or Vickers pyramid) is the same whatever the size of indentation, as long as the material is uniform and homogeneous. He finds, for fully work-hardened metals, the semiempirical relation

$$p_m = \frac{2P}{D^2} = CY \tag{3}$$

where

Y = the yield stress of the tested materials,
D = the diagonal of the Vickers indentation, and
C = a dependent constant of the indenter ($k = 3.2 - 3.3$ for a Vickers pyramid and work-hardened specimens).

Lockett [9] gives theoretical values of the proportionality factor, C, for the variable apical angle, θ, of cones. In this theory, there is no friction on the indenter.

So, we can say that the normal load, P, during loading is proportional to the square depth of the indentation, h_d

$$P = K_d(Y, \theta)h_d^2 \tag{4}$$

If, as in wedge indentation [10], the height of the ridge, h_b, due to the pileup effect is proportional to the penetration depth, h, which is different from the indentation depth, h_d, we can write the following for plastic materials such as elastic materials during loading

$$P = K_p(Y, \theta)h^2 \tag{5}$$

where $K_p(Y, \theta)$ is a proportionality factor depending on the mechanical and geometrical parameters.

In the course of unloading, since there is no elastic strain, the strength becomes instantly nil (Fig. 1*b*).

Experimental Technique

Figure 2 shows a schematic diagram of the equipment used. The Vickers Indenter B penetrates Specimen C to a controlled indentation depth. During the test, Distance $\alpha\beta$ stays constant, while Distance $\alpha\gamma$ increases linearly with time by means of the dilatation of Frame A. The indentation depth is measured by an induction coil transducer, D, which monitors the distance be-

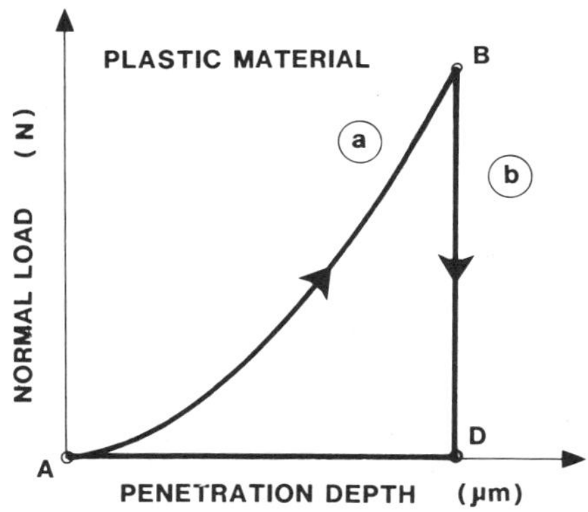

FIG. 1b—*Indentation curve of plastic materials:* ⓐ *loading curve;* ⓑ *unloading curve.*

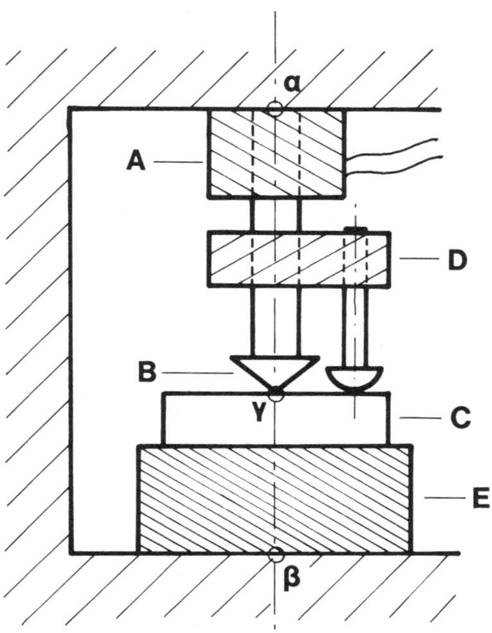

FIG. 2—*Schematic diagram of the equipment:* (A) *dilation system;* (B) *Vickers indenter;* (C) *specimen;* (D) *displacement transducer;* (E) *load transducer.*

tween the indenter and the surface of the specimen. The positioning of the transducer ensures that even if the rigidity between α and β is not infinite (about 5×10^7 N/m in our case), no error is introduced into the indentation depth measurement. The test Specimen, C, is mounted on a piezoelectric transducer, E, which measures the resulting load, P.

Inelastic relaxation is assumed to be negligible compared to elastic relaxation, so the indenter is withdrawn as soon as maximum penetration is reached. The maximum normal load, P, varies from 0.5 to 20 N with a resolution of 10^{-4} N. The maximum penetration depth, h, is 40 μm with a resolution of displacement of 4×10^{-4} μm. Three controlled speeds, V, can be used: 0.1 μm/s, 0.05 μm/s, and 0.01 μm/s. The Vickers indenter used is a square-based pyramid with a 136° angle between its faces. A small defect of 0.1 μm height and 0.7 μm breadth was detected by a scanning electron microscope (SEM) observation on the pyramid tip. The indenter was cleaned with acetone (analytical reagent) before each test. All the indentation tests were conducted in a clean room, at room temperature (20°C). The diamond indenter is assumed to be infinitely rigid.

Two materials are tested, first, a hardened Type 52100 steel plane. It is polished with diamond paste and then ultrasonically cleaned in hexane (analytical reagent). The height of the asperities measured with a Talystep instrument does not exceed 0.1 μm. Its superficial hardness is HV_{500} = 850 to 900 *da* N/mm^2. Its Young's modulus is evaluated as 210 GPa. Second, an annealed aluminum plane is tested. It is 99.9% pure. The main impurities are iron, silicon, copper, and titanium. The plane is ultrasonically cleaned in a hexane (analytical reagent) bath before the experiments. The height of its asperities does not exceed 0.5 μm. Its superficial hardness, obtained with a classic apparatus, is HV_{500} = 46 to 47 *da* N/mm^2. Its Young's modulus is evaluated as 65 GPa.

Description of an Experimental Curve

The study of the indentation curves is an experimental method recently developed [*11–14*].

The experimental Vickers indentation curves are similar to those in Figs. 3*a* and 3*b*. Parts AB and BC are called, respectively, the "loading curve" and the "unloading curve."

Point A corresponds to the first contact between the Vickers pyramid, and the specimen surface. Point B is the maximum load and penetration reached. Point C corresponds to the last contact between the indenter and the specimen.

We define P_{max} as the maximum load of the test, h_T as the maximum penetration depth (Length AD) of the pyramid, and h_R as the remanent depth of the indentation (Length AC).

From the tangent to the unloading curve, we determine the distance $h_{R'}$

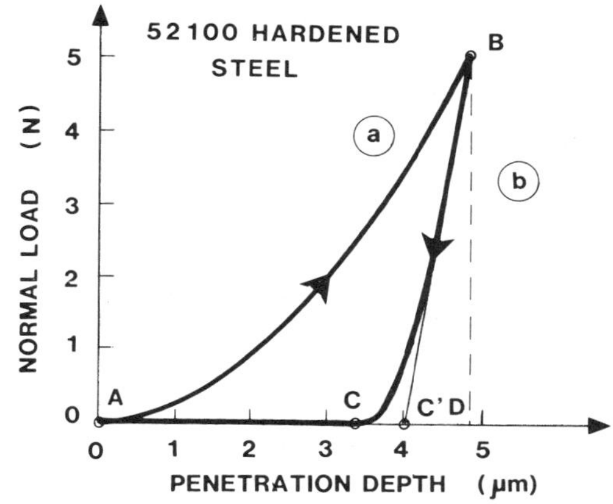

FIG. 3a—*Indentation curve of 52100 hardened steel:* ⓐ *loading curve;* ⓑ *unloading curve.*

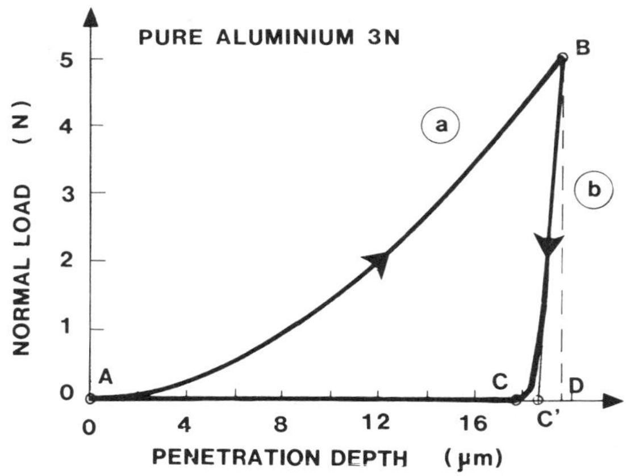

FIG. 3b—*Indentation of aluminum 3 N:* ⓐ *loading curve;* ⓑ *unloading curve.*

(Length AC′). The latter is very useful for the determination of Young's modulus, as will be shown later. Because of Point A, the three length parameters are measured from the nondeformed surface of the specimen.

The Loading Curve Study

The study of experimental loading curves shows that, for the limited regions of testing loads, we can write [*15*]

$$P = Kh^n \tag{6}$$

where n is a constant, not necessarily an integer. Because P is proportional to h^2, if the contact is purely elastic or rigid-plastic, we propose, for a uniform and homogeneous elastoplastic material with perfect plasticity, the relation

$$P = K_{ep}h^2 \tag{7}$$

where K_{ep} is a proportionality factor function of the elastic parameters, (E, ν), the plastic parameter, Y, and the geometrical parameter (semiapical angle θ of the Vickers pyramid or the conical indenter).

On the experimental curves, we can verify the validity of this relation (Fig. 3) with the aid of different methods.

1. Fröhlich et al [*4*] proposes to plot P/h against h (Fig. 4*a*).
2. Pollock suggests the representation $\sqrt{P}$ against h (Fig. 4*b*).
3. We can plot log P as a function of log h (Fig. 4*c*).
4. With the derivative of the load, we can plot dP/dh against h (Fig. 4*d*).

Each of these four methods has its own advantages and disadvantages, in particular the first and the third methods give, respectively, the proportionality factor, K_{ep}, and the exponent, n, of the curve, but they are both very sensitive to the experimental determination of Point A.

The second and the fourth plotting systems guarantee the best verification or deviation around the proposed law, but the second method is not directly related to the parameters, and the fourth is very sensitive to noise in the load and displacement signals.

We can see in Fig. 4*c* that aluminum 3 N and 52100 hardened steel verify the hypothesis in the main part of the experimental curve, but the exponent, n, is not 2. We think this is attributable both to a harder layer near the surface and to the pyramid's blunt point. The purpose of this paper is not to explain this deviation, however, so we shall not go into this matter further.

The determination of the relationship between K_{ep} and the mechanical parameters (E, ν, Y, or HV) usually used is a difficult problem. As the theory does not exist, several approximations were proposed [*6*]. We describe here the one that gives us the best results (Fig. 5).

First, we assume that plastic and elastic behavior of the material can be disassociated, and we write

$$P = K_p h_p^2 \tag{8}$$

where K_p is a function of Y or HV for the plastic part and

$$P = K_e h_e^2 \tag{9}$$

where K_e is a function of E, ν, and θ for the elastic part.

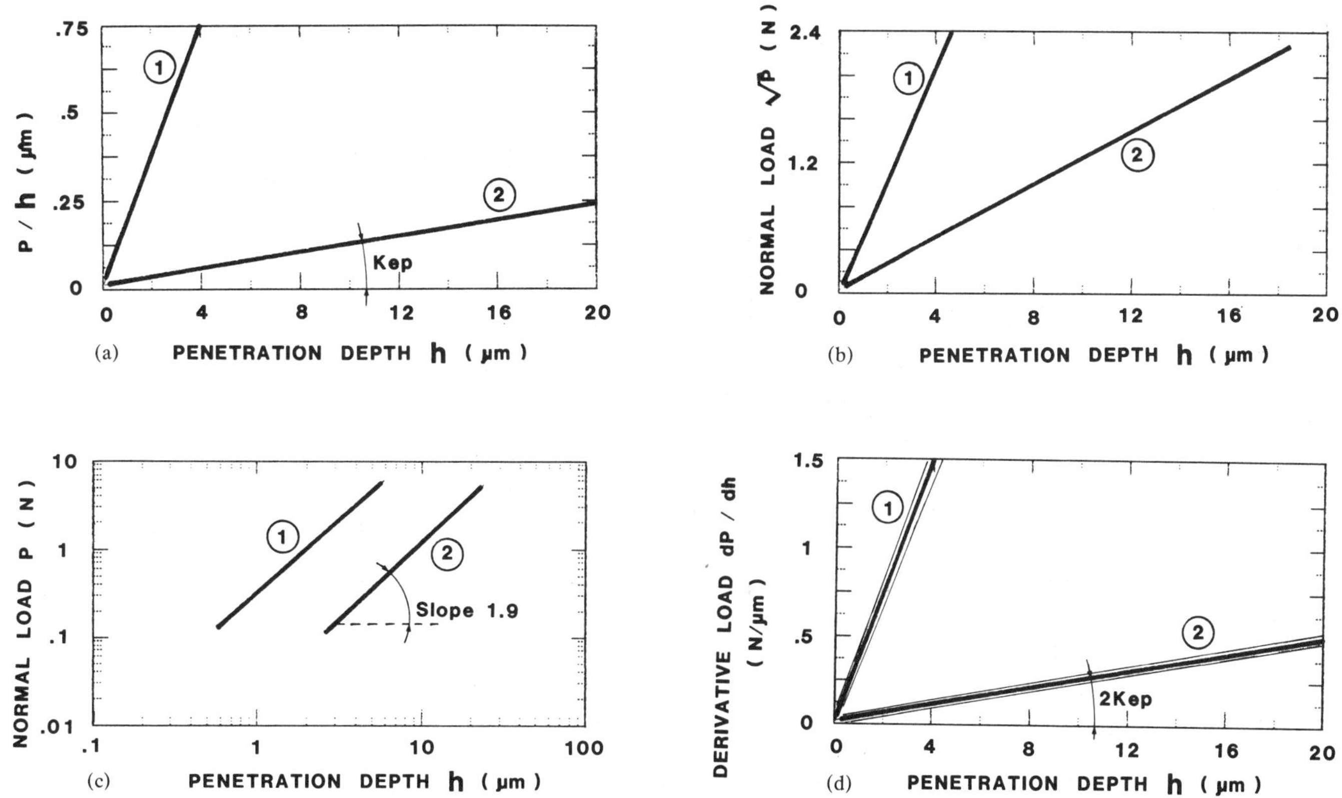

FIG. 4—*Graphical examination of the loading curve (observed deviation around the mean curves drawn are less than 5%):* (a) P/h *as a function of* h; (b) $\sqrt{P}$ *as a function of* h; (c) *log* P *as a function of log* h; (d) dP/dh *as a function of* h. *Index* ① *is 52100 hardened steel and Index* ② *is aluminum 3 N.*

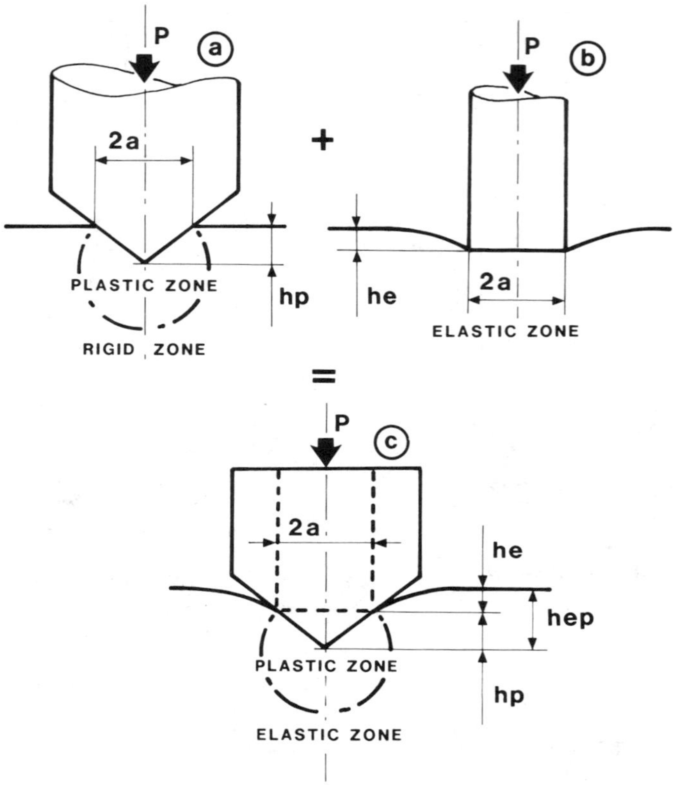

FIG. 5—*Model of the behavior of an elastoplastic indentation:* (a) *plastic contact;* (b) *equivalent elastic contact;* (c) *elastoplastic contact. (The terms are defined in the text.)*

Second, we said that Eqs 8 and 9 are for two springs. For the plastic law, there is a problem with the return, but we can replace it with an elastic spring that has the same behavior in the loading part.

Third, because we have supposed a behavior described in the theoretical Eq 7 for the loading curve, we can say that the two springs are in series, so that $h_{ep} = h_e + h_p$. The springs are both representative of the elastoplastic behavior, therefore

$$K_{ep} = [(K_e)^{-1/2} + (K_p)^{-1/2}]^{-2} \tag{10}$$

Fourth, we have to select the plastic and elastic proportionality factor. For the plastic factor, K_p, we pick out the theoretical relation between the Vickers hardness value (HV), load, P, and plastic penetration depth, h_p.

$$\mathrm{HV} = \frac{\cos^2\theta}{4\sin\theta}\frac{P}{h_p^2} = 3.78 \times 10^{-2}\frac{P}{h_p^2} \tag{11}$$

therefore

$$K_p = \mathrm{HV}\frac{4\sin\theta}{\cos^2\theta} = 26.4\ \mathrm{HV} \tag{12}$$

For the elastic proportionality factor, K_e, we suppose there is no elastic penetration in the contact but an elastic deflection, h_e, around it. The penetration is done only by the plastic part of the model. The elastic behavior does not create an increase in the contact area.

We assume that the elastic deflection can be calculated with the aid of the theory of an expansible flat punch, which means an imaginary punch with a variable base area.

Sneddon [*1*] gives the relation between the load, P, and the deflection, h_e, in the case of a nonadhesive, rigid, cylindrical flat punch normally loaded on the plane surface of the smooth elastic body

$$P = \frac{2E}{1-\nu^2}ah_e \tag{13}$$

where a is the radius of the contact area.

The contact area must always be the same as the one given by the plastic penetration, so that

$$\pi a^2 = \frac{D^2}{2} \rightarrow a = \frac{D}{\sqrt{2\pi}} \tag{14}$$

where D is the diagonal of the Vickers indentation squared. It is given by the Vickers relation

$$D = \left(1.854\frac{P}{\mathrm{HV}}\right)^{1/2} \tag{15}$$

We can now write the elastic relation as

$$P = \left(\frac{E}{1-\nu^2}\right)^2\frac{1.18}{\mathrm{HV}}h_e^2 \tag{16}$$

so that

$$K_e = \left(\frac{E}{1-\nu^2}\right)^2 \frac{1.18}{\text{HV}} \tag{17}$$

The elastoplastic coefficient can be expressed as

$$K_{ep} = \left[0.92\left(\frac{1-\nu^2}{E}\right)\sqrt{\text{HV}} + \frac{0.194}{\sqrt{\text{HV}}}\right]^{-2} \tag{18}$$

The comparison between the model and the experimental loading curves (Fig. 6) on the 52100 hardened steel shows only a 10% deviation between them.

The Unloading Curve Study

During the unloading stage (Curve BC in Fig. 7), two effects can be observed. (1) the elastic recovery of the zone adjacent to the contact zone and (2) the elastic recovery of the plastic edge of the remanent print (Fig. 8).

As pointed out by Shorshorov et al [*16*], during the unloading stage, the relation $P(h)$ is not linear. The derivative, dP/dh (Curve BC′ in Fig. 7), for the initial stage of the unloading can be given by the elastic flat punch theory. This punch has a base area equal to the projected area of the indentation. This is the elastic recovery of the elastic deflection created during the loading stage. Therefore, the equation of Line BC′ (Fig. 7) is

$$P = \frac{2E}{1-\nu^2}\,\frac{D}{\sqrt{2\pi}}(h - h_{R'}) \tag{19}$$

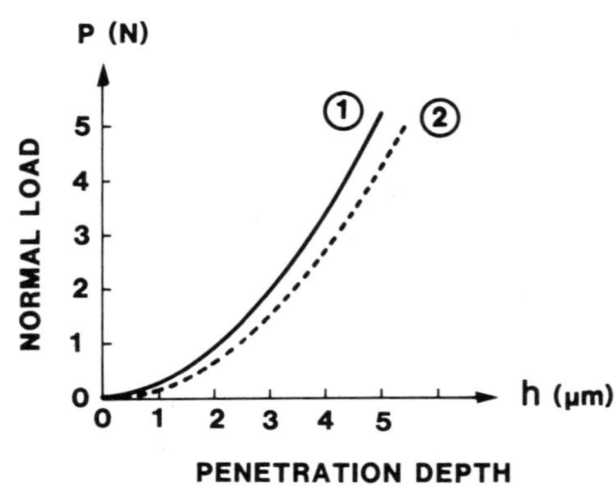

FIG. 6—*Comparison between the experiment loading curve of 52100 hardened steel (Index ①) and the theoretical curve corresponding to the elastoplastic calculation developed (Index ②).*

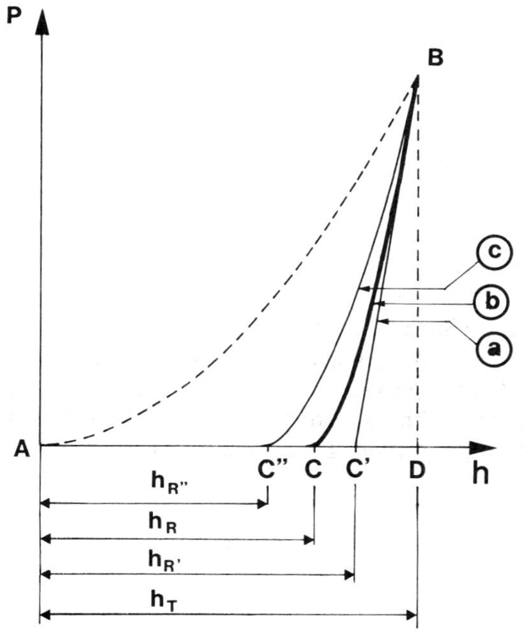

FIG. 7—*Comparison between an experimental unloading curve, ⓑ, and the two possible extreme theoretical positions (52100 hardened steel): Curve ⓐ is the rigid flat punch law on an elastic plane; Curve ⓒ is the rigid conical punch law on an elastic plane. (The terms are defined in the text.)*

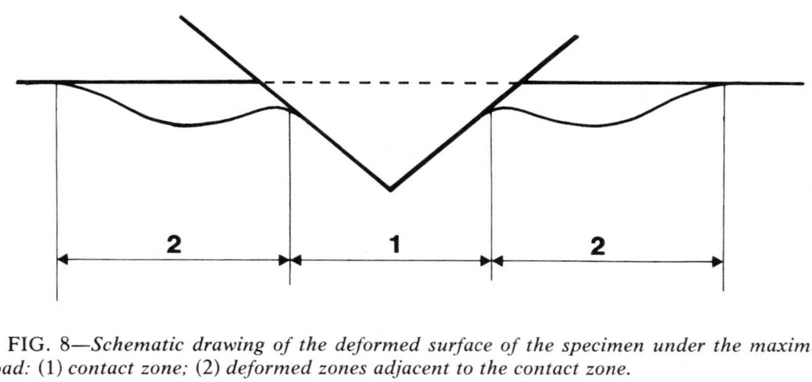

FIG. 8—*Schematic drawing of the deformed surface of the specimen under the maximum load:* (1) *contact zone;* (2) *deformed zones adjacent to the contact zone.*

This equation allows us to calculate the Young's modulus of the material. Using Point B (Fig. 7), we find

$$\frac{E}{1 - \nu^2} = \frac{P_{\max}\sqrt{2\pi}}{2D(h_T - h_{R'})} \tag{20}$$

For the 52100 hardened steel, the experimental value is

$$\frac{E}{1 - \nu^2} = 218 \text{ GPa}$$

This agrees with the literature data.

The hardness value is obtained by optical measurement of the remanent print diagonal. On the curve, the penetration depth, $h_{R'}$, is the indentation depth measured from the nondeformed surface of the specimen with respect to the geometrical properties of the Vickers indenter. If the pileup effect is negligible

$$D = 7h_{R'} \tag{21}$$

The experimental data show $D > 7\ h_{R'}$. For example, on the 52100 steel used, we found $D = 7.9\ h_{R'}$. The proportionality factor depends on the material tested (aluminum $= 3$ N; $D = 7.7\ h_{R'}$; in a single crystal of magnesium oxide [MgO], $D = 8.95\ h_{R'}$ [*17*]). This experimental parameter between D and $h_{R'}$ allows us to measure the height of the ridge, h_b, resulting from the pileup effect by the formula

$$h_b = \frac{D}{7} - h_{R'} \tag{22}$$

Or, if we say that this parameter is roughly equal to 7, we can estimate the "optical" hardness value, using Formula 11, with $h_p = h_{R'}$.

The convexity of the unloading curve exhibits the second elastic effect: the elastic recovery of the plastic edge of the remanent print by the opening of its apical angle. The remanent indentation depth, h_R (Fig. 7), allows us to obtain the experimental value of the remanent semiapical angle, θ'. If we assume there is practically no change in the diagonal of indentation during the unloading process [*18*], we obtain a geometrical relation between θ and θ'

$$\tan\theta' = \frac{h_{R'}}{h_R}\tan\theta \tag{23}$$

where $\theta = 68°$ for a Vickers pyramid.

On the tested specimens, we found $\theta' = 70°$ for the 52100 hardened steel and $\theta' = 68.5°$ for the pure aluminum.

Curve c (Fig. 7) corresponds to a translation of the theoretical curve of an elastic material indented by a conical punch. Its contact area (at Point B) is the same as that of the experimental remanent print. The Young's modulus of the calculation is assumed to be the same as that of the tested specimen. Using this hypothesis, we note that Curve a is also tangent to Curve c. Together, they form the external envelope of the possible experimental unloading curves. This law can be used, as proposed by Lawn [*19*], to calculate the Young's modulus of the tested material by measuring the recovery of the indentation. The mathematical expression of the experimental unloading curve could be given by

$$P = K_e^d(h - h_R)^{nd} \tag{24}$$

with $1 \leq nd \leq 2$

$$h_{R''} \leq h_R \leq h_R'$$

where $h_{R''}$ is the theoretical elastic conical punch law translation value. For example, on the 52100 hardened steel, we found the experimental values

$$K_e^d = 3.19\ \text{N}/(\mu\text{m})^{nd}; \qquad nd = 1.5$$

for

$$P_{\max} = 5\ \text{N}; \quad h_{R''} = 3.15\ \mu\text{m}; \quad h_R = 3.6\ \mu\text{m}; \quad h_{R'} = 4.05\ \mu\text{m}$$

Total Interpretation of the Curves

The experiments are made at a very slow penetration speed (maximum 0.1 μm/s). The load and unload processes are a succession of states of equilibrium (Fig. 9); it is a quasi-static evolution. There are no dynamic or viscous effects. The recorded energies are identical to the strain energy of the specimen, if friction energy is assumed to be negligible.

Area ABDA under the loading curve is the necessary work for the creation of the indentation, W_T. Surface BCDB is the work released by the specimen, W_E. Area ABCA is the retained work, W_R. These parameters are, by convention, positive so that

$$W_R = W_T - W_E \tag{25}$$

W_T is an elastoplastic work, W_E is an elastic work, and W_R is mainly a plastic work [*20*].

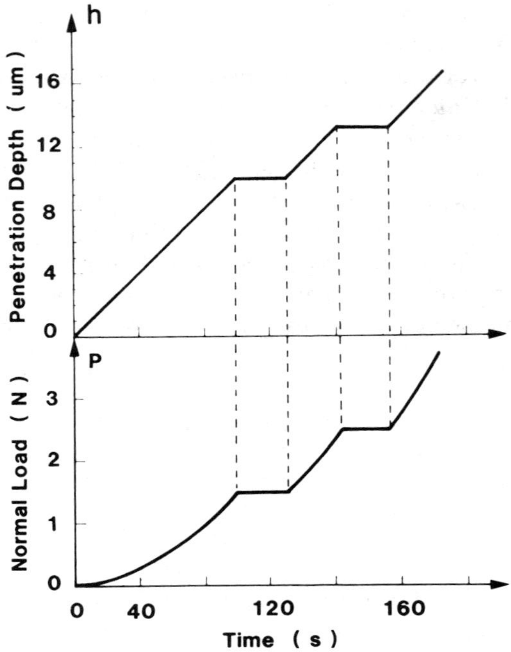

FIG. 9—*The independence of the imposed penetration depth and the resultant normal load on aluminum 3 N. A stop in the indenter course promotes an instantaneous stabilization of the load.*

The ratios W_R/W_T and h_R/h_T are descriptors for the elastoplastic indentation behavior. For an elastic material, $W_R = h_R = 0$, so that

$$\frac{W_R}{W_T} = \frac{h_R}{h_T} = 0$$

For a rigid plastic material, $W_R = W_T$ and $h_R = h_T$, so that

$$\frac{W_R}{W_T} = \frac{h_R}{h_T} = 1$$

(See Fig. 10.) For an elastoplastic material, we find that the experimental value of W_R/W_T is not very different from the h_R/h_T value. This relation agrees with the experimental law proposed, as we can see in the following.

On the loading curve we have found a behavior described by Eq 6, and on the unloading curve the mathematical expression has already been written in Eq 24. So, at the maximum loading Point B (Fig. 7), we obtain

$$Kh_T^n = K_e^d(h_T - h_R)^{nd} \tag{26}$$

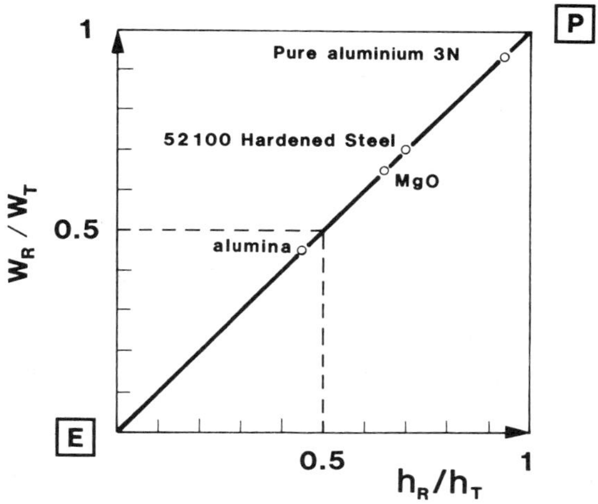

FIG. 10—*Classification function of parameters* W_R/W_T *and* h_R/h_T *of different materials tested (alumina* [17]*).*

The expression of the total work is

$$W_T = \int_0^{h_T} P dh; \qquad \text{so that } W_T = \frac{1}{n+1} K h_T^{n+1} \tag{27}$$

The expression of the elastic work is

$$W_E = \int_{h_R}^{h_T} P dh; \qquad \text{so that } W_E = \frac{K_e^d}{n_d + 1} (h_T - h_R)^{nd+1} \tag{28}$$

As $W_R/W_T = 1 - W_E/W_T$, we find the relation between W_R/W_T and h_R/h_T to be

$$\frac{W_R}{W_T} = 1 - \left(\frac{n+1}{nd+1}\right)\left(\frac{1 - h_R}{h_T}\right) \tag{29}$$

On the 52100 hardened steel, we have the experimental data $n = 1.9$; $nd = 1.5$; and $h_R/h_T = 0.7$. Therefore

$$\frac{W_R}{W_T} = 0.93 \frac{h_R}{h_T}$$

which proves the point in question.

Conclusion

The analysis of the Vickers indentation curves of elastoplastic materials demonstrates four main points.

1. In the main part of the experimental loading curve, for plastic and elastic materials, the load is roughly proportional to the square of the penetration depth, h^2.

2. Present theories do not allow the correct calculation of the proportionality factor, K_{ep}, between P and h^2. The approximate calculation proposed can explain the part played by the elasticity behavior in the indentation process.

3. The unloading curve allows the calculation of (*a*) the Young's modulus of the tested material, (*b*) the angular aperture of the indentation apical angle, and (*c*) the "optical" hardness number if the pileup effect is negligible.

4. The total curve exhibits the different works brought into play in the indentation process. The retained work, W_R, and the supplied work, W_T, define an elastoplastic index, (W_R/W_T), which is related to the ratio h_R/h_T (remanent indentation depth/maximum loaded penetration depth).

Acknowledgments

The authors are indebted to the French Direction des Recherches, Etudes et Techniques. They wish to thank Drs. Kapsa, Sidoroff, and Pollock for their helpful discussions and advice.

References

[*1*] Sneddon, I. N., "The Relation between Load and Penetration in the Axisymmetric Boussinesq Problem for a Punch of Arbitrary Profile," *International Journal of Engineering Science*, Vol. 3, 1965, pp. 47–57.
[*2*] Tabor, D., *The Hardness of Metals*, Clarendon Press, Oxford, 1951.
[*3*] Johnson, K. L., "The Correlation of Indentation Experiments," *Journal of the Mechanics and Physics of Solids*, Vol. 18, 1970, pp. 115–126.
[*4*] Fröhlich, F., Grau, P., and Grellmann, W., "Performance and Analysis of Recording of Microhardness Tests," *Physica Status Solidi (A)*, Vol. 42, 1977, pp. 79–82.
[*5*] Newey, D., Wilkins, M. A., and Pollock, H. M., "An Ultra-Low Load Penetration Hardness Tester," *Journal of Physics E: Scientific Instruments*, Vol. 15, 1982, pp. 119–122.
[*6*] Loubet, J. L., "Etude des Courbes d'Indentation, Cas du Poinçon Conique et de la pyramide Vickers," Thèse 3rd cycle, Université Lyon I, No. 1332, 1983.
[*7*] Sneddon, I. N., *Proceedings of the Cambridge Philosophical Society*, Vol. 44, 1948, p. 429.
[*8*] Tabor, D., "The Hardness of Solids," *Review of Physics in Technology*, Vol. 1, 1970, pp. 145–179.
[*9*] Lockett, F. J., *Journal of the Mechanics and Physics of Solids*, Vol. 11, 1963, p. 345.
[*10*] Hill, F., Lee, E. H., and Tupper, S. J., "The Theory of Wedge Indentation of Ductile Materials," *Proceedings of the Royal Society, Series A*, Vol. 188, Plate 4, 1947, pp. 273–290.
[*11*] Lee, C. H., Masaki, S., and Kobayashi, S., "Analysis of Ball Indentation," *International Journal of Mechanical Science*, Vol. 14, 1972, pp. 417–426.
[*12*] Newey, D., "An Ultra Load Penetration Hardness Tester and Its Uses," Ph.D. thesis, Lancaster University, Lancaster, England, 1981.
[*13*] Gasilin, V. V. and Kunchenko, V. V., "Application of an Indenter Penetration Recording

Device to the Study of Mechanical Characteristics of Materials," *Zavodskaya Laboratoriya*, Vol. 46, No. 4, 1980, pp. 353-356.

[*14*] Berkovich, E. S., Matveebskii, R. M., Karpova, T. M., Skvortsov, V. N., and Emel'yanov, N. M., "Determination of Rubbing Surface Mechanical Characteristics from Indentation Diagram," *Trenie i Iznos*, Vol. 2, No. 3, 1981, pp. 556-560.

[*15*] Bückle, H., *Mikroharteprufung*, Berliner Union Verlag, Stuttgart, 1965.

[*16*] Bulychev, S. I., Alekhin, V. P., and Shorshorov, M. Kh., *Zavodskaya Laboratoriya*, No. 9, 1975, pp. 1137-1140.

[*17*] Loubet, J. L., Georges, J. M., Marchesini, O., and Meille, G., "Vickers Indentation Curves of Magnesium Oxide (MgO)," *Transactions of the American Society of Mechanical Engineers: Journal of Tribology*, Vol. 106, January 1984, pp. 43-48.

[*18*] Stiwell, N. A. and Tabor, D., "Elastic Recovery of Conical Indentations," *Proceedings of the Physical Society*, Vol. 78, 1961, pp. 169-179.

[*19*] Lawn, B. R. and Howes, V. R., "Elastic Recovery at Hardness Indentations," *Journal of Materials Science*, Vol. 16, 1981, pp. 2745-2752.

[*20*] Shorshorov, M. Kh., Bulychev, S. I., and Alcklin, V. P., "Work of Plastic and Elastic Deformation during Indenter Indentation," *Soviet Physics Doklady*, Vol. 26, 1981, pp. 769-770.

W. C. Oliver,[1] *R. Hutchings,*[2] *and J. B. Pethica*[3]

Measurement of Hardness at Indentation Depths as Low as 20 Nanometres

REFERENCE: Oliver, W. C., Hutchings, R., and Pethica, J. B., **"Measurement of Hardness at Indentation Depths as Low as 20 Nanometres,"** *Microindentation Techniques in Materials Science and Engineering, ASTM STP 889,* P. J. Blau and B. R. Lawn, Eds., American Society for Testing and Materials, Philadelphia, 1986, pp. 90–108.

ABSTRACT: Hardness as a function of depth is examined over a wide range of depths in gold, nickel, lithium fluoride (LiF), and silicon. To examine hardness in very shallow indents (100 nm), a special hardness tester with 2.5-μN load resolution and 0.2-nm indenter displacement resolution was used. The tester monitors the load continuously as the load is applied and removed. The hardness is calculated using the depth measurements. Images of indents in the scanning electron microscope (SEM) and transmission electron microscope (TEM) are used to check that the expected contact areas are achieved. It is shown that the contact area can be calculated if elastic effects are accounted for. The loading and unloading portions of the curves are discussed in detail. Attention is paid to the mechanisms of hardness changes with depth and the effects of elastic recovery on contact area measurement.

KEY WORDS: microhardness, surfaces, mechanical properties, elastic recovery, gold, nickel, lithium fluoride, silicon, microindentation, hardness testing

The data obtained through controlled indentation of materials can be divided into two areas of interest. They are the loading and unloading portions of the load-contact area data. The loading data are usually influenced more by the plastic properties of the material and the unloading data by the elastic

[1]Research scientist, Metals and Ceramics Division, Oak Ridge National Laboratory, Oak Ridge, TN 37831.

[2]Staff member, Department of Metallurgy and Materials Science, University of Cape Town, Cape Town, South Africa.

[3]Staff member, Cavendish Laboratory, University of Cambridge, Cambridge, United Kingdom.

properties. Clearly there are overlaps that must be considered; however, each section of this paper will be divided along these lines.

Loading

The loading portion of the data is that from which hardness is calculated. In particular, we will be discussing hardness as a function of the depth or diameter of the indentation, with emphasis on the hardness of very small volumes of material.

Hardness is a measurement that is often used because of the ease with which the data can be obtained. Often, hardness measurements can be made when it is difficult to do any other sort of mechanical test. The most direct correlation between hardness and other mechanical properties is the relationship between the hardness, as measured with an indenter of a specific geometry, and the flow strength of the material in a uniaxial test at a specific strain. The strain at which this correlation may be made is characteristic of the geometry of the indenter. The definition for hardness we choose to use is the contact force divided by the projected contact area. Using this definition, Tabor [*1*] showed that when the indenter is blunt, one can write $H \cong 3\sigma$, where H is hardness and σ is the flow strength. Thus, if the flow strength remains constant as a function of deformation volume and the indentation geometry scales directly with the depth, as is the case with a pyramid or cone, the hardness should be independent of the depth. It is now well established that this is seldom the case. This fact leads to the conclusion that the yield processes and the stresses required to drive them vary as the deformed volume changes.

Interest in this area has been rekindled by the need for mechanical testing on a miniature scale. Many new coating techniques, such as ion implantation, involve very thin layers (hundreds of nanometres). Miniaturization is an ever-present force in the electronics industry, and the need for mechanical property measurements of such structures is now apparent. In addition, if the hardness of materials on a local scale can be well understood, hardness could be used as a strength microprobe and would give unique experimental insight into many aspects of the relationship between the microstructure and the mechanical properties of materials.

The methods for measuring hardness can be divided into two types. These are techniques in which the indent is imaged and its area measured and those in which the depth of the indent is measured and, with knowledge of the indenter's geometry, the contact area is calculated. The method of imaging depends on the resolution required. The higher the required resolution, the more sophisticated and difficult the measurement becomes. The optical, scanning electron (SEM), and transmission electron microscopes (TEM) have all been applied to this problem. Except for the difference in resolution, these experiments are similar and have the following advantages and disadvantages.

The most important advantage of the first technique is that it gives a direct measure of one of the variables needed to determine hardness—the contact area. In addition, because there is no need for accurate displacement measurements, the experiment is fairly simple. This advantage applies more to the case of optical measurements than to those accomplished using the SEM and TEM.

The disadvantages include a resolution limit. If frictional effects are to be avoided, the depth of an indent must be at least four times smaller than the diameter. If replication and the TEM are used, it is very difficult to image indents less than 100 nm deep. The imaging technique also gives very little information about relaxation effects unless special techniques are used. Finally, imaging the indent can be done only at high resolution for one depth per indent; hence, the accumulation of data is very time-consuming.

The use of displacement measurements to calculate contact areas avoids some of these problems but has some disadvantages of its own. First of all, because no imaging is required, the hardness as a function of depth can be obtained from a single indent. Because very small displacements (nanometres) can be measured without great difficulty, this type of measurement can be extended to shallower depths and therefore smaller deformed volumes. The depth-sensing apparatus can be used during loading and unloading; thus, elastic effects can be directly measured and accounted for. The final advantage of a depth-sensing system is that it is easily automated. This is an important consideration when hardness as a function of depth is required at very shallow depths at which it is critical that a statistical approach be employed. Unless the system can be automated, the time needed to generate sufficient data on a single material becomes prohibitive.

The most important disadvantage of the depth-sensing technique is that the quantity measured (the depth) must be squared to calculate the hardness. Thus, the error is $2d\Delta d$, where d = depth and Δd = error in the depth. This is not usually an important problem if the depth-sensing system is sensitive enough. To calculate the hardness from the depth, the geometry of the indenter must be known to the accuracy required for the depths of interest. This is difficult if one desires to get results from indents tens of nanometres deep.

The imaging technique is conclusive but time-consuming. The depth-measuring technique is very efficient but must be properly calibrated (the indenter shape determined). In this work, the two techniques are combined to obtain hardness data from the smallest volumes of material ever tested in this fashion. These data will extend the range over which the dependence of hardness on depth, or deformed volume, is known.

Early work in the area of the depth, or load, dependence of microhardness was confused by experimental error in the load applied or the area measured [*2*]. Once these problems were realized, the resolution of the optical microscope limited the range of depth that could be sampled.

Gane and co-workers [*3–6*] used SEM and TEM for imaging and produced a series of papers on the deformation of small volumes of materials, in particular gold, which extended the range of study tremendously. Some of their conclusions are reviewed here. They found that, on very clean surfaces, increases of hardness of three to four times the bulk values were apparent at shallow depths. This result was corroborated by Pethica [*7*]. In addition, the effects of very thin contaminant layers were found to be significant. With contaminants between the surface and the indenter, strengths approaching the theoretical limits were obtained. Gane [*3*] correlated the results of hardness tests with the results of sharp tip blunting, filament bending, and small sphere deformation experiments and found general agreement. Gane and Cox [*4*] drew several conclusions concerning the work hardening of small volumes of material and the additive nature of bulk work hardening and hardness increases at shallow depths. This series of papers also considered possible mechanisms for hardness increases at shallow depths. Some of these conclusions will be reconsidered in the light of the data presented in this work.

Upit and Varchenya [*8,9*], using a special hardness tester, explored the behavior of metals, semiconductors, and ionic solids in a range of depths similar to those Gane and co-workers used in their experiments. They found that all the materials they tested satisfied the modified Meyer's equation first put forward by Onitsch and co-workers [*10,11*]: $H_m = 1.854\ ad^{n-2}$, where n and a are characteristic of the material being tested and d is the diameter of the indent. Upit and Varchenya showed that, over the range of depths investigated, n was the same for all the test materials with the sodium chloride (NaCl) crystal structure, leading to the conclusion that this parameter is structure dependent. They also showed that very fine grained material did not show any increase ($n = 2$). This result led to the conclusion that the size effect was related to the size of dislocation loops formed during indentation—a conclusion which is hard to reconcile with the additive increase found by Gane and Cox.

In this work, the validity of the modified Meyer's law will be reexamined, and deviations will be noted. We will examine the hardness as a function of depth or diameter for four materials: gold, nickel, silicon, and lithium fluoride (LiF). For nickel and gold, the data cover four orders of magnitude in depth. Gane and Cox's data for gold and the additive nature of hardness effects will be reexamined in terms of the modified Meyer's law.

Unloading

Clearly, if one hopes to measure the hardness of materials by measuring indenter displacement and calculating contact areas from the indenter shape, the amount and type of elastic accommodation of the indenter displacement are important. The important distinction is between the displacement recovered through shape changes of the indent and that accounted for by an elastic field outside of the borders of the indent. If the entire elastic recovery was due to changes in the shape of the indent, then the depth under load would be used to calculate the contact area. However, if the recovery were entirely due to a general motion of the area surrounding the indent, the depth under load minus the recovered depth or the residual depth would give the best approximation of the contact area.

Two interesting works on this subject are those of Stilwell and Tabor [*12*] and of Lawn and Howes [*13*]. Stilwell and Tabor use the elastic solution for a rigid cone contacting a flat elastic body to reach the conclusion that 36% of the recovery is outside of the indent, and 64% is due to changes in the shape of the indent. The recovery curves they predicted are in good agreement with their experimental results; however, they tested a limited number of materials, all with similar values of E/H, where E is the elastic modulus. The metals used have small elastic recoveries.

Lawn and Howes found fault with the Stilwell and Tabor treatment in that no direct account of the effect of the loading process on elastic recovery was taken. Lawn and Howes put forward a model that accounted for the effects of the loading process. They also measured the change in shape of indents in a variety of materials with a range of values for E/H. The researchers did not address the question of the amount of elastic accommodation outside the indent except through a material-independent constant, γ, the value of which is used as a fitting parameter. They discuss the problems with this approach and point out that γ may change with the material.

In this paper we will discuss relaxation data from a range of materials and present a model of the relaxation process.

Experimental

The experimental technique used to prepare the specimens and take the data presented in this work have been described elsewhere [*14*]. However, some important points are noted here. The surfaces of the nickel were electropolished, the gold was deformed against glass, the silicon was mechanically polished, and the LiF was cleaved. The experimental hardness tester at Brown-Boveri and Company Research Center used for the hardness tests has a load resolution of $\sim 2.5\ \mu$N and an indenter displacement resolution of 0.2 nm. Proper steps were taken to eliminate impact effects (approach speeds < 10 nm/s $\rightarrow$ impact energy $< 10^{-18}$ J) and vibrational effects. Vibrational effects were reduced by locating the system on the ground level on a 700-kg stone slab, which was isolated from the floor by an air suspension system. The microscopy was done on a JOEL 200 CX scanning transmission electron microscope (STEM) using conventional replication techniques.

Results

Loading

The only way to be certain that the hardness measurements calculated from depth measurements were as accurate as possible was to image indents of various sizes and measure their areas. To accomplish this, arrays of indents like the one shown in Fig. 1 were made on each material. This process was facilitated by using the automated indentation and high precision x–y tables. The number, depth, and position of each indent is preprogrammed, and the whole series is done in a single computer run. The map shows the position and depth of each indent. Figures 2 and 3 show indents with nominal depths of 500 and 10 nm, respectively, replicated from the surfaces of electropolished nickel. The measurements of indents from nickel were used to generate a depth-versus-area function for the diamond used. The deviation of this area curve from that calculated from the intended geometry of the indenter can be accounted for by a 1.2° error in the face angle and a 1.5-nm truncation of the tip.

The material characteristics that may affect this function are the amount of pileup or sink-in around the indent and the amount of elastic recovery of the indent depth. Gold, nickel, and LiF showed similar amounts of elastic recovery, and the edges of the indents appeared convex, indicating some pileup material. Figure 4 shows the load versus depth and the depth versus hardness (as calculated using a depth versus area function) for nickel. These curves are the average of twelve indentations. They are representative of the results for gold and LiF. The error bars on the hardness curve and unloading curves are two standard deviations long.

The silicon behaved quite differently, as shown in Fig. 5. When silicon is tested, 54% of the indenter displacement is elastically recovered, and the indent edges appear concave. These effects will be discussed in greater detail in the section concerning elastic recovery.

The importance of the piled up material around the indent is questionable. The constraint of this material is not the same as that of the material beneath the indenter and therefore cannot support the same stresses. This was shown experimentally by Bergsman [*15*]. He showed that when piled up material was removed and the indentation reloaded, the resulting increase in indent size was very small. The reduced area of the silicon indents due to elastic recovery and convex edges must, of course, be accounted for when hardness is calculated from depth.

Figures 6 through 9 show the hardness of nickel, gold, LiF, and silicon over the range studied in this work. The data for gold and nickel cover four orders of magnitude in depth. The data for LiF and silicon cover two orders of magnitude. The data are plotted on log-log plots; thus, if the modified Meyer's law is obeyed, a straight line will result.

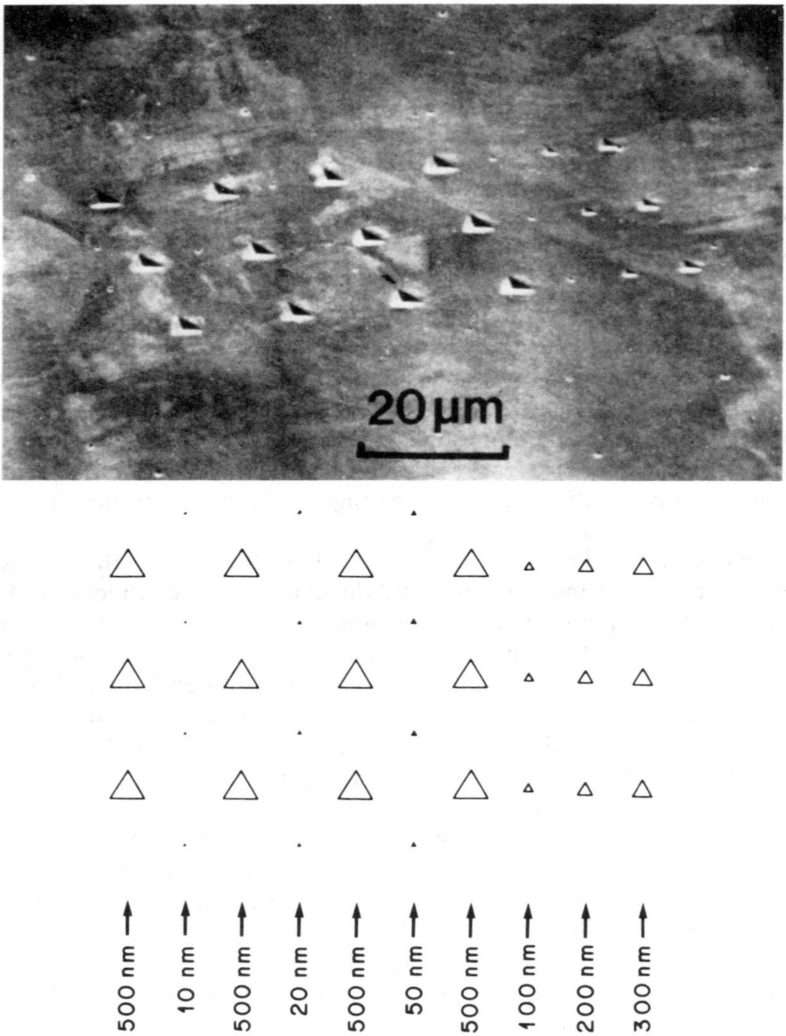

FIG. 1—*Array of hardness indents in electropolished nickel imaged directly in the SEM and a schematic showing nominal depths.*

The data of Gane and co-workers for gold are included on Fig. 6. To convert their data from diameter to depth, it was assumed that the depth would be that associated with a triangular indent of the geometry used in this work with a center-to-tip distance equal to the radius of the indent originally reported. The $d/D = 0.4$ (D = indenter diameter) data with a characteristic strain of 8% were used as this factor most closely corresponds to the indenter

1 μm

FIG. 2—*TEM image of a replicated indent from electropolished nickel,* d = *500 nm.*

used here. The best method for such a conversion is certainly debatable, but any reasonable method would simply result in some constant factor times the depth plotted and would not affect the shape of the curves. The range of slopes for log hardness versus log depth relationships, as determined by Upit and Varchenya for the NaCl structure, are shown with the LiF data in Fig. 8.

Figure 9 shows the data for silicon. No cracking was apparent from the load–depth curves or the replicas examined. To correct for elastic relaxation in the depth, the hardness value calculated from depth measurements was corrected by multiplying the hardness by the factor 0.54^{-2}. The 0.54 factor is the ratio of the elastic relaxation factor obtained for nickel divided by that

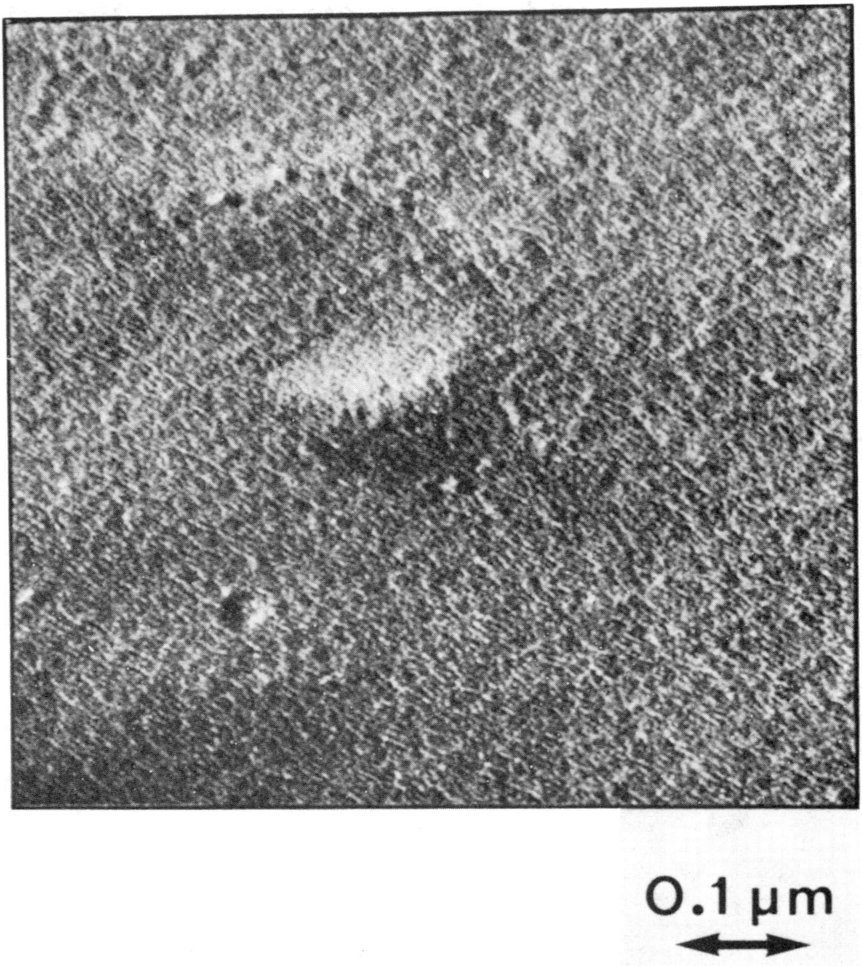

FIG. 3—*TEM image of a replicated indent from electropolished nickel:* d $= 10$ *nm; the tilt is 50°.*

obtained for silicon. The elastic relaxation factor is obtained by taking the ratio of the depth obtained by projecting the relaxation curve to zero load to the depth under load. Except for this correction, the area curve is the same as that used for the nickel data. Clearly this is a crude correction and ignores the concave nature of the indent edges as well as the shape changes of the impression that might occur during unloading; however, it allows us to reach some approximation of the hardness in a region where it is impossible to image indents (25 to 100-nm depth).

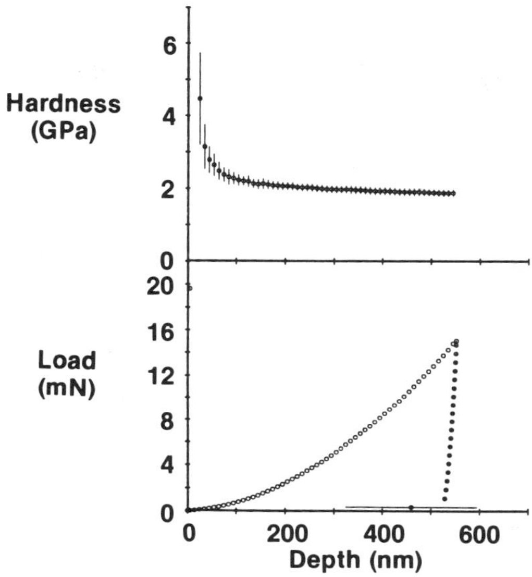

FIG. 4—*Load and hardness versus depth for electropolished nickel; the error bars are two standard deviations in length.*

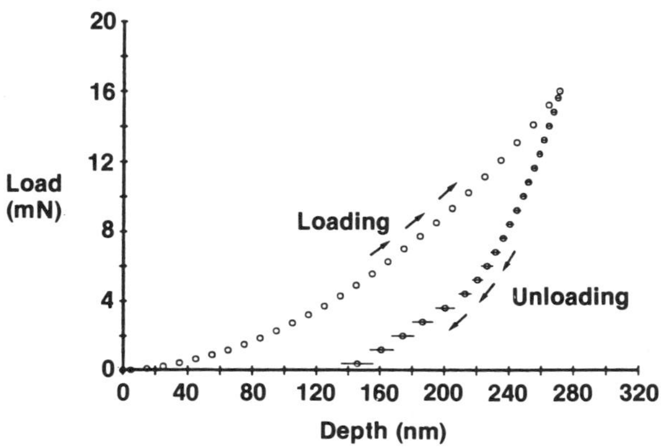

FIG. 5—*Load versus depth for silicon; the error bars on the relaxation part of the curve are two standard deviations in length.*

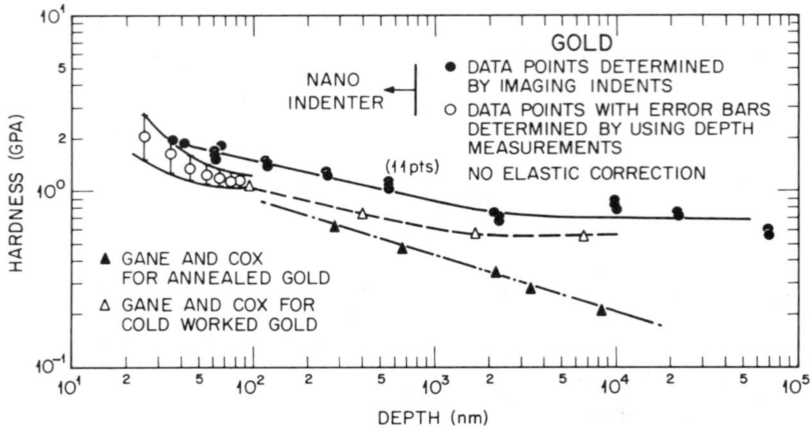

FIG. 6—*Hardness versus indent depth for gold.*

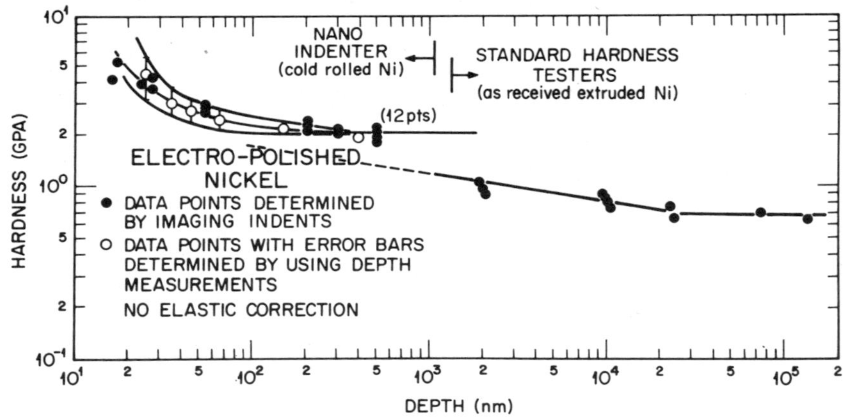

FIG. 7—*Hardness versus indent depth for nickel.*

Unloading

From the data for indenter penetration versus load, it can be seen that the indenter relaxes away from the surface as the load is decreased (Figs. 4 and 5). This unloading part of the curve is almost exactly reversible, implying that the movement involves the release of elastic strains under the indenter. It is possible that some small amount of nonelastic backflow is occurring; however, it is not a significant effect. In what follows, we will take the recovery to be fully elastic; any error thereby introduced is small.

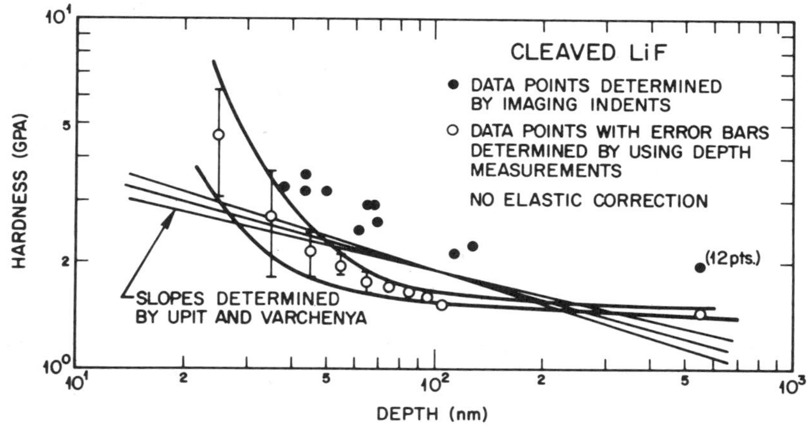

FIG. 8—*Hardness versus indent depth for LiF.*

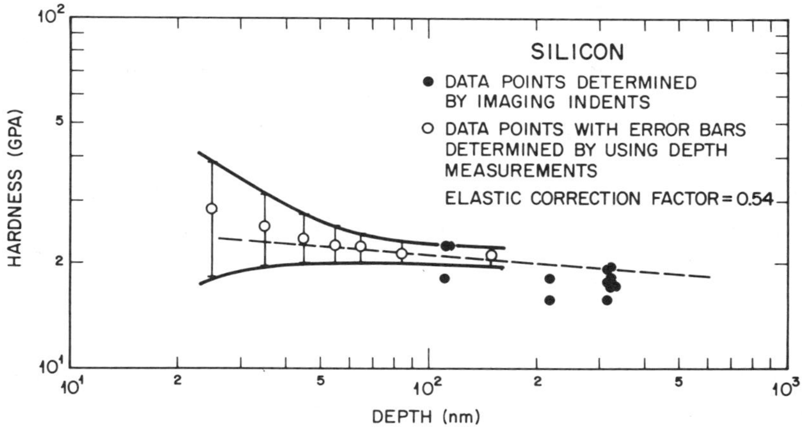

FIG. 9—*Hardness versus indent depth for silicon.*

Discussion

Loading

Several precautions must be taken to ensure the quality of data when very light load hardness testing is undertaken. The most important problems are the dangers of impact loads, vibrations, and the geometry of the indenter. The impact loads and vibrations were discussed briefly in the experimental section of this paper.

One important consideration concerning the geometry of the indenter is that it is sharp enough that the assumption of geometric similarity is satisfied. This means that the load required to develop a fully plastic zone is small enough that it does not affect the measurements. As previously discussed, the area curve determined for nickel indicates a 1.5-nm blunting of the diamond tip. If the extreme case of a flat end is considered, this implies a tip with an area of 16.87 nm^2. The hardest material we measured was silicon, $H \approx 25$ GPa. Thus, the force required to make an initial indent covering the tip area is 4.2×10^{-1} μN. This is less than one fifth of the force resolution of the hardness tester (2.5 μN).

The data for gold presented in Fig. 6 extends the range of the depth at which measurements have been made by a factor of three in the shallower direction. The differences between the data generated through imaging indents and depth measurements are most likely caused by the use of the area function determined for nickel.

The slopes of the straight-line sloping portions of all of the data, including Gane and Cox's, are similar. The depth at which the data for the cold-worked gold becomes horizontal is approximately 1 to 2 μm. Gane and Cox concluded that the hardness increase at shallow depths was additive to the increase of the bulk hardness due to cold working. This conclusion was reached by plotting the data on a linear scale. Such a plot makes it difficult to see the effects at shallow depths. The log–log plot in Fig. 6 shows these effects more clearly. We feel that the replotting of Gane and Cox's data taken together with our results show that work hardening causes the hardness to become invariant at depths greater than a particular one. That particular depth is determined by the intersection of a horizontal line through the bulk hardness and a sloping line representative of the modified Meyer's law. The modified Meyer's law line does not seem to be dependent on prior deformation history.

This result suggests that a possible cause for the increase in hardness at depths at which the modified Meyer's law is obeyed ($n \neq 2$) is a decreasing dislocation source size. Clearly, if the increase is due to changes in dislocation source size, when the size of the indent approaches the subgrain size, the grain size, or some other barrier to slip, the depth dependence would change. As these other potential barriers in the material become more closely spaced, the plateau of hardness occurs at higher hardness values. Thus, for a recrystallized material, as the grain size decreases, the bulk hardness should increase. This effect is verified by Upit and Varenchya in their work on fine-grained materials and by various authors who have shown the grain-size dependence of hardness. One interesting question is the identity of the potential barrier causing the hardness in the gold tested in this work, and in the work by Gane and Cox, to become invariant. Considering the characteristic depth of 1 μm, where the transition occurs, the most likely obstacle would be the cell size, caused by the large strain deformation used to cold work the gold. The grain size is certainly too large, and the dislocation intersection length is probably too small. Further experimentation is planned to sort out this question.

The nickel data shown in Fig. 7 also indicate that the modified Meyer's law could explain the data. If one interprets the data in the same way as described previously for gold, one must assume that the bulk hardness for nickel changes rapidly with deformation. This would be expected with a material that has a high work-hardening coefficient, as nickel does. For the higher bulk hardness nickel, which was used for very shallow testing (the material had been cold worked), the bulk hardness is so high that the mechanism giving rise to the sloping behavior shown by the less worked nickel does not control the strength of the material at any depth.

The nickel data show that there is a clear deviation from the straight-line portion of the curve at very shallow depths ($d < 50$ nm). This trend is shown in both the data taken by imaging indents and the hardness calculated using depth data. Figure 10 shows that the hardness is inversely dependent on the depth in this region. However, the relationship between hardness and depth does not pass through zero but is of the form $H = C_1 d^{-1} + C_2$. It is not clear if such a region is shown in the gold data; if so, it occurs at much shallower depths.

The LiF data shown in Figs. 8 and 11 also show the region in which the hardness has a d^{-1} dependence. The silicon data in Fig. 9 are inconclusive on this point. The form of the equation $H = C_1 d^{-1} + C_2$ strongly suggests two additive strengthening effects. In fact, the equation could be written as $H = C_1 d^{-1} + 1.854\ a d^{n-2}$. The second term of this equation changes slowly enough to be approximately constant over the range that the d^{-1} term has been observed.

The LiF data show the largest discrepancy between the areas imaged and

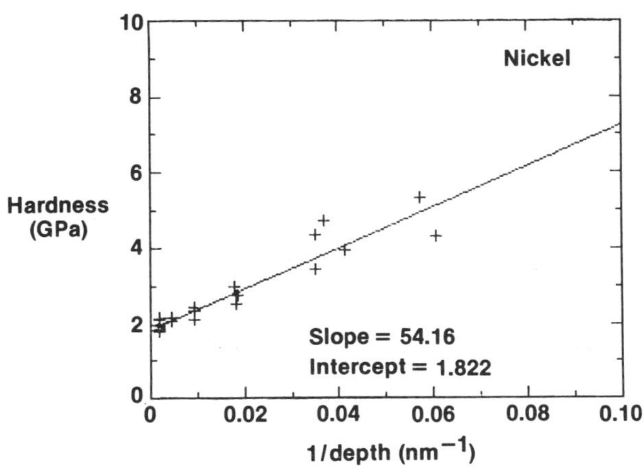

FIG. 10—*Hardness versus* d^{-1} *for nickel.*

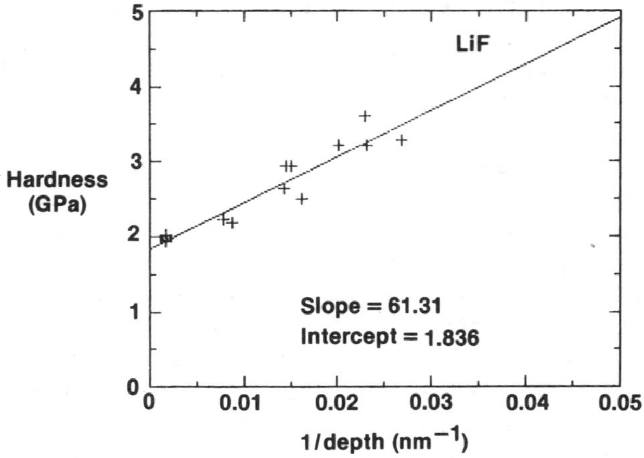

FIG. 11—*Hardness versus* d^{-1} *for LiF.*

the calculated areas. This could be due, in part, to the use of an area function determined for nickel; however, the 30% difference is large and the elastic effects and indent shapes are very similar to those of nickel. It is likely that some systematic experimental error occurred in determining the depths of the indents in LiF. The shapes of the curves defined by each set of data are very similar, but there is an effect in absolute hardness. It is also difficult to explain the shape of the curves when the slopes determined by Upit and Varchenya are considered. Figure 11 shows that the LiF data fits a d^{-1} dependence better than the modified Meyer's law in the region investigated by this work.

The data for silicon are shown in Fig. 9. These data are again well represented by the modified Meyer's law. One interesting result of the large elastic effect in silicon is that it makes the imaging of indents more difficult, but makes the depth calculation data capable of measuring indents roughly half as deep. In fact, Fig. 9 shows that the depth-sensing technique can be used to take hardness measurements at depths an order of magnitude shallower than the smallest indents that could be imaged with the TEM.

Although the data shown in Figs. 6 through 11 are not completely conclusive, three important features of the functional dependence of hardness and depth are indicated. These three regions of behavior can be characterized as follows, in order of reducing indent size. The first is a region in which the volume of deformed material caused by the hardness indent is as large as or larger than all the important barriers to slip in the material. In this region the hardness is invariant as the depth is changed. The second is a region where the modified Meyer's law, with $n \neq 2$, is representative of the hardness. The limits of this region depend on the prior deformation of the material; however, the exponent and constant in the modified Meyer's law do not. There are indications that this behavior is a result of the reduced volume of deformation, perhaps through reduced dislocation source size. Calculations show that the depth at which direct surface effects will be observed is much shallower. The third region is apparent within a few tens of nanometres of the surface. Here, the surface could possibly directly affect hardness. Image forces, contamination layers, and electric fields are some possibilities of surface phenomena that could affect hardness. The functional dependence of hardness on depth in this region is close to

$$H = C_1 d^{-1} + C_2$$

These suggestions are supported by the apparent way in which the separate hardness contributions combine. The hardness increase due to bulk work hardening is not additive with the hardening resulting from reducing the dislocation source size, but it does limit the depth range over which such a dependence occurs. The hardness increase from the surface effects should be, and appears to be, additive with other contributions.

The functional dependence of hardness on depth described earlier is somewhat speculative considering the available data. More experiments on materials with carefully controlled microstructures are now being undertaken in an attempt to better understand and separate those hardening effects that are the result of the reduced volume being deformed and those that are directly related to the presence of a free surface.

Unloading

A problem in analyzing indenter relaxation occurs because the location and intensity of stresses during contact are determined by the loading plastic process, and hence it is unclear which elastic case, for which a solution exists, would most closely resemble the final state obtained. However, from the relation between the indenter displacement, d, and the load, L, in the relaxation portion of a curve, it is possible to discuss qualitatively the nearest appropriate elastic contact. Indenters of various degrees of sharpness can be treated in the following way. If the indenter is taken to be a rigid axisymmetric solid of revolution about the z-axis of the curve $z = r^{2m}$, and such an indenter is brought in contact with a flat surface with all strains remaining elastic, then L and d are related by [*16*]

$$L \propto d^{(2m + 1)/2m}$$

If $m = 1/2$, a cone results. As m increases, the indenter becomes blunter. The corresponding interfacial pressure distribution for the cases $m = 1$ and $m = 2$ are given in Fig. 12 [*16*]. Clearly the amount of elastic recovery depends heavily on the ratios of the elastic modulus to the hardness. Some extreme

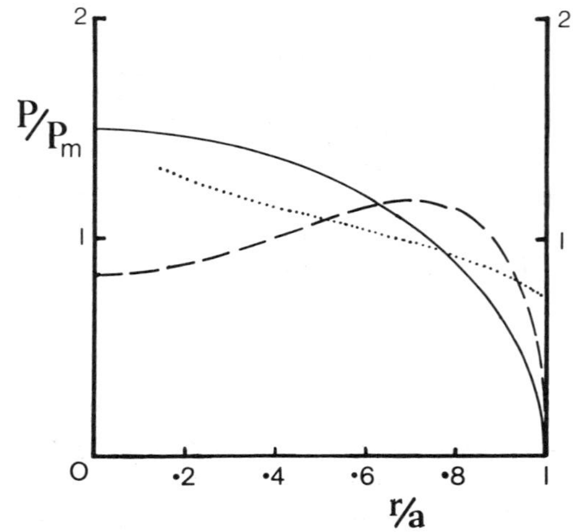

FIG. 12—*Pressure,* P*, at radius,* r*, within a contact of the radius,* a*. The mean pressure is* $P_m = L/\pi a^2$*. The solid line is elastic loading* m = *1 (sphere). The broken line is elastic loading* m = *2. The dotted line is the ideal plastic deformation under a cone.*

cases that we tested were gold, chromium, and silicon, for which E/H is 138, 25, and 7.9, respectively. For the wide range of materials we have tested, the relaxation data may be fitted to the aforementioned power law, with values of m ranging between 1 and 2. Thus, the interfacial pressure is likely to be between the two curves $m = 1$ and $m = 2$. The actual pressure distribution under our (pyramidal) indenter is likely to be similar to that obtained for rigid ideal plastic indentation with a cone [*17*], shown as a dotted line in Fig. 12. Interestingly, this distribution is in quite good agreement with that inferred from the preceding simple elastic analysis. The model seems reasonable. Note that this result implies that the conical/pyramidal indenter actually behaves upon unloading as if it were considerably rounder and flatter (m is greater than the $m = 1/2$ of a cone). This is in agreement with the expanding cavity models of the indentation process [*18*] in which a cap of "dead" material is taken to exist around the indenter tip, effectively blunting the tip.

Thus, the model suggested here predicts that the elastic motion of the indenter is behaving like a blunt tip. This conclusion has consequences for the important question of where the largest elastic effects are encountered. In the case of a flat punch, all of the elastic accommodations must be outside of the indent shape.

Stilwell and Tabor's model does a good job of predicting the elastic displacement of a conical indenter for materials with small elastic effects (high values of E/H). However, the prediction of their model is that the contact area would decrease continuously with load. This is in disagreement with the results of Pethica [*7*], which showed that the contact resistance stayed constant as the load decreased until most of the load was removed. In addition, their prediction that 64% of the relaxation occurs within the boundaries of the indent seems unreasonable in the cases of materials with large elastic recoveries.

The data for silicon shown in Fig. 5 are interesting. The large elastic effect is a result of its low value of E/H. Images of the indents, published elsewhere [*14*], show very noticeable concavity of the edges of the indent. This indicates that a significant amount of the elastic effect was outside of the indent area. The same effect, to a lesser degree, can be observed in some of the images of indents published by Lawn and Howes [*13*]. In addition, if the residual depth (depth under load minus the elastically recovered depth) is used to calculate the area, the values agree reasonably with the areas measured in the TEM, again indicating large elastic effects outside of the indent area.

The available data are not conclusive on the question of how indenter displacement is elastically recovered; however, it seems that when large fractions of the indenter motion are elastically recovered, the majority of the evidence indicates that the best approximation of the contact area is obtained using the residual depth measurement. The model presented in this work is capable of handling a wide range of material behaviors (values of E/H) and can be used to model relaxation curves, but it does not really answer the question of where the elastic strain is recovered.

Conclusions

1. Hardness becomes invariant with depth above a certain size indent. The value of the hardness in this invariant region is strongly dependent on the prior deformation history, that is, the microstructure of the material being tested.

2. When indents are made shallower than the depth at which the hardness becomes invariant, the hardness can obey the modified Meyer's law

$$H = 1.85ad^{n-2}$$

where d = indent diameter and a and n are material constants.

3. At still shallower depths ($d < 50$ nm), other effects become apparent and the hardness can vary inversely with the depth, that is

$$H = C_1 d^{-1} + C_2$$

4. The mechanism that gives rise to the modified Meyer's law behavior is not additive with the bulk hardness. The larger of the two gives the apparent hardness.

5. The unloading portion of the load displacement curve is well modeled by

$$L \propto d^{(2m + 1)/2m}$$

where $1 < m < 2$, which is the elastic solution for a restricted range of indenter shapes.

6. When large elastic recoveries are observed, a large portion of the relaxation is due to strain outside the area of contact.

Acknowledgment

This research was carried out in part at Brown-Boveri Research Center, Baden, Switzerland.

References

[*1*] Tabor, D., *The Hardness of Metals,* Clarendon Press, Oxford, 1951, p. 31.
[*2*] Mott, B. W., *Microindentation Hardness Testing,* Butterworth, London, 1956, p. 101.
[*3*] Gane, N., *Proceedings of the Royal Society (London),* Series A, Vol. 317, 1970, pp. 367–391.
[*4*] Gane, N. and Cox, J. M., *Philosophical Magazine,* Vol. 22, 1970, pp. 881–891.
[*5*] Macmillan, N. H. and Gane, N., *Journal of Applied Physics,* Vol. 41, No. 2, February 1970, pp. 672–673.
[*6*] Gane, N. and Bowden, F. P., *Journal of Applied Physics,* Vol. 39, 1968, p. 1432.
[*7*] Pethica, J. B. and Tabor, D., "Contact of Characterized Metal Surfaces at Very Low Loads: Deformation and Adhesion," *Surface Science,* Vol. 89, 1979, pp. 182–190.
[*8*] Upit, G. P. and Varchenya, S. A., *Physica Status Solidi,* Vol. 17, 1966, p. 83.
[*9*] Upit, G. P. and Varchenya, S. A., in *The Science of Hardness Testing and Its Research Applications,* J. H. Westbrook and H. Conrad, Eds. American Society for Metals, 1973, pp. 135–144.
[*10*] Onitsch, E. M., *Microskopie,* Vol. 2, 1947, p. 345.
[*11*] Onitsch, E. M., "Microhardness, Its Theory and Practice with the Reichert Microhardness Tester," Reichert, Vienna, Austria, 1950.
[*12*] Stilwell, N. A. and Tabor, D., *Proceedings of the Physical Society, London,* Vol. 78, No. 2, 1961, pp. 169–179.
[*13*] Lawn, B. R. and Howes, V. R., *Journal of Materials Science,* Vol. 16, 1981, pp. 2745–2752.
[*14*] Pethica, J. B., Hutchings, R., and Oliver, W. C., *Philosophical Magazine A,* Vol. 48, No. 4, 1983, pp. 593–606.
[*15*] Bergsman, E. B., *Sheet Metal Industries,* Vol. 31, 1954, p. 31.
[*16*] Lur'e, A. I., *Three-Dimensional Problems of the Theory of Elasticity,* Interscience, New York, 1964, especially Chapter 5.
[*17*] Lockett, F. J., "Indentation of a Rigid/Plastic Material by a Conical Indenter," *Journal of the Mechanics and Physics of Solids,* Vol. 11, 1963, pp. 345–355.
[*18*] Johnson, K. L., "The Correlation of Indentation Experiments," *Journal of the Mechanics and Physics of Solids,* Vol. 18, 1970, pp. 115–126.

Ronald W. Armstrong[1] *and Wayne L. Elban*[2]

Dislocation Aspects of Plastic Flow and Cracking at Indentations in Magnesium Oxide and Cyclotrimethylenetrinitramine Explosive Crystals

REFERENCE: Armstrong, R. W. and Elban, W. L., "**Dislocation Aspects of Plastic Flow and Cracking at Indentations in Magnesium Oxide and Cyclotrimethylenetrinitramine Explosive Crystals,**" *Microindentation Techniques in Materials Science and Engineering, ASTM STP 889,* P. J. Blau and B. R. Lawn, Eds., American Society for Testing and Materials, Philadelphia, 1986, pp. 109–126.

ABSTRACT: Unique information about the hardness properties of crystalline materials can be obtained from studying the dislocation mechanisms involved in microindentation testing. The dislocation description provides information in addition to the knowledge of indentation processes gained from earlier continuum mechanics analyses and from models of the orientation dependence of hardness based on the geometrical arrangement of slip systems. For example, (1) mapping of the residual dislocation distributions within microindentation plastic zones can be used to characterize the deformation field and (2) specification of the dislocation interactions that occur among the crystal deformation systems is useful in evaluating the extent to which the hardness is affected by work hardening and cracking mechanisms. An interesting application of this information is to microindentation hardness testing of explosive crystals.

KEY WORDS: microindentation hardness, dislocations, pileups, cracking, scanning electron microscopy, X-ray topography, chemical etch pitting, magnesium oxide, cyclotrimethylenetrinitramine (RDX), energy dissipation, hot spots, microindentation hardness testing

Elementary Considerations

Multiple dislocation processes are generally involved in the plastic strain fields surrounding hardness indentations—particularly when cracking oc-

[1]Professor, Department of Mechanical Engineering, University of Maryland, College Park, MD 20742.

[2]Materials research engineer, Naval Surface Weapons Center, Silver Spring, MD 20910; now associate professor, Department of Engineering Science, Loyola College, Baltimore, MD 21210.

curs. In several ways the various dislocation aspects of plastic flow and cracking contribute to the complicated meaning of the hardness property. However, the theoretical and experimental properties of dislocations are now understood reasonably well for simple deformations and for well-defined states of stress. The positions of dislocations and the way they interact can be observed using a number of techniques. It is advantageous to apply this knowledge and the observational techniques, also, to obtain a better understanding of the hardness property. It is gratifying when such understanding can be combined with obtaining new information on the plastic deformation and cracking properties of any material. In this investigation, results relating to these considerations are reported for single crystals of magnesium oxide (MgO) and cyclotrimethylenetrinitramine ($C_3H_6N_6O_6$ or RDX) explosive.

Figure 1 shows in several simple representations the surface and (sectioned) subsurface geometry of a diamond pyramid indentation put into an otherwise flat material surface. To accommodate the indenter force without any change in volume, material has been moved from the central lower terrace structures to the surrounding ledge structure framing the indentation. The three-dimensional description leads to a relatively small size in the sectioned two-dimensional view of the raised ledge in comparison with the depressed region. Of greater interest are the mechanisms involved in moving material in accor-

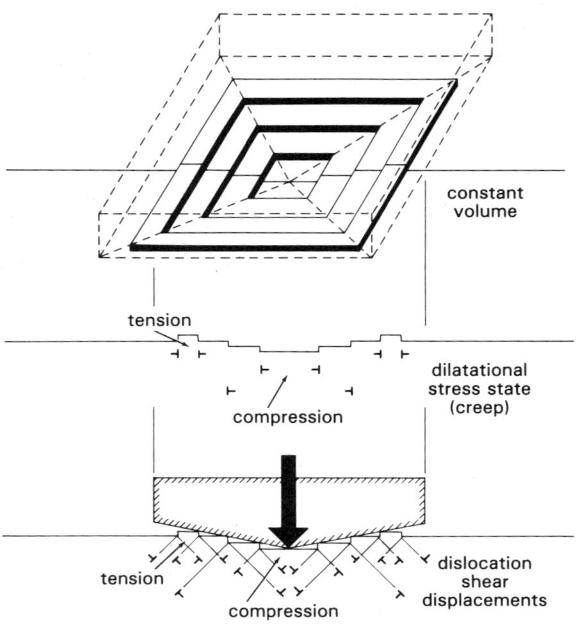

FIG. 1—*Rudimentary surface geometry and dislocation arrangements at an impression made with a diamond pyramid indenter (dashed lines).*

dance with the two dislocation configurations shown in the section views. The simplest dislocation configuration of prismatic loops might produce the required deformation under creep-type conditions—for example, if a small indenter force is slowly applied at a high temperature so that dislocation motion occurs predominantly by means of diffusional climb. The internal stress field is largely a dilatational one, exhibiting the compressive and tensile regions shown in Fig. 1. In general, there should be a volume change. No special considerations arise between the two-dimensional and three-dimensional views of the problem. Alternatively, the second case of accounting for the deformation by dislocations slipping on specific crystallographic planes is more realistic for the conventional hardness testing of single crystals. Although dilatational components of the residual stress field occur, as before, the dislocation interactions are more complex. The three-dimensional strain field is more complicated.

Dislocation Description of Indented MgO

The combined techniques of dislocation etch pitting, X-ray diffraction topography, and scanning electron microscopy have been applied to determining the distribution of dislocations around microindentation diamond pyramid hardness impressions put into the (001) cleavage surface of MgO crystals [*1,2*]. Figure 2 shows the three-dimensional distribution of dislocations which are produced on ⟨101⟩{10$\bar{1}$} slip systems to satisfy the imposed displacements of an indentation with its diagonal edges aligned along the ±[110] and ±[1$\bar{1}$0] directions. The observation of troughs spreading along the ±[100]

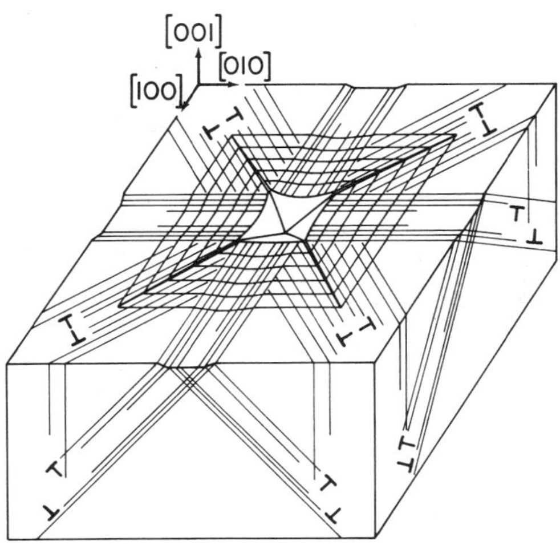

FIG. 2—*Primary dislocation distribution surrounding a diamond pyramid indentation on a (001) MgO crystal surface* [2].

and ±[010] directions in bands defined by the width of the indentation seems a surprising result because of the expectation that the microindentation will be totally surrounded by raised material to minimize any volume change (compare Fig. 1). However, the result in Fig. 2 is a natural consequence of the requirement in three dimensions that dislocations producing the primary displacements for forming the microindentation facets, say, with edges along ±[100] directions by slip on $[01\bar{1}]$ (011) and $[0\bar{1}\bar{1}]$ $(0\bar{1}1)$ systems, must carry their downward displacement vectors along the orthogonal ±[100] direction in a screw orientation, as the dislocations are forced out from the immediate vicinity of the indentation by others following behind.

Figure 3 is a scanning electron microscope (SEM) image obtained with backscattered electrons from a thin gold film put onto the surface of an MgO crystal indented in the manner just described [*2,3*]. The troughs are easily recognized even though the slip structure producing them is not. This is because of the cross-slip required of the particular screw dislocations involved. Stereoscopic SEM images were obtained to study in detail the surface relief of the slip band structures and the microindentation shapes so as to assess the extent to which any recovery has occurred during unloading of the indenter [*3*]. In general, the recovery effects are small, but they show an important crystallographic dependence. In Fig. 3, for example, the indentation has recovered to a greater extent along the centerline ±[100] and ±[010] directions than in the ±[110] and $\pm[1\bar{1}0]$ directions. The effect is attributed to the dislocation reactions on adjacent secondary ⟨101⟩ $\{10\bar{1}\}$ slip systems whose slip band traces are well defined in Fig. 3. Dislocations on these systems are

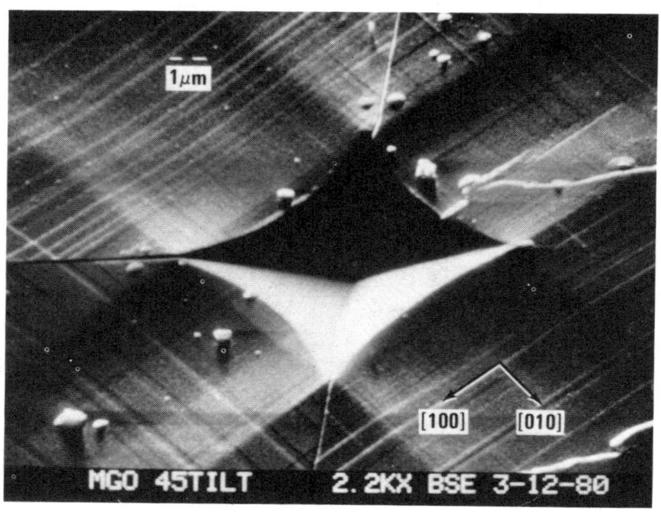

FIG. 3—*Scanning electron photomicrograph of a diamond pyramid indentation on a MgO (001) cleavage surface* [3].

responsible for the formation of shallow $\{1\bar{1}0\}$ cracks observed along the $\pm[110]$ and $\pm[1\bar{1}0]$ indenter diagonals. Such cracks are formed when the indenter diagonals are placed along the $\langle 100 \rangle$ family of directions and when a spherical indenter is used; this demonstrates the significant conclusion that the cracking is crystallographically determined.

Pileup Description of $\{1\bar{1}0\}$ Cracking

The square array of $\langle 101 \rangle$ $\{10\bar{1}\}$ slip systems surrounding indentations made on the (001) plane operate to produce volume-accommodating outward surface displacements. In Fig. 3, the material is raised along the $\pm[110]$ and $\pm[1\bar{1}0]$ directions, beginning from the tips of the indentation diagonals, where the $\{1\bar{1}0\}$ cracks are observed. Such cracking was first attributed to the reaction of dislocations, for example, on adjacent $[10\bar{1}]$ (101) and $[0\bar{1}1]$ (011) slip systems from dislocation etch pit results [*4*]. Figure 4 gives a schematic view of the dislocation features involved. Dislocation half-loops are shown to move from central positions on the respective slip planes to the adjacent diagonal $\langle \bar{1}\bar{1}1 \rangle$ lines of intersections of the planes [*5*]. The Burgers vectors, ($\vec{b}$), and line vectors, ($\vec{\xi}$), for the dislocations were chosen according to the self-consistent start-to-finish, righthand (SF/RH) vector system [*6*]. On the basis indicated in Fig. 4, the combination of $(a/2)$ $[10\bar{1}]$ and $(a/2)$ $[0\bar{1}1]$ Burgers vectors at the intersection of (101) and (011) slip planes, respectively, gives an $(a/2)$ $[1\bar{1}0]$ Burgers vector for a dislocation with a line vector along the $[11\bar{1}]$

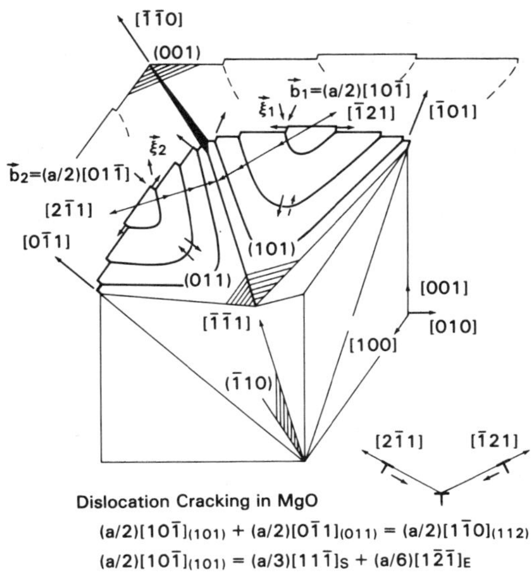

FIG. 4—*Dislocation pileup model for $(1\bar{1}0)$ cracking in MgO* [7].

direction, and is thus contained within the (112) plane. The resultant dislocation is sessile both on a crystallographic and an applied stress basis. By consideration of the screw and edge components of each dislocation, for example, $(a/3)\ [11\bar{1}]_S$ and $(a/6)\ [1\bar{2}\bar{1}]_E$ of the unit $(a/2)\ [10\bar{1}]$ dislocation, the annihilation of screw components is shown to make the overall reaction favorable [*7*]. Subsequent dislocations following behind the first ones are proposed to be blocked until a sufficiently large pileup stress is generated to force a crack nucleus to form by dislocation coalescence [*8*].

The dislocation explanation for $\{1\bar{1}0\}$ cracking at indentations on {001} surfaces of MgO and other alkali halide crystals with the rock salt structure is important because such cracking is unexpected relative to easier {100} cleavage cracking, which also occurs according to a dislocation reaction [*2*]. The full explanation of $\{1\bar{1}0\}$ cracking depends on the capability of the limited deformation systems available in these structures to satisfy the imposed indenter strain fields. This consideration relates directly to the operation of the adjacent ⟨101⟩ $\{10\bar{1}\}$ slip systems in Fig. 4. Slip along the $[\bar{1}01]$ and $[0\bar{1}1]$ directions is required to allow a resolved extensional strain along the [001] direction. However, the components of these slip vectors contained in the (001) plane are orthogonal and produce a corresponding tensile strain in the (001) plane, as given by the reacted $(a/2)\ [1\bar{1}0]$ dislocation Burgers vector. Continued forcing of the indenter strain requirement along the [001] direction gives an accompanying larger tensile strain in the (001) plane, which must eventually produce cracking unless some other deformation mechanism or system is activated to alleviate the situation. The six ⟨101⟩ $\{10\bar{1}\}$ slip systems of the rock salt structure give only two independent deformation systems, which, when taken with the three independent ⟨101⟩ {010} systems, satisfy the minimum condition of allowing the five independent strains required for an arbitrary deformation [*9*]. The ⟨101⟩ {010} systems are not capable of giving an extensional strain along the [001] direction, as shown previously [*2*]. Consequently, either cross-slip of the pileup dislocations or slip on an alternative deformation system such as ⟨101⟩ $\{11\bar{1}\}$ appears to be required to prevent $\{1\bar{1}0\}$ cracking. The major point is that the character of the dislocations and their interactions are connected directly with the properties of the deformational field imposed by indenters put into MgO crystals. Such dislocation interactions were proposed to be a major factor influencing the anisotropy of Knoop hardness measurements in MgO [*2*].

Hardness Dependence on Indentation Size and Crack Size

A number of investigations have been made of the effective indenter force required to produce different indentation sizes and, also, diametral crack sizes in MgO crystals. Some particular results are shown in Fig. 5 [*1,10–12*]. The data cover a range between essentially static, crack-free, diamond pyramid microindentation hardness measurements made at a force as small as

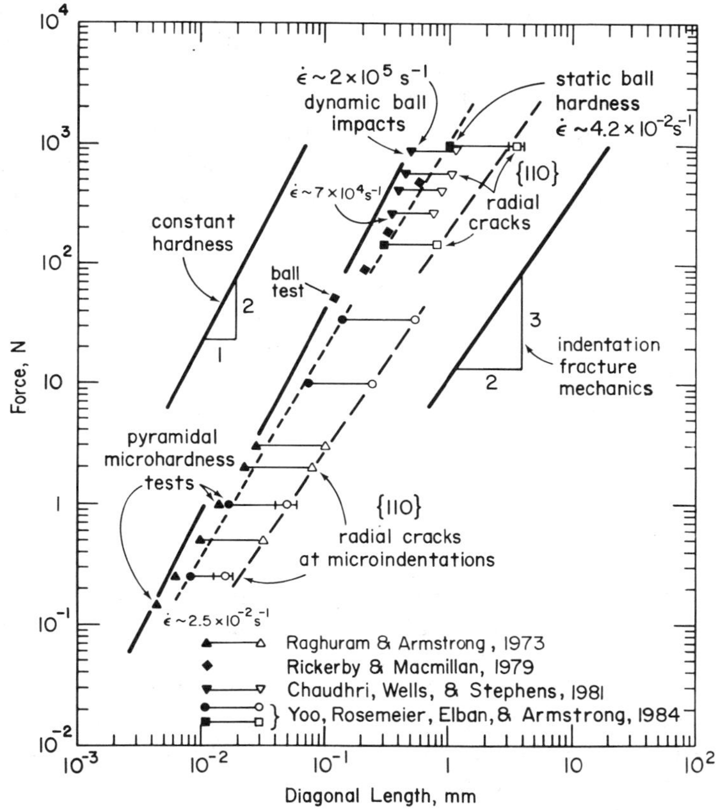

FIG. 5—*Logarithmic variation of indenter forces against indentation sizes and crack lengths for MgO* [12], *the closed symbols are indentation sizes and the open symbols are diagonal crack lengths.*

0.15 N [*1*], on the one hand, and macroscopically cracked, either static [*12*] or dynamic [*11*] ball indentation tests made with forces as large as 1.0 kN, on the other hand. The measured force dependence on indentation size and crack size is compared with an exponential slope of 2.0 for constant hardness [*1*] and an exponential slope of 1.5 for constant work of fracture on an indentation fracture mechanics basis [*13–15*]. The hardness strain rate, $\dot{\epsilon}$, values which are shown for the essentially static results were estimated by dividing the projected ball indentation-to-diameter ratio, or its effective strain equivalent of 0.375 for the diamond pyramid tests, by the 15-s dwell time of the indenter. The individual hardness pressures corresponding to these data range between 15 kN/mm^2 for the crack-free 0.15-N indentation and 1.3 kN/mm^2 for the slowly loaded but cracked 1.0-kN indentation.

The total data in Fig. 5 are consistent with the observation that reduced hardness occurs with an increased extent of $\{1\bar{1}0\}$ radial cracking for MgO crystals. The comparison of diamond pyramid microindentation hardness values at 0.25 N shows clearly that the smallest amount of cracking is associated with a reduced hardness [*1,12*]. The indentation diagonals and the associated crack lengths show connected dependences in Fig. 5. The overall diagonal lengths of the cracks, which form during loading but otherwise are similar to Palmqvist cracks, are very nearly in agreement with a ($^3/_2$) dependence of the applied force [*16*]. The macroscopic static ball results show a similar crack length dependence of the indenter force. Although the dynamic ball impact results give larger hardness values than are obtained from the static results, even the highest rate dynamic results give lower hardness values than are obtained for the static, crack-free microindentation results. The dynamic results in Fig. 5 do show that an increasing hardness is obtained with increasing strain rate even if cracking occurs. No doubt, the positive rate effect is due to the rate dependence of the dislocation processes involved in the indentation plastic flow and cracking events.

Energy Dissipation

One way to interpret the reduction in hardness pressures accompanying an increased extent of cracking in MgO at larger applied forces is that the cracking reduces the amount of work which is otherwise necessary to produce an increasingly larger indentation. It might be expected that cracks emanating from the indentation site act as stress raisers for enhancing the effective stress field. The dislocation flow depicted on (110) and $(1\bar{1}0)$ slip systems in Fig. 2 gives evidence of that. Also, the cracks are capable of dissipating stored energy in the otherwise work-hardened material, and this dissipation process has been proposed on the basis of X-ray diffraction topography evidence to be intimately involved in contributing to the measured reduction in hardness values [*12*].

Figure 6 shows a matched set of optical and X-ray topographic pictures of macroindentations put into a MgO (001) cleavage surface with forces of 147 and 981 N acting on a 1.59-mm spherical steel indenter in a Rockwell tester. Stable support of the indenter load was not reached for either indentation, but the rate of indenter penetration had decreased very appreciably by the end of the 15-s dwell time. Sequential cracking noises were heard during settling of the indenter. Such cracking has been observed to produce load drops in autographic records of load versus time as obtained during a continuously monitored indentation test [*10*]. In Fig. 6 (*top*), the significant extent of cracking along the $[1\bar{1}0]$ direction can be recognized at the larger indentation. The same crack is also recognizable by its relatively minor effect on the X-ray diffraction contrast in Fig. 6 (*bottom*). From the results of previous investigations [*1,2*], the central white area of absent diffracted X-ray intensity should cover the physical area included within the extent of cracking while an

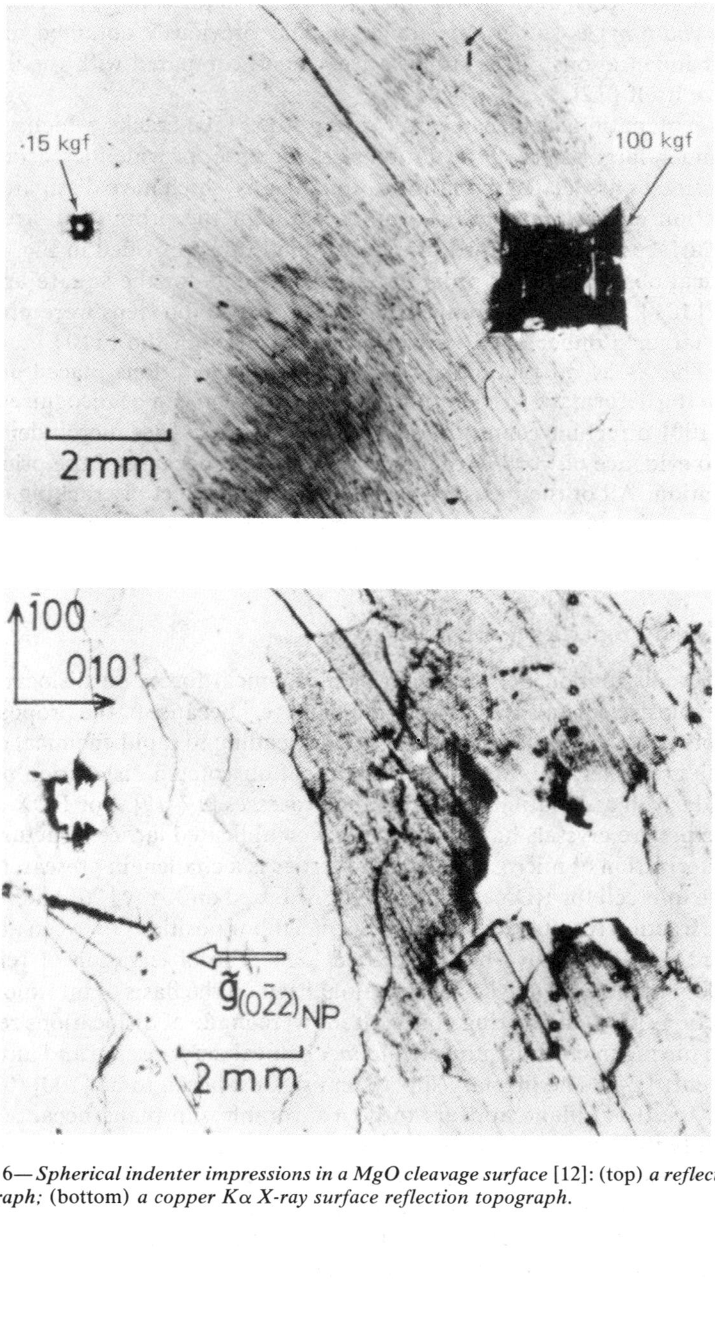

FIG. 6—*Spherical indenter impressions in a MgO cleavage surface* [12]: (top) *a reflected light photograph;* (bottom) *a copper Kα X-ray surface reflection topograph.*

outer region of enhanced intensity should be expected because of the cumulative residual dislocation strain fields locked in at the indentation site. The diffraction contrast at the X-ray image centered on the smaller indentation in Fig. 6 (*bottom*) is somewhat similar to that previously obtained at much smaller indentations but is still relatively small compared with the indentation size itself [*12*].

The explanation of the new results is that the $\{1\bar{1}0\}$ cracks, which are relatively much larger as the indentation sizes increase, provide increasing relief of the stresses associated with dislocation pileups which have dissipated their interaction energy, mainly by running through the prominent stress-free crack surfaces. Evidence for this explanation is also provided in Fig. 6 (*top*) by optical observations of pronounced step heights for the square arrays of $\langle 101 \rangle \{10\bar{1}\}$ volume-accommodating slip, because the steps were produced by the larger number of dislocations moving through the $\{1\bar{1}0\}$ crack surfaces. The X-ray diffraction contrast at microindentations placed at 1.0-N force in the deformation field of the large indentation can be recognized along its $\pm[100]$ direction centerline in Fig. 6 (*bottom*). The microindentations gave no evidence of work hardening being present because of the prior large indentation. All of the measurements confirm the effect of cracking on promoting strain energy release by the mechanism of dislocations escaping through the $\{1\bar{1}0\}$ crack surfaces [*12*].

Dislocation Properties in RDX

Energy dissipation by the action of mechanical forces on dislocations in RDX explosive crystals is of considerable interest because of the proposal that hot spots (small, high-temperature regions) leading to rapid chemical decomposition are generated during the collapse of obstructed dislocation pileups, particularly in association with cracking processes [*17–19*]. For RDX and related explosive crystals having reasonably complicated lattice structures, the characterization of microstructural properties is a challenging research activity. The unit cell for RDX is shown in Fig. 7 based on X-ray [*20*] and neutron [*21*] diffraction results [*22*]. The nearest neighbor positions between atoms in adjacent molecules have been specified [*21*]. The observation of relatively easy (001) plane cleavage has been rationalized on the basis of intermolecular interactions [*23*]. Pioneering observations were made of dislocations revealed by etch pitting in crystals subjected to mechanical deformation and annealing treatments [*24*]. The primary slip system was proposed to be [100] (010). In Fig. 7, the (010) plane appears to be a favorable slip plane because of the possibility that shear translation can occur (for example, across the rhombohedral bases which are shown) without severely affecting the molecular units. X-ray topography by the Lang transmission method was applied to observing dislocations introduced during crystal growth from solution [*25,26*]. Results were reported on the indentation fracture mechanics testing of RDX [*27*],

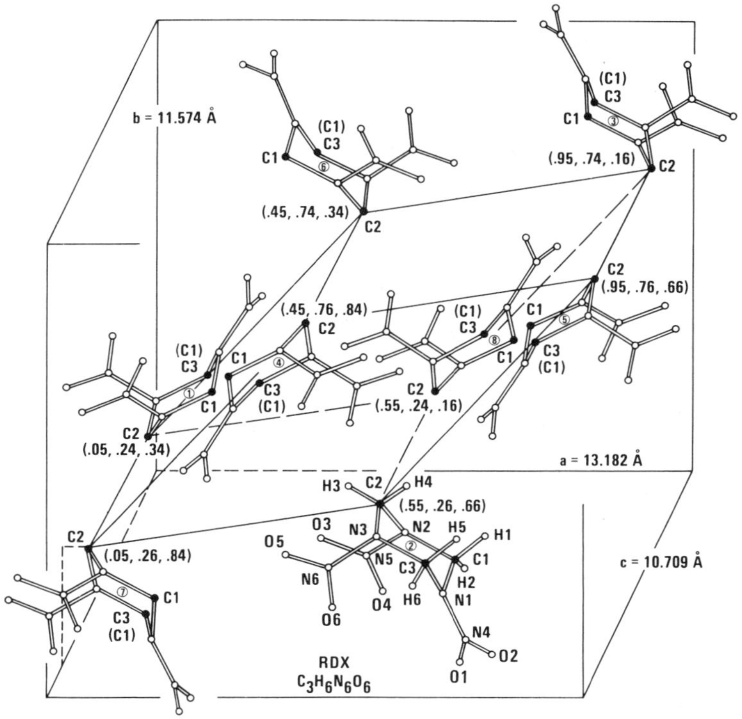

FIG. 7—*Orthorhombic unit cell for RDX (cyclotrimethylenetrinitramine)* [22].

and the derived fracture surface energy was compared with the thermodynamic value [*28*]. Combined hardness and surface reflection topography results have been reported for RDX [*29*].

Dislocation etch pitting at microindentation hardness impressions in RDX crystals was employed to investigate the nature of the dislocation behavior possibly relating to hot spot formation [*18*]. Figure 8 (*top*) shows a typical example of the extreme localization of plastic flow, which was observed, in this case, around a Knoop impression aligned at a small angle to the $[00\bar{1}]$ direction. Similar results have been reported in a later study [*30*]. An example of localized flow being associated with cracking on the {241} family of planes around a (Vickers) diamond pyramid microindentation is shown in Fig. 8 (*bottom*). The long cleavage cracks on the $(00\bar{1})$ plane occurred during subsequent handling of the crystal; however, the etch pit traces along the $[\bar{1}\bar{2}\bar{4}]$ and $[\bar{1}\bar{2}4]$ directions were attributed to slip on the $(02\bar{1})$ and (021) planes at the microindentation [*18*]. On the basis of the unit cell construction in Fig. 7, $[\bar{1}00]$ slip on either the $(02\bar{1})$ or (010) plane was proposed to be an energetically favorable possibility because the relatively large $[\bar{1}00]$ Burgers

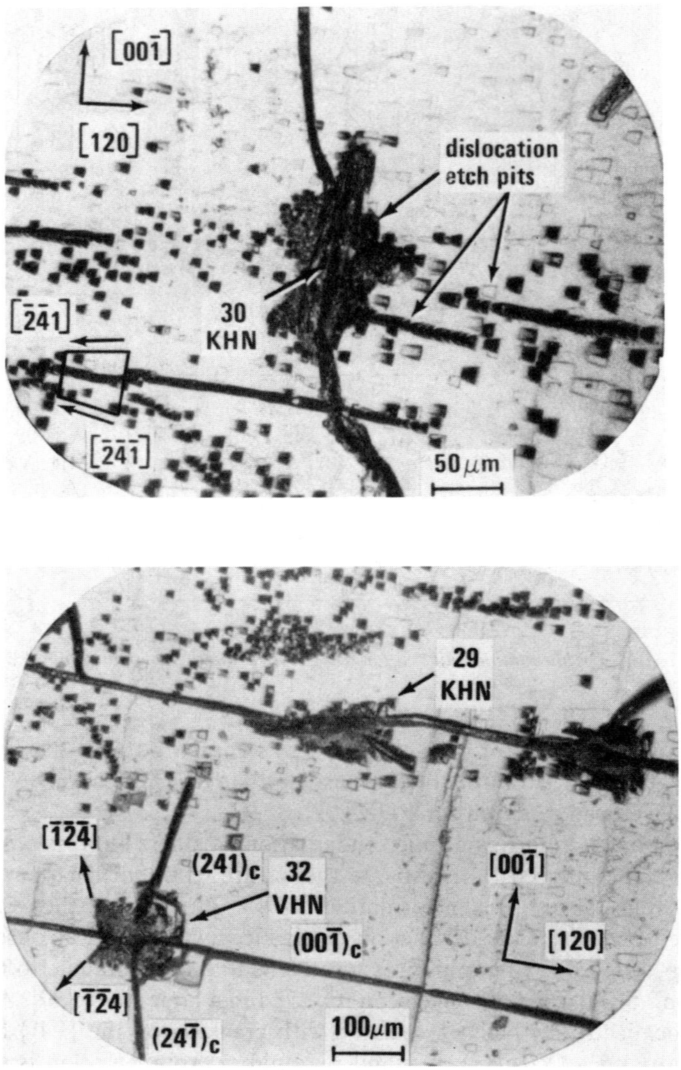

FIG. 8—*Dislocation etch pits centered on microindentations in the ($\bar{2}$10) plane growth surface of RDX* [18]: (top) *Knoop indentation parallel to the [00$\bar{1}$] direction;* (bottom) *Vickers diamond pyramid indentation with one diagonal parallel to the [00$\bar{1}$] direction.*

vector dislocation can be dissociated into lesser energy partial dislocations [*18,22*]. Partial Burgers vectors along the [$\bar{2}12$] and [$\bar{2}1\bar{2}$] directions were described in the former case, in addition to the [$\bar{1}00$] direction, and in the latter case, the [$\bar{1}01$], [$\bar{1}0\bar{1}$], and [$\bar{1}00$] directions would be involved. Such considerations should be important for molecular crystals in general because the large Burgers vectors contribute to the occurrence of relatively large self-energies, line tensions, and interaction energies for the dislocations—relating also to the energy dissipation to be expected from collapsing dislocation pileups associated with cracking.

Microhardness Anisotropy of RDX

An initial assessment of the anisotropy of Knoop microindentation hardness measurements was previously reported [*18,22*]. Figure 9 relates to more extensive measurements [*29*] on a solution-grown RDX crystal having reasonable microstructural perfection.[3] The crystallographic features of the indented crystal are shown in Fig. 9 (*top*) and a surface reflection X-ray diffraction topograph, taken with a ($\bar{6}3\bar{2}$) plane normal as the diffraction vector, is shown in Fig. 9 (*bottom*). The diffraction contrast at Knoop indentations placed at various angles to the [$00\bar{1}$] direction on the ($\bar{2}10$) plane is shown in the topograph. In agreement with previously reported etch pit results [*18,30*], the diffraction contrast in Fig. 9 (*bottom*) is extremely localized at the indentation sites.

The Knoop hardness number (HK) dependence on orientation is shown in Fig. 10. Significant cracking occurred at two indenter orientations: on the ($00\bar{1}$) plane for the indenter axis aligned near the [$00\bar{1}$] direction, and on the ($24\bar{1}$) or (241) plane, or both, for the indenter axis positioned on either side of alignment along the [$\bar{1}\bar{2}0$] direction. Lower HK values were measured for both orientations where significant cracking occurred. The result agrees with the description for the influence of cracking on the MgO crystal hardness indicated in Fig. 5. The measured hardness anisotropy and the cracking observations for RDX, particularly the {241} cracking, are consistent with the proposition that large internal stresses are generated within local regions and subsequently relieved by the cracking that occurs.

The Knoop hardness test can be understood on a crystallographic basis by use of the stereographic projection shown in Fig. 11. The figure is centered on the ($\bar{2}10$) surface of the indented crystal and includes the (010), (021), ($02\bar{1}$), ($00\bar{1}$), ($24\bar{1}$), and (241) slip or cleavage systems, which have been discussed. A reference orientation is shown for a Knoop indenter aligned along the [$\bar{1}\bar{2}0$] direction. In accordance with one model [*31*] of the effective force system for explaining the HK orientation dependence on a critical resolved shear stress basis, the directions of forces, F, are shown at positions contained in the respective planes of the four indenter facets. The model has been improved and

[3] R. Y. Yee, Naval Weapons Center, China Lake, CA, personal communication, March 1983.

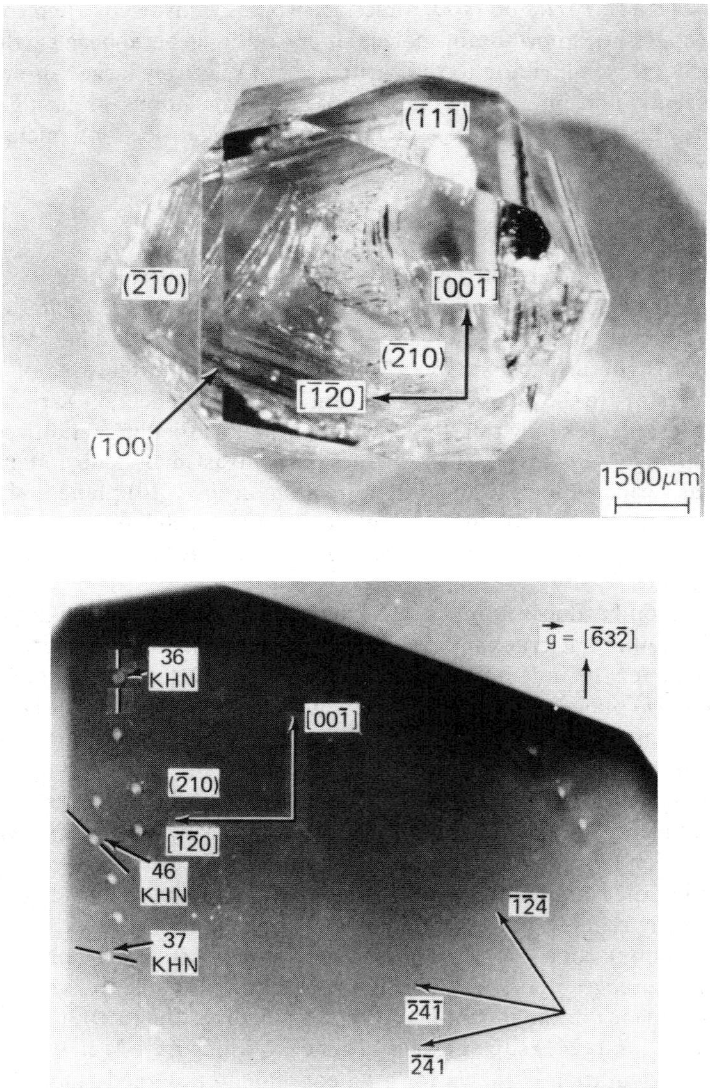

FIG. 9—*RDX crystal grown at the Naval Weapons Center by R. Y. Yee* [29]: (top) *a reflected light photograph;* (bottom) *a copper Kα X-ray surface reflection topograph.*

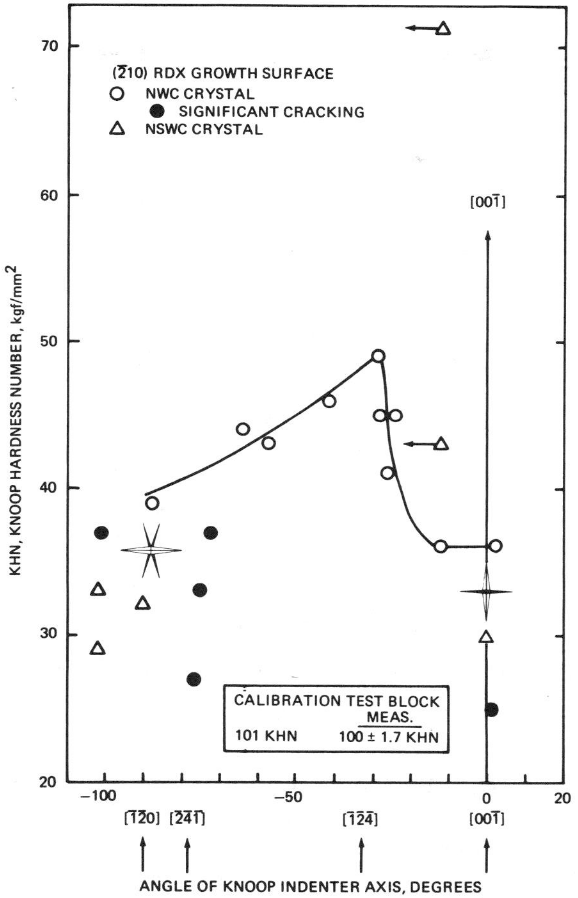

FIG. 10—*Knoop hardness anisotropy for a Naval Weapons Center-grown RDX crystal* [29].

applied to microhardness anisotropy measurements for a wide range of crystalline materials [*32*]. Most recently the model has been used to explain the hardness anisotropy measurements for RDX on the basis that the (010) and {021} slip systems control the hardness property.[4] The model appears to be in agreement with the measurements in Fig. 10; for example, low values of the effective resolved shear stress factor are obtained for the indenter axis aligned along the $[\bar{1}\bar{2}\bar{4}]$ direction and, hence, a high hardness should be measured, as observed. However, the $[\bar{1}00]$ and $[00\bar{1}]$ slip directions in the (010) plane and the $[\bar{1}00]$ slip direction in the $(02\bar{1})$ and (021) planes give fewer than the requi-

[4]J. N. Sherwood, University of Strathclyde, Glasgow, UK, personal communication, April 1984.

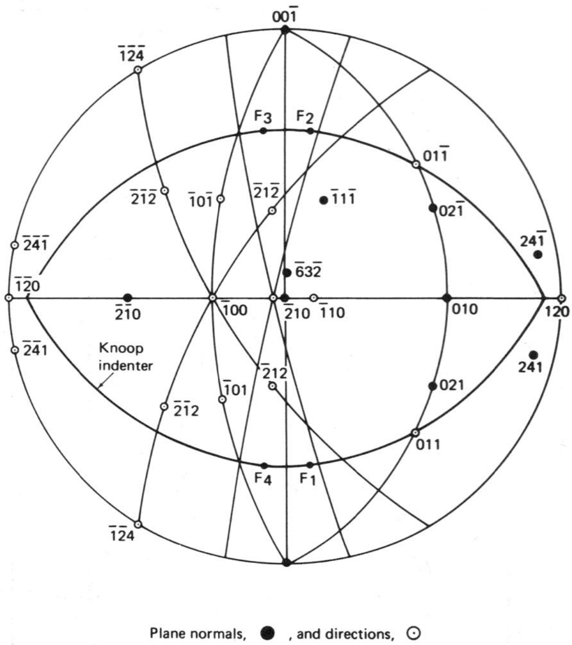

FIG. 11—*A ($\bar{2}10$) plane stereographic projection for slip and cracking around Knoop indentations in RDX.*

site five deformation systems needed for an (arbitrary) indentation strain. Unless more easy cracking or other deformation systems operate to give the required strains, the hardness anisotropy must be affected by additional factors as well. Further study is being made of this consideration of the availability of deformation systems and of the influences that work hardening [*1,2*], cracking [*2,12*], and strain energy release mechanisms [*12,19,29*] should have on the total material properties of RDX.

Summary

An analysis of dislocation behavior within the plastic deformation zones surrounding indentation sites is necessary for a fuller understanding of the hardness property of single crystals. The extent of plastic flow, the potential onset of cracking, and the anisotropy of microindentation hardness measurements are examples of instances in which the hardness properties should be connected with the dislocation behavior. Investigating the nature of hot spot generation in inert and explosive crystals by hardness testing provides a method of assessing the problem in terms of the fundamental dislocation characteristics involved.

Acknowledgments

This research activity has been supported primarily by the Office of Naval Research (ONR)—Contract N00014-84-WR24086 at the Naval Surface Weapons Center (NSWC), and Contract N00014-82-K0263 at the University of Maryland—and by Dr. R. S. Miller, program manager at ONR. Drs. R. G. Rosemeier (Brimrose Corporation of America), R. Y. Yee (Naval Weapons Center), and K.-C. Yoo (Solarex Corporation) contributed to the work. In addition to the authors' helpful discussions with these persons and colleagues at NSWC and the University of Maryland, one of the authors (RWA) has benefited both from a sabbatical study period at the Department of Metallurgy and Materials Science, University of Cambridge, the United Kingdom, and from numerous contacts with researchers while on leave at ONR, London.

References

[*1*] Armstrong, R. W. and Raghuram, A. C. in *The Science of Hardness Testing and Its Research Applications,* J. H. Westbrook and H. Conrad, Eds., American Society for Metals, Metals Park, OH, 1973, Chapter 13, pp. 174-186.

[*2*] Armstrong, R. W. and Wu, C. C., *Journal of the American Ceramic Society,* Vol. 61, Nos. 3-4, 1978, pp. 102-106.

[*3*] Hammond, B. L., "SEM Studies of Diamond Pyramid (Vickers) Indentations in an MgO Single Crystal," *Electron Microscope Central Facility Newsletter,* No. 9, University of Maryland, College Park, MD, 1981, pp. 16-17.

[*4*] Keh, A. S., *Journal of Applied Physics,* Vol. 31, No. 9, 1960, pp. 1538-1545.

[*5*] Pande, D. R. and Murty, T. S., *Journal of Physics D. Applied Physics,* Vol. 7, No. 3, 1974, pp. 403-406.

[*6*] Hirth, J. P. and Lothe, J., *Theory of Dislocations,* McGraw-Hill, New York, 1967, pp. 19-22.

[*7*] Keh, A. S., Li, J. C. M., and Chou, Y. T., *Acta Metallurgica,* Vol. 7, No. 10, 1959, pp. 694-696.

[*8*] Cottrell, A. H., *Transactions of the American Institute of Mining and Metallurgical Engineers,* Vol. 212, No. 2, 1958, pp. 192-203.

[*9*] Groves, G. W. and Kelly, A., *Philosophical Magazine,* Vol. 8, No. 89, 1963, pp. 877-887.

[*10*] Rickerby, D. G. and Macmillan, N. H., *Materials Science and Engineering,* Vol. 40, No. 2, 1979, pp. 251-259.

[*11*] Chaudhri, M. M., Wells, J. K., and Stephens, A., *Philosophical Magazine,* Vol. A43, No. 3, 1981, pp. 643-664.

[*12*] Yoo, K.-C., Rosemeier, R. G., Elban, W. L., and Armstrong, R. W., *Journal of Materials Science Letters,* Vol. 3, No. 6, 1984, pp. 560-562.

[*13*] Frank, F. C. and Lawn, B. R., *Proceedings of the Royal Society* (London), Vol. A299, No. 1458, 1967, pp. 291-306.

[*14*] Lawn, B. R. and Fuller, E. R., *Journal of Materials Science,* Vol. 10, No. 12, 1975, pp. 2016-2024.

[*15*] Lankford, J., *Journal of Materials Science Letters,* Vol. 1, No. 11, 1982, pp. 493-495.

[*16*] Niihara, K., *Journal of Materials Science Letters,* Vol. 2, No. 5, 1983, pp. 221-223.

[*17*] Coffey, C. S. and Armstrong, R. W., in *Shock Waves and High Strain Rate Phenomena in Metals,* M. A. Meyers and L. E. Murr, Eds., Plenum, New York, 1981, pp. 313-324.

[*18*] Elban, W. L. and Armstrong, R. W. in *Proceedings,* Seventh Symposium (International) on Detonation, 16-19 June 1981, NSWC MP 82-334, Naval Surface Weapons Center, Silver Spring, MD, pp. 976-985.

[*19*] Armstrong, R. W., Coffey, C. S., and Elban, W. L., *Acta Metallurgica,* Vol. 30, No. 12, 1982, pp. 2111–2116.
[*20*] Reed, P. T. in "Structures of Trinitro-Aromatic Crystals and Related Substances," P. M. Harris, Ed., Technical Report AFOSR-TR-59-165, Ohio State University, Columbus, OH, 1959.
[*21*] Choi, C. S. and Prince, E., *Acta Crystallographica,* Vol. B28, No. 9, 1972, pp. 2857–2862.
[*22*] Elban, W. L., Hoffsommer, J. C., and Armstrong, R. W., *Journal of Materials Science,* Vol. 19, No. 2, 1984, pp. 552–566.
[*23*] Stals, J. *Australian Journal of Chemistry,* Vol. 22, No. 12, 1969, pp, 2505–2514.
[*24*] Connick, W. and May, F. G. J., *Journal of Crystal Growth,* Vol. 5, No. 1, 1969, pp. 65–69.
[*25*] Lang, A. R. in *Diffraction and Imaging Techniques in Material Science,* Vol. II, S. Amelinckx, R. Gevers, and J. Van Landuyt, Eds., North-Holland, Amsterdam, 1978, p. 678.
[*26*] McDermott, I. T. and Phakey, P. P., *Physica Status Solidi,* Vol. (a)8, No. 2, 1971, pp. 505–511.
[*27*] Hagan, J. T. and Chaudhri, M. M., *Journal of Materials Science,* Vol. 12, No. 5, 1977, pp. 1055–1058.
[*28*] Elban, W. L., *Journal of Materials Science,* Vol. 14, No. 4, 1979, pp. 1008–1011.
[*29*] Elban, W. L., Coffey, C. S., Armstrong, R. W., Yoo, K.-C., and Rosemeier, R. G., "Microstructural Origins of Hot Spots in RDX Explosive and Several Reference Inert Materials," NSWC MP 83-116, Naval Surface Weapons Center, Silver Spring, MD, 1983.
[*30*] Halfpenny, P. J., Roberts, K. J., and Sherwood, J. N., *Journal of Materials Science,* Vol. 19, No. 5, 1984, pp. 1629–1637.
[*31*] Daniels, F. W. and Dunn, C. G., *Transactions of the American Society for Metals,* Vol. 41, 1949, pp. 419–442.
[*32*] O'Neill, J. B., Redfern, B. A. W., and Brookes, C. A., *Journal of Materials Science,* Vol. 8, No. 1, 1973, pp. 47–58.

Techniques and Measurement

David Tabor[1]

Indentation Hardness and Its Measurement: Some Cautionary Comments

REFERENCE: Tabor, D., "**Indentation Hardness and Its Measurement: Some Cautionary Comments,**" *Microindentation Techniques in Materials Science and Engineering, ASTM STP 889*, P. J. Blau and B. R. Lawn, Eds., American Society for Testing and Materials, Philadelphia, 1986, pp. 129–159.

ABSTRACT: This paper discusses some of the main developments in our understanding of indentation hardness during the last 20 years and indicates the limitations of several current models. In particular, it deals critically with the expanding spherical cavity as applied to elastic-plastic solids and to solids with marked creep. It also discusses the problem of load application, the indentation behavior of single crystals, the hardness of semiconductors, and recent developments in crack formation by indentations in plastic-brittle solids. Finally, it deals with the more specific problem of hardness at extremely small loads and the practical as well as theoretical difficulties involved in characterizing microhardness in terms of depth of penetration.

KEY WORDS: indentation hardness, elastic-plastic solids, semiconductors, plastic-brittle solids, creep, microhardness, microindentation hardness testing

The hardness of a solid is usually understood to mean its resistance to local deformation [*1–6*]. The simplest method of quantifying it is to press a hard indenter of specified geometry into the body, divide the load by the area of the indentation formed, and express the answer in units of kilograms per square millimetres or Pascals (1 kg $mm^{-2} \simeq 10^7$ Pa). In what follows, we shall use the projected area of the indentation, since the deduced contact pressure or hardness is physically more meaningful. In practical hardness tests the superficial indentation area is often used and gives a slightly smaller value for the hardness. Since the unit of hardness has the dimensions of a pressure or stress, it is natural to attempt to correlate the indentation hardness with some stress property of the solid, for example, its elastic modulus, E, or its uniaxial

[1] Professor, University of Cambridge, Cavendish Laboratory, Cambridge CB3 OHE, United Kingdom.

yield stress, Y. A large part of the hardness literature since the time of Brinell has been concerned with establishing correlations of this type. Indentation hardness has also been used to study creep and the fracture of brittle materials.

The type of indentation formed depends on the nature of the solid. With ideal elastic solids, the deformation is reversible and determined by classic elasticity. In some cases the stress field will initiate cracks, particularly at the periphery of the indentation. Most solids, however, show some ductility, especially under compression. With such materials irreversible flow occurs beneath the indenter, the displaced material being accommodated by elastic strains in the hinterland. These strains impose a constraint on the flow, and the behavior can be described by various versions of the expanding spherical cavity model. If the constraint is too large, it becomes energetically more favorable for the displaced material to flow upward around the boundary of the indentation, and the behavior approximates that described by classic slip-line field theory for rigid-plastic solids. However, with real solids plastic indentation is always accompanied by elastic strains in the hinterland, and when the indenter is removed, elastic stresses are partially released. These change the shape and, to some extent, the size of the residual indentation. With plastic-brittle solids, released elastic stresses may induce additional crack features, analyzed in a recent model which also allows for possible compaction of the material during indentation.

These ideas are developed in the following review, which covers a broad range of selected topics in indentation hardness measurements. Some of the concepts are applicable, with suitable modification, to microindentation. The main themes are as follows:

1. *Indentation hardness of various types of solids and a critical account of current theoretical models:*

(*a*) elastic solids,
(*b*) rigid-plastic solids and the slip-line field approach,
(*c*) elastic-plastic solids and the expanding cavity model,
(*d*) work hardening produced by the indentation process itself,
(*e*) solids which creep under indentation; a different expanding cavity model,
(*f*) elastic-brittle and plastic-brittle solids; a recent model allowing for compaction,
(*g*) elastic, rigid-plastic, elastoplastic solids analyzed by a nonphysical model of indentation,
(*h*) single crystals and the relationship between hardness and shear modulus, and
(*i*) semiconductor single crystals and the relationship between hardness and band gap.

2. *Some practical problems encountered in indentation hardness measurements:*

(*a*) loading of the indenter,
(*b*) elastic recovery of the indentation in depth and diameter, and
(*c*) measurement of the effective area of the indentation.

3. *Microindentation hardness:*

(*a*) load range: the increase in hardness at small loads, and
(*b*) specific experimental difficulties, the indenter, loading device, the specimen, and the areal or depth measurements of indentation.

4. *Conclusions, future trends, and challenges.*

Indentation Hardness of Various Types of Solids

Elastic Solids

With elastic solids such as rubber the indentation pressure is a direct measure of the elastic properties of the material. If E is Young's modulus, ν is Poisson's ratio, and the indenter is a cone of semiapical angle, θ, the indentation pressure, p (assuming zero interfacial friction), is

$$p = H = \frac{E \cot \theta}{2(1 - \nu^2)} \tag{1}$$

It can be seen that H is independent of the applied load, a result which follows from the principle of geometric similarity—for a conical or pyramidal indenter, the indentation process is identical whatever the size of the indentation. By contrast, if the indenter is a sphere, H will depend on the load, W, and the radius of curvature, R, of the sphere as well as on E and ν

$$H = \frac{W^{1/3}}{\pi} \left| \frac{4E}{3R(1 - \nu^2)} \right|^{2/3} \tag{2}$$

assuming that any elastic deformation of the indenter may be ignored. If there is friction between the indenter and the rubber, the measured hardness will be increased. We also note that the indentation vanishes when the indenter is removed, so that measurements must be made while the load is applied. For this reason it is often easier to measure the depth of the indentation, h_0, corresponding to a (circular) indentation of radius, a_0. For a conical indenter, $h_0 = a_0 \cot \theta$. However, the indenter produces a further depression of the rubber around the indentation, so that the total penetration distance moved by the indenter into the rubber is h (Fig. 1*a*), where

$$h = \frac{\pi}{2}(h_0) \tag{3}$$

From this we may deduce that if the applied load is W

$$H = \frac{W}{h^2}\frac{\pi \cot^2 \theta}{4} \tag{4}$$

This assumes ideal elastic behavior of the rubber. With real systems, h is never exactly $(\pi/2)h_0$, and there is always a certain amount of creep in the rubber as well as some interfacial friction between the indenter and the specimen. In addition, the cone invariably has a rounded tip. Nevertheless, Eq 4 provides the basic relation for penetrometer tests of the "hardness" of rubber. Relations similar to Eq 4 can be derived for spherical indenters, although in this case $h = 2h_0$ (Fig. 1*b*). Equations 1, 2, and 3 may also be used to estimate the elastic recovery of plastically formed indentations (described later). The generation of cracks around and beneath elastic indentations will be discussed in a later section.

With polymers, the indentation process involves viscoelastic effects. The deformation is time dependent and often markedly temperature dependent (described later). When the load is removed, the indentation relaxes. At first only shallowing occurs, but over prolonged periods the diameter of the indentation shrinks and, with spherical indenters, the indentation finally disappears entirely. With pyramidal indenters, an impression of the diagonals may remain for very long periods.

Metals: Rigid-Plastic Solids

Consider, first, an ideal rigid-plastic solid (Fig. 2*a*). Under uniaxial tension or compression, no deformation occurs until a stress, Y, is reached, when plastic flow takes place; this is known as the yield or flow stress. To a close approximation, hydrostatic pressure does not change the behavior. Because the only part of a stress system that is unaffected by hydrostatic pressure is a shear stress, we deduce that yield is associated with a critical shear stress, k, a conclusion that is consistent with the atomic process of shear or slip along specific crystallographic planes. The relationship between k and Y depends on the yield criterion. The critical shear stress criterion of Tresca gives $k = (1/2)Y$; the critical shear energy criterion (von Mises) gives $k = 1/\sqrt{3}Y$.

If we now consider the indentation of such a material by a hard punch, we can construct the stress field within the solid, indicating only the trajectories at which the shear stress reaches the critical value, k. This, then, defines the regions where plastic flow occurs. Certain boundary conditions and "velocity" criteria must also be incorporated into the solution [7].

The "slip-line field" for a flat two-dimensional punch, assuming friction-

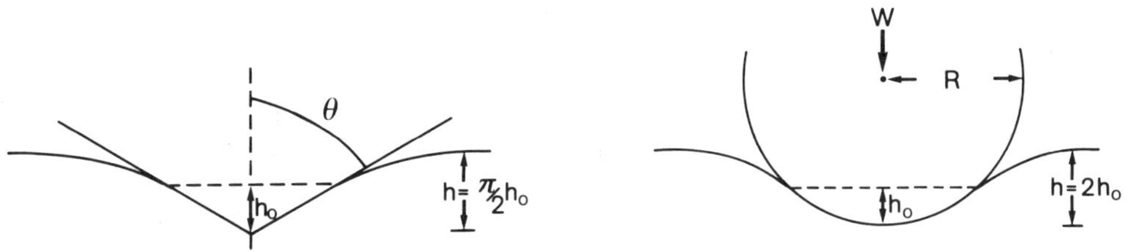

$$H = \frac{E \cot \theta}{1 - \nu^2} \qquad H = \frac{W^{1/3}}{\pi}\left[\frac{4E}{3R(1-\nu^2)}\right]^{2/3}$$

FIG. 1—*Indentation of ideally elastic solids by conical and spherical indenters.*

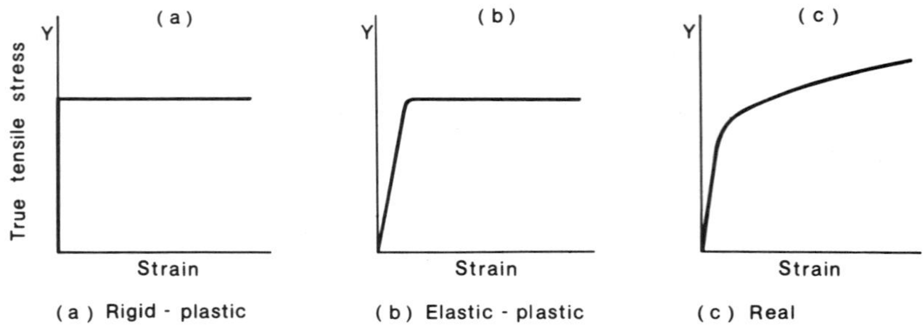

FIG. 2—*Stress-strain curves for* (a) *rigid-plastic,* (b) *elastic-plastic, and* (c) *real work-hardening solids.*

less contact, is shown in Fig. 3*a*. The pressure, p, across the punch is uniform and has the value

$$p = H = 2k\left(1 + \frac{\pi}{2}\right) \tag{5}$$

Thus

$$H \simeq 2.6Y \text{ (Tresca)} \tag{6}$$

$$H \approx 3Y \text{ (von Mises)} \tag{7}$$

In general we may write

$$H = cY \tag{8}$$

where c is the appropriate constraint factor: this depends on the geometry of the indenter and on the effect of interfacial friction, which always increases c. The slip-line field for a two-dimensional wedge of semiapical angle θ is shown in Fig. 4, and the pressure across the face of the indenter for frictionless indentations is [*8*]

$$p = H = 2k(1 + \alpha) \tag{9}$$

where

$$\cos(2\theta - \alpha) = \frac{\cos\alpha}{1 + \sin\alpha} \tag{10}$$

For axially symmetrical indenters (circular punches or conical indenters), analogous solutions are available. There is also a solution for the plastic indentation of a plastic-rigid solid by a spherical indenter.

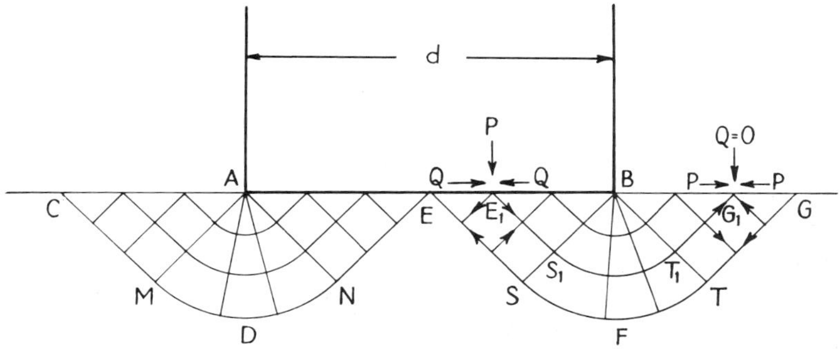

FIG. 3—*Slip-line field solution for the indentation of a rigid-plastic solid by a frictionless punch (two-dimensional).*

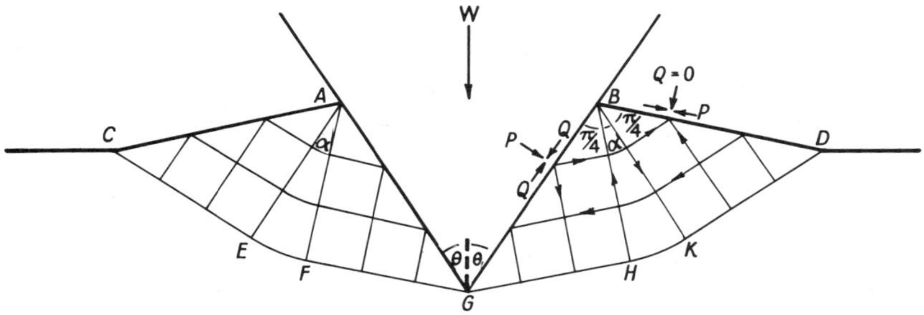

FIG. 4—*Slip-line field solution for the indentation of a rigid-plastic solid by a frictionless wedge of semiapical angle, θ.*

The slip-line field approach to the plastic deformation of metals is a well-established technique which has been widely applied to problems of metal working and metal forming as well as to the specific problem of indentation hardness. This approach is elegantly described in Hill's seminal book published in 1950 on the mathematical theory of plasticity [*8*]. Under its influence, Honeycombe and I[2] carried out some indentation experiments at that time using a spherical or pyramidal indenter and a split specimen with a grid inscribed on one face. We were never able to observe the deformation pattern predicted by slip-line field theory and attributed this to a wedging apart of the specimen or some other experimental inadequacy. Evidently, we were observing a different mode of indentation, later referred to by Samuels and Mulhearn [*9*] as a "compression mode." This mode arises because metals are not rigid-plastic but elastic-plastic materials.

[2] J. Honeycombe and D. Tabor, Cambridge University, Cambridge, United Kingdom, unpublished work, 1950.

Metals: Elastic-Plastic Solids

We first deal with an "ideal" elastic-plastic solid such as that in Fig. 2*b*: it deforms elastically according to its Young's modulus and then, at tensile stress, Y, yields plastically at a constant yield or flow stress. A good approximation to such behavior is a fully work-hardened polycrystalline specimen of copper or iron.

The effect of elasticity is especially marked with spherical indenters. At small loads, the specimen deforms elastically; then, at a higher load, the critical shear stress is first exceeded at a region below the center of the contact zone (Fig. 5*a*) [*3,11*]. This corresponds to the onset of plastic indentation, and it occurs for a mean contact pressure [*3,7*].

$$p = H \simeq 1.1Y \tag{11}$$

As the load, W, is increased further, the indentation becomes larger and the plastic zone grows until the whole of the material surrounding the indenter undergoes plastic deformation [*3,10*]. The contact pressure is not quite uniform over the indentation, but the details are not certain. On the other hand, the mean contact pressure (see Fig. 5*b*) is fairly well defined and is given by

$$H \simeq 3Y \tag{12}$$

The increase of H with W is shown in Fig. 5*c* [*10*].

With a conical or pyramidal indenter, the smallest load will produce infinite stresses at the tip and plastic flow will occur. The indentations formed will, in principle, be geometrically similar whatever their size, so that the indentation pressure will be independent of load. However, only indenters with a semiapical angle less than about 50° show the plastic displacement pattern represented by Fig. 3. For shallower or blunter indenters there is practically no outward or upward flow of metal. The behavior, as pointed out by Samuels and Mulhearn in 1957 [*9*], resembles what would be expected with *radial flow* of the metal away from the indenter. This led Marsh [*12*] to suggest in 1964 that the behavior was dominated by elastic yielding of the hinterland, the deformation resembling the expansion of a spherical cavity into an elastic-plastic solid by an internal hydrostatic pressure. The problem had already been analyzed in 1945 by Bishop, Hill, and Mott [*13*], who had shown that the pressure, p, at which the cavity expands is

$$\frac{p}{Y} = \frac{2}{3}\left(1 + \ell n \frac{E}{3(1 - \nu)Y}\right) \tag{13}$$

The simplest model assumes that, if the hardness indentation has a diameter $d = 2a$, there is a hemispherical core of material of radius a below the

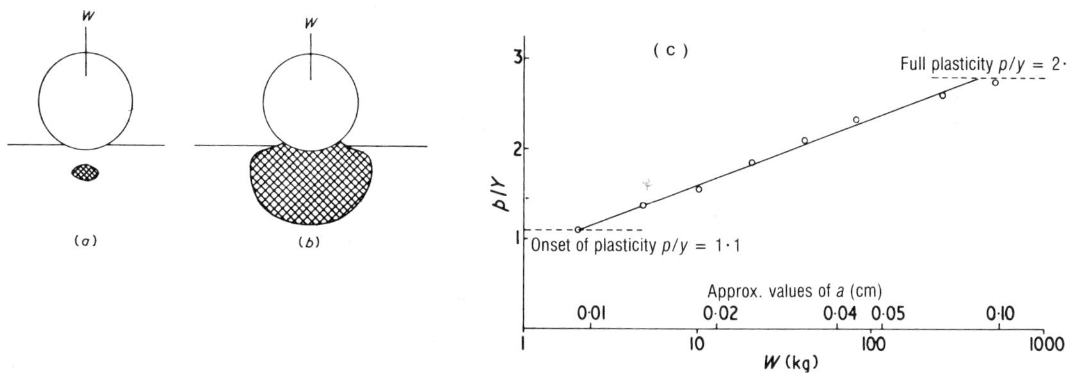

FIG. 5—*Indentation of a rigid-plastic solid by a spherical indenter:* (a) *onset of plasticity below the surface at a contact pressure,* p, *of approximately 1.1* Y; (b) *at a higher load, "full" plasticity is reached and the plastic flow extends to the free surface (at this stage* p $\approx$ *3*Y*),* (c) *mean indentation pressure,* p, *as a function of load,* W, *plotted as* p/Y *against log* W. *The figure also shows that the radius,* a, *of the indentation increases by a factor of about ten as the deformation passes from the onset of plasticity to full plasticity. (The results are for a hard steel sphere 10 mm in diameter pressed into the surface of a fully work-hardened mild steel specimen of yield stress* Y = *77 kg/mm*$^{-2}$ [3,11]*).*

indenter which is under hydrostatic pressure equal to the indentation pressure, p (Fig. 6*a*). From this boundary, plastic flow spreads into the surrounding material, the plastic strains gradually diminishing until they match the elastic strains in the hinterland at some radius, c; this marks the plastic-elastic boundary. Clearly, on this model the behavior depends little on the shape of the indenter itself. Figure 6*b* shows results obtained by Marsh [*12*] for a Vickers indenter. The most interesting feature is that the hardness depends on the ratio E/Y. For solids such as polymers, for which E/Y is of the order of 10, H is $<1.5Y$. For most metals the ratio E/Y is greater than 100, and H is about $3Y$ but could be appreciably more were it not that the rigid-plastic solution now takes over (see later).

An improved analysis comes from Johnson [*14*]. The radial expansion model is the "bottom half" of a spherical cavity and therefore does not provide for any piling up of any displaced material outside the indentation. Evidently, the volume of the indentation is ultimately taken up in the elastic hinterland, so that the shape of the indenter must be involved. For a cone of semiapical angle, θ, the critical parameter now becomes $(E/Y) \cot \theta$, while for a spherical indenter the corresponding quantity is $(E/Y)(a/R)$. The final relation for cones or pyramids is then

$$\frac{H}{Y} = \frac{2}{3}\left(1 + \ln \frac{E \cot \theta}{3Y}\right) \tag{14}$$

while for spherical indenters $\cot \theta$ is replaced by a/R. This treatment assumes that the material has a constant yield stress, Y, that is, that no work hardening is produced by the indentation process itself. The result is shown in Fig. 7.

Limitations of the Expanding Cavity Model

According to Hill's original solution, the pressure in the cavity when the plastic-elastic boundary is at a distance, c, from the center is

$$\frac{P}{Y} = \frac{2}{3} + 2 \ln \frac{c}{a} \tag{15}$$

This implies that the elastic-plastic boundary coincides with the boundary of the cavity itself ($c = a$) at $P = \frac{2}{3}Y$, and below this contact pressure the analysis fails and no plastic flow can occur. However, there is no case of plastic indentation occurring in any system for an indentation pressure less than Y. On the other hand, in the region where $p \simeq 3Y$, Eq 15 shows that $c = 3.2a$, and we must assume that the elastic yielding of the hinterland no longer influences the plastic flow of the material. The contact pressure now corresponds to the classic theory for a rigid-plastic solid. However, there is nothing in the expanding cavity model to indicate that the indentation pressure has an

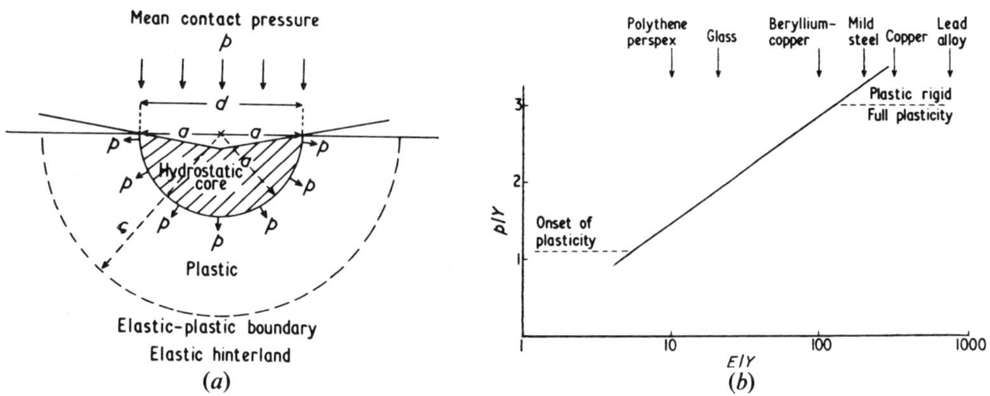

FIG. 6—(a) *Indentation in an ideal elastic-plastic solid treating the process as resembling the expansion of a hemispherical core;* (b) *variation of indentation pressure,* p, *with the ratio* E/Y, *where* E *is Young's modulus and* Y *the yield stress. The results were deduced by March* [11] *for a Vickers indenter. The horizontal broken lines also represent the approximate limits for a spherical indenter.*

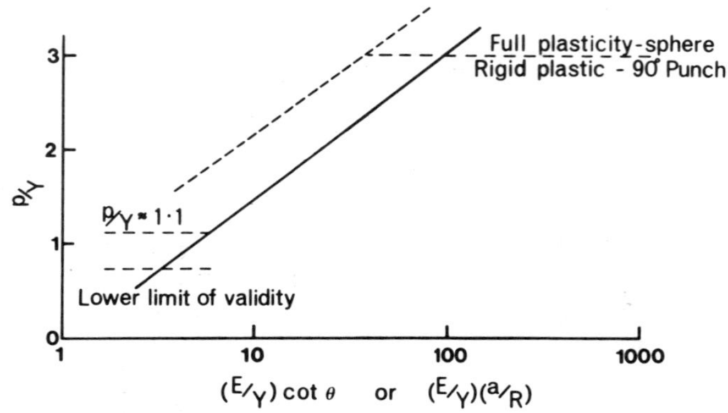

FIG. 7—*Variation of indentation pressure,* p, *with (*E/Y*) cot θ for a conical indenter where θ is the semiapical angle of the cone. The full line is the theoretical curve from Johnson* [20]. *The same theoretical result is obtained for a spherical indenter of radius* R *if the horizontal ordinate is replaced by (*E/Y*)(*a/R*), where* a *is the radius of the indentation. The dotted line represents the experimental results of Marsh* [11] *for a Vickers indenter; this line is the same as the full line in Fig. 6, displaced to the left by a factor of approximately 0.4 (cot 70°). The discrepancy between theory and experiment is greatly reduced if the theoretical values of* p *are all increased by 2 Y/3. (See the text for details.)*

upper limit of $3Y$. One must accept the conclusion that the rigid-plastic model involves less plastic work than the expanding cavity model for $c/a > 3.2$, and this then becomes the preferred mode of deformation.

Figure 7 from Ref *10* is helpful in indicating why conical or pyramidal indenters are likely to give the radial expansion behavior for large values of θ, for example, the Vickers indenter for which $\theta \simeq 70°$ and $\cot \theta \approx 0.4$, since this gives a low value of $(E/Y) \cot \theta$. By contrast, for indenters with $\theta \leqq 45°$, $\cot \theta > 1$ and the corresponding value of $(E/Y) \cot \theta$ is greater and moves toward the rigid-plastic solution. However, although the trend is right, the factor $\cot \theta$ is not a particularly large one when viewed on a logarithmic scale. Figure 7 also indicates how the indentation pressure for a spherical indenter increases from $H = 1.1Y$ to $H \simeq 3Y$ as the size of the indentation (a/R) increases. At the lower end of the horizontal ordinate, where $(E/Y)(a/R)$ has a value of ~ 10, the hardness is about $1.3Y$; for a tenfold increase in a/R the hardness is $\sim 3Y$. It is, however, worth noting that the early part of the indentation process involves a small plastic zone below the indenter, which gradually grows as the size of the indentation increases. This in no way resembles the expanding cavity model, although the experimental results fit Fig. 7 rather well (compare Fig. 5*c*). The model is thus not too discriminating. The later part of the indentation process, in which the plastic zone has reached the indenter surface, is much closer to the expanding cavity model.

Another factor is the uncertain nature of the hemispherical zone beneath the indenter where the stresses are assumed to be hydrostatic. If they were truly hydrostatic, this material could not yield plastically and would have to be considered "dead" metal. In that case, the major plastic deformation would occur well away from the surface. Does this imply that microhardness measurements can never sample the outermost surface layers unless the indentations are less than, say, 20 Å in diameter?

It seems that the stress situation in the hemispherical zone includes shear stresses capable of producing plastic flow, but these stresses do not greatly distort the assumption of a hydrostatic pressure acting on the surrounding material [*15*]. Indeed, Johnson [*15*] suggests that it might be more reasonable to make the pressure, p, under the indenter greater than the hydrostatic pressure around the core by an amount $\frac{2}{3}Y$ with a radial component of $Y/3$ (so that plasticity is achieved in the core). Thus, the line in Fig. 7 would be everywhere raised by an amount $p/Y = \frac{2}{3}$. As the figure shows, this gives better agreement with experimental data but, of course, introduces other inconsistencies. The basic difficulty is that the assumption of a true hydrostatic pressure is not valid.

Finally, we return to the basic problem of using the flow equations of a complete spherical cavity to describe the behavior of the lower half of such a cavity. Clearly, this description will leave the free surface of the half-space in a condition of nonequilibrium. There must be additional surface stresses. This point has recently been discussed and analyzed in some detail by Chiang, Marshall, and Evans [*16*]; they have extended the treatment to deal particularly with residual stresses and their role in the fracture of plastic-brittle solids.

In summary, the expanding cavity model in its simple form is an extremely helpful and fairly realistic description of the indentation process in the "compression" or radial flow mode. It must, however, not be pushed too far. Attempts have been made to improve it to give better numerical agreement with experiment. A more satisfactory approach involves finite element analysis and the newer technique of boundary element analysis, but even these techniques have limitations imposed by the theoretical assumptions involved.

Work Hardening

The previous sections have treated the specimen as a material of constant yield stress. With most real materials, the yield or flow stress increases as a result of plastic deformation or work hardening (Fig. 2*c*). Consequently, when an indentation is made in such a material, the indentation process itself produces an increase in the flow stress. However, the plastic strains will vary over the deformed region so that the amount of work hardening will vary from point to point. A detailed study of this problem using slip-line field theory has been described by Lee, Masaki, and Kobayashi [*17*] for a material of specified flow stress-plastic strain characteristics. It involves a finite-element analysis at incremental stages of the indentation process. A far simpler approach is that of Tabor [10], who suggested that there is an "effective" strain such that the indentation hardness is

$$H = cY_R \tag{16}$$

where Y_R is the uniaxial flow stress at this increased value of strain. For a spherical indenter of diameter D, producing an indentation of chordal diameter, d, the "effective" strain must be a function of the dimensionless ratio, d/D. Experiments show that *under conditions of full plasticity* the hardness increases with the size of the indentation in a manner consistent with the empirical assumption that the effective strain is approximately equal to 0.2 (d/D). This agrees rather well with the theoretical analysis of Lee et al mentioned earlier.

Similarly, with the Vickers indenter, for which geometric similarity applies, the average effective strain produced by the indenter is between 0.08 and 0.10; the indentation hardness is then c times the yield stress at this augmented value of strain [3].

Using the empirical result for spherical indenters, it is possible to reconstruct the flow stress-plastic strain curve of a material over the first 20% of strain by making indentations of increasing size. One must, of course, be sure that the whole series of measurements is within the regimen of full plasticity, where the constraint factor, c, is a constant. A similar reconstruction can be carried out using conical indenters of different apical angles which produce different amounts of strain [18]. However, this is complicated by the fact that the factor c itself depends on the apical angle and on the friction at the indenter–specimen interface.

Indentation Hardness and Creep

Indentation hardness decreases if the loading time is increased or if the temperature is raised. Such effects are small at room temperature with high-melting-point metals but become appreciable at temperatures exceeding about 0.4 T_m, where T_m is the absolute melting point. This observation is due to the creep of the metal. The detailed model is naturally very complicated, but a simple model (from Hill[3]), once again involving radial flow, provides a useful approach. The indentation is assumed to correspond to the plastic movement of a series of shells, concentric with a hemispherical core surrounding the indentation, into the bulk of the specimen. The elastic properties of the hinterland are ignored. Instead, it is assumed that at some distance, r, from the center of the core, the rate of flow of material is determined by the creep properties of the material at that distance. The shear strain rate,

[3] R. Hill, Department of Applied Mathematics and Theoretical Physics, Cambridge, United Kingdom, personal communication, 1964.

$\dot{\gamma}$, is proportional to $1/r^3$, so that it decreases fairly rapidly into the bulk of the specimen. By applying a convenient creep equation and integrating over the whole half-space surrounding the indentation, the indentation hardness may be deduced as a function of loading time, temperature, and the stress power index of the appropriate creep process. One may even deduce the activation energy for the creep process. In some cases it agrees with the activation energy for self diffusion [*19*].

The model is rather crude. The final equation turns out to be the same whether transient or steady-state (viscous) creep is assumed. It is, however, useful for comparative studies, and, as Johnson has shown [*20*], it provides a simple explanation as to why under creep conditions the hardness for conical and pyramidal indenters depends so little on the apical angle [of the order $(\cot \theta)^{1/9}$]. This is, of course, for "blunt" indenters for which radial flow dominates the indentation process. For more "pointed" indenters (say $\theta < 45°$) a different behavior might apply. It should be noted that this model, even when radial flow applies, is quite inapplicable to the indentation behavior of non-creeping solids. For such materials, the role of the elastic hinterland is crucial.

Indentation of Elastic-Brittle Solids

In this section we discuss briefly the indentation behavior of hard brittle solids such as glasses, ceramics, and rocks. Some of these materials undergo brittle fracture within the elastic range of loading. With spherical indenters, crack formation is then explicable in terms of the Hertzian solution of the stress field; cracking is produced by a critical *tensile* stress. However, an energy balance must also be satisfied, that is, the released elastic strain energy must be sufficient to provide for the surface energy involved in the production of the crack surfaces [*21*]. This is the classic approach of fracture mechanics, and one of its consequences is a scaling effect: the released energy involves volume and is proportional to L^3, where L is a linear dimension related to the size of the circle of contact, whereas the area of the crack is proportional to L^2. Consequently, the smaller the indentation, the larger the stress (this determines the strain energy per unit volume) required to produce cracking [*22,23*]. The formation and growth of the crack are also influenced by interfacial friction, differences in modulus between the indenter and specimen [*24*], and the presence of surface flaws [*25*].

Many solids undergo some plastic flow involving true shear deformation (Hagan [*26*]) before cracking occurs, partly because the high hydrostatic pressures beneath the indenter inhibit crack formation. In many cases, apart from cracks which are formed when the load is applied, additional families of cracks may form when the load is removed. This behavior is generally attributed to a mismatch between the strains in the plastically deformed zone and those in the elastic hinterland, which produces residual tensile stresses when the load is removed. Recently, Yoffe [27] has shown how this may be tackled analytically to include the loss of volume due to compaction which occurs with many glasses when they are subjected to indentation.

The main ideas are presented in Fig. 8. It is assumed that at distances, r, from the indenter, the elastic field is given by the classic Boussinesq solution for a point load on the flat surface of a semi-infinite half-space. The stresses fall off as r^{-2}, and they are sufficient to support the load from below. The problem of the plastically deformed material beneath the indenter is handled in an original way. The material is treated as a small volume element, which is subjected to a normal compressive stress and equal outward radial stresses. This treatment leaves the surface free of stress. Thus, the small cylinder shown in Fig. 8 changes its shape (and maybe its volume) as indicated. At distances, r, from the plastically deformed material, the local stresses fall off as r^{-3}, so that they contribute nothing to the support of the load at large distances. (The stress field is similar to that of a prismatic dislocation loop and is referred to by the author as a "blister field.") The displacements are particularly interesting; we quote only the radial displacements, U_r, which are given by

$$U_r = \frac{B}{G_r^2} [2(1 - \nu) - (5 - 4\nu) \cos^2 \theta] \tag{17}$$

where ν is Poisson's ratio, G the elastic shear modulus, and B the "strength of the field." These displacements are sketched in Fig. 8, where it may be seen

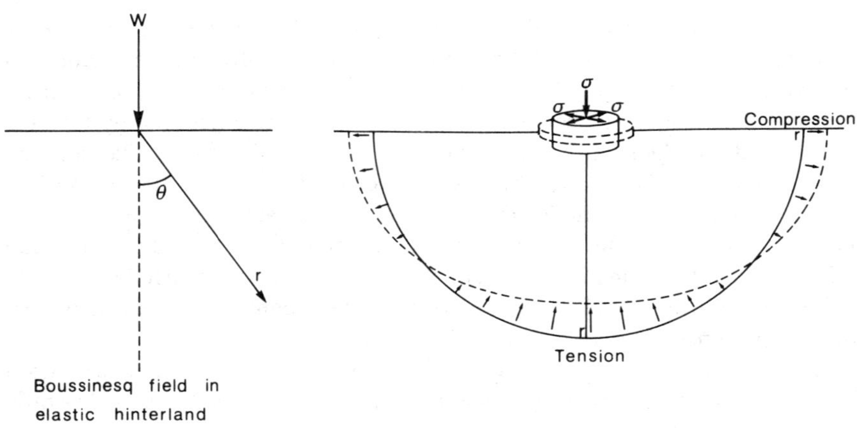

FIG. 8—*Indentation of a solid allowing for volume changes due to compaction, using the analysis of Yoffe* [27], *which treats the elastic field in the hinterland as though it were produced by a point load:* (left) *the elastic field and* (right) *the plastic displacements of a surface "blister," due to a normal pressure and horizontal orthogonal tensions. At a radius,* r, *there is a mismatch in displacements, producing tensions below the indentation and horizontal compressions near the free surface.*

that a hemisphere of arbitrary radius, r, expands radially near the surface, giving rise to a compressive stress, and contracts beneath the indenter, resulting in a tensile stress.

The stresses produced in the elastic region depend critically on Factor B. For a value $B = 0$, the value of U_r is everywhere zero. There is zero compaction of the deformed material, and no volume change occurs. This result is also shown by calculating the volume change, ΔV, of a hemisphere of any radius, r: it turns out that ΔV is independent of r and is given by

$$\Delta V = \frac{2\pi B(1 - 2\nu)}{3G} \tag{20}$$

The analysis shows how the stresses on and beneath the surface vary with B and predicts the cracking behavior to be expected both when the load is applied and when it is removed.

This original treatment is still at an early stage; in particular, it is not predictive since Factor B cannot yet be derived *ab initio*.

A Modelless Model

We have already discussed a number of models which have been developed to explain the indentation behavior of various types of solids. They all involve different ways of tackling the flow of displaced material, the stresses and strains in the hinterland, and the criteria of yield. They are all approximations to a rigorous solution.

It is therefore of particular interest to observe that Matthews has recently presented an extremely neat approximate analysis of indentation by a sphere without involving a physical model [*28*]. He assumes that the yield of the material in uniaxial tension or compression may be written in the form

$$\frac{\epsilon}{\epsilon_0} = \left(\frac{Y}{Y_0}\right)^n \tag{19}$$

where ϵ is the strain, Y is the stress, and ϵ_0 and Y_0 are appropriate material constants which may be found by fitting the data to a simple tension or compression test. We note that if $n = 1$, Eq 19 represents a linear *elastic* solid whose Young's modulus is $E = Y/\epsilon = Y_0/\epsilon_0$. At the other extreme, when $n \to \infty$, Eq 19 represents an ideal rigid-plastic solid of yield stress $Y = Y_0$. For any value of n between 1 and ∞, Eq 21 represents an annealed plastic material which work hardens according to a power law. Matthews then proposes a contact pressure distribution of the form

$$p(r) = p_m \frac{2n + 1}{2n}\left(1 - \frac{r^2}{a^2}\right)^{1/2n} \tag{20}$$

where the mean pressure, p_m, over the circle of contact is

$$p_m = \frac{W}{\pi a^2} = \frac{6nY_0}{2n+1}\left(\frac{8a}{9\pi R \epsilon_0}\right)^{1/n} \tag{21}$$

and where a is the radius of the circle of contact, r is the distance from the center, and R is the radius of curvature of the sphere. It may be seen that when $n = 1$, Eqs 20 and 21 reduce to the classic Hertzian expressions for elastic contact; when $n \to \infty$, the pressure distribution becomes uniform, of magnitude $3Y_0$, which is close to the behavior of an ideal rigid-plastic solid according to slip-line field theory. For n between 1 and ∞, the solution applies surprisingly well to work-hardening metals.

There are further extensions to this. We may estimate the "representative" yield stress, Y_R, of the solid by using the old semiempirical relation

$$Y_R = \frac{1}{3}H = \frac{1}{3}p_m = \frac{2nY_0}{2n+1}\left(\frac{8}{9\pi\epsilon_0}\frac{a}{R}\right)^{1/n} \tag{22}$$

Combining this with Eq 19, we may deduce a "representative" strain, ϵ_R. It may be written

$$\epsilon_R = \frac{8}{9\pi}\left(\frac{2n}{2n+1}\right)^n \frac{a}{R} \tag{23}$$

The strain, ϵ_R, has a value which ranges from 0.188 (a/R) to 0.171 (a/R) as n varies from 1 to ∞. This may be compared with the empirical value of 0.2 (a/R) given 30 years earlier by Tabor [*3,10*].

By analogy with Eqs 22 and 23, Matthews proposes an empirical expression for the penetration, δ, of the indenter into the half-space, namely

$$\delta = \left(\frac{2n}{2n+1}\right)^{2(n-1)} \frac{a^2}{R} \tag{24}$$

Thus, δ varies from the elastic value a^2/R for $n = 1$ to 0.37 (a^2/R) for a rigid-plastic material when $n \to \infty$. This is in good agreement with the ideal plastic analysis of Richmond et al (see Ref *31*) in which the pileup of material around the indentation is allowed for. If there were no pileup, the simple geometry of indentation would give a value of $\delta = a^2/2R$. If $\delta < a^2/2R$, pileup must occur, and this takes place when $n > 3.8$. Conversely, when $\delta > a^2/2R$, the material around the periphery of the indentation sinks in. This occurs for low values of n, that is, for annealed materials. Matthews successfully applies a similar approach to materials which creep according to a simple power law.

These extraordinary results are all derived from a nonphysical model using a pressure distribution which appears to be entirely intuitive, although it may have been inspired by some earlier approximate solutions of indentation problems involving nonlinear creep [*29,30*]. This is discussed more explicitly by Johnson [*31*].

Hardness of Single Crystals

If hardness indentations are made in the surface of a single crystal using a Knoop indenter (which produces indentations seven times as long as they are wide), it is found that the hardness is often markedly anisotropic [*31*]. Anisotropies of up to a factor of two have been observed. Anisotropy is not surprising since various slip systems are involved and the resolved shear stress will depend on the orientation of the indenter relative to the crystal, as Brookes [*32*] and his colleagues have shown. However, a further complication is that glide on the relevant slip planes is probably accompanied by bending of the slip planes themselves [*33*].

We shall not discuss hardness anisotropy further. Instead, we shall consider the *magnitude* of the hardness in terms of the yield properties of the crystal. With polycrystalline specimens, as discussed previously, the indentation hardness, H, is of the order 1.5 to 3 Y. With single crystals, there is a very large discrepancy between H and the yield stress. Westbrook [*34*] found that for many crystals the ratio of H to Y was of the order of 35 instead of 1.5, but the strains involved in determining H and Y are probably very different indeed. The lowest hardness is observed with crystals of high purity. A good deal of effort has gone into a study of the indentation behavior in terms of dislocation mechanisms; the analysis is very difficult. A much simpler approach is that of Gerk [*35*], who suggests that indentation behavior can be understood in terms of the work-hardening characteristics of the material.

It turns out that with very pure single crystals of many crystal types other than hexagonal close-packed, the initial shear stress is very small and does not greatly increase with strain. This is the regime of easy glide (Stage I) in which slip occurs on favorably oriented slip planes. Beyond a certain level of strain, other slip planes become involved: there is a rapid multiplication of dislocations and the shear stress, τ, increases almost linearly with strain (Stage II). If we express the strain, γ, as a tensorial shear strain (where γ is one half the engineering strain), we find that for face-centered cubic metals and rock-salt structures [*35*]

$$\tau \simeq \left(\frac{G}{125}\right)\gamma \tag{25}$$

where G is the elastic shear modulus of the crystal. We thus have a material for which the stress is roughly proportional to the tensorial strain, although the strain is plastic. We may take over the elastic solution for the indentation of a material of elastic shear modulus, S, by a cone of semiapical angle, θ. This gives

$$P = \frac{E \cot \theta}{2(1 - \nu^2)} = \frac{S \cot \theta}{(1 - \nu)} \tag{26}$$

since the shear modulus $S = E/2(1 + \nu)$. We now replace S with $G/125$, assume $\nu = 1/2$ for plastic deformation, and take θ to be about 70° for a Vickers indenter. Then

$$p = H = 7 \times 10^{-3} G \tag{27}$$

A plot of average hardness against G for a number of solids is given in Fig. 9. It is seen that for many single crystals the relation

$$H \approx 10^{-2} G \tag{28}$$

is fairly well obeyed. The hardness values are thus about 40% higher than those predicted by Eq 27—possibly because planes other than those of easy slip are involved—but the general correlation is well satisfied. We may well ask if correlations really validate a particular explanation. The main point of Gerk's work is that the hardness of pure single crystals is largely determined by their work-hardening characteristics.

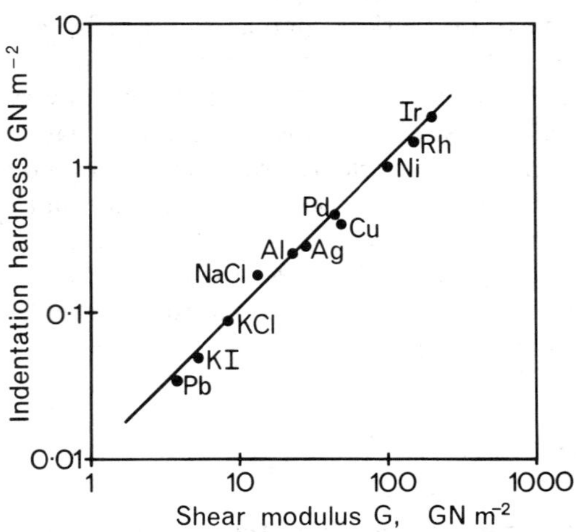

FIG. 9—*Vickers indentation hardness of pure single crystals plotted against their elastic shear modulus,* G *(from Gerk* [34]*). Note that 100 kg mm^{-2} $\simeq$ 1 GPa.*

Hardness of Semiconductor Single Crystals

Plastic hardness indentations may be made at low loads in single crystals of germanium and silicon and even in diamond, and there is convincing evidence for the generation and movement of dislocations around the indentations. With germanium and silicon at very small loads, no cracks are observed. With these materials, the indentation hardness is of the order of 0.1 to 0.2 times the elastic shear modulus, compared with metals for which the ratio is nearer 10^{-2}. The reason for this, as pointed out by Gilman [*36*] and other workers, is that with these essentially covalent materials, plastic flow involves the breaking and remaking of electron-pair bonds. Consequently, one would expect some connection between the energy of the band gap and the hardness. Since the latter has the dimensions of pressure, the energy must be divided by volume to give similar dimensions. Gilman has indeed shown that there is reasonably good correlation between the hardness and band gap/unit cell volume. Of course, it is possible to argue that other parameters are more directly involved, for example, the homopolar part of the band gap, or the bond strength [*37*]. There is one further point. It is well known that under a hydrostatic pressure of about 120 kbar, germanium becomes metallic, that is, electrons are able to move freely in the material and it acquires some ductility. This pressure is close to the indentation hardness of germanium (800 kg $mm^{-2} \approx$ 80 kbar). Similarly, with silicon the transition pressure is 190 kbar, while the indentation hardness is ~120 kbar. From this it would be reasonable to deduce that with diamond, for which the hardness is ~1000 kbar, a metallic transition should occur at a pressure somewhat greater than 1000 kbar [*38*]. Whether or not the hardness is, indeed, related to some phase change is still to be established.

Although this correlation is crude, it has interesting ramifications in an associated line of research—the effect of doping. If plastic flow involves the breaking and reformation of bonds, we should expect the presence of donor or acceptor levels in the band gap to enhance the mobility of the dislocations. This has, indeed, been observed in silicon [*39*]. Apparently, the stress required to generate dislocations is unchanged, that is, the hardness is the same for doped and undoped material, but the size of the rosettes exposed by etching increases with the concentration of dopant. Similar effects are produced by irradiation with light or electrons, but it is probable that different mechanisms, for example, multiphonon processes, are involved.

Some Practical Problems Encountered in Indentation Hardness Measurement

The Problem of Loading

If a static hardness indentation is formed, it is assumed implicitly but not explicitly that the load is applied incrementally, so that the indentation is in equilibrium with the load at all times. Consider a conical (or pyramidal) indenter which, under a final load of W, produces an indentation of diameter $d = 2a$. The Meyer hardness may be defined as

$$H = \frac{W}{\pi a^2} \quad \text{or} \quad \pi a^2 = \frac{W}{H} \tag{29}$$

If, however, the indenter is placed on the surface of the specimen and the full load, W, is applied instantaneously, the force opposing the load will initially be zero when $a = 0$, so that the load will accelerate as the indenter penetrates the specimen. It will finally come to rest, producing an indentation of chordal diameter $d_0 = 2a_0$ when the work done by W is equal to the plastic work of forming the indentation. The latter is equal to H times the volume of the indentation $= H^1/_3\pi a_0^2 h$, where h is the depth of the indentation. Ignoring piling up or sinking in, h is also the distance moved by the load, W. Hence

$$Wh = H\frac{1}{3}\pi a_0^2 h$$

or (30)

$$\pi a_0^2 = \frac{3W}{H}$$

Evidently, the area of the indentation will be three times larger than that obtained under conditions of incremental loading. If the hardness were calculated from this area, the deduced hardness would be one third of the true equilibrium value. For a spherical indenter, instantaneous loading gives an apparent hardness value of one half the true equilibrium value. Of course, energy may be dissipated in other ways, for example, by wave propagation, but this simple calculation indicates the importance of avoiding step loading. This must be especially important in microhardness measurements. For example, if an induction period is recorded before creep is observed, it could be due to some initial steplike application of the load.

Elastic Recovery of the Indentation

When the indenting load, W, is removed, a certain amount of recovery in the elastic hinterland occurs. As a result, the indentation changes its shape. With spherical indenters, there appears to be little change in the chordal diameter, d, of the indentation but a marked increase in its radius of curvature. It is shallower than if it were of the same curvature as the indenter. Assuming that the recovered indentation is spherical, we may calculate its curvature by arguing that if we were to replace the indenter and reapply the original load, W, the surfaces would deform elastically until the contact occurred just over the initial diameter, d [10]. Thus, the amount of elastic recovery can be calculated and, from this, the elastic energy released on removing the indenting load. This procedure provides a means of correlating rebound hardness with the elastic and the plastic properties of the solid [3,10]. Similar calculations may be carried out for conical indenters, making the assumption that the recovered indentation is also conical but of a larger semiapical angle [40].

Although this approach gives fairly good quantitative agreement with the experimental results, it leaves a number of questions. First, the recovered shape is not exactly spherical for the spherical indenter nor conical for the conical indenter. Associated with this problem is the presence of residual stresses. Furthermore, these may be large enough to produce reversed plastic flow when the load is removed, particularly if the metal shows a marked Bauschinger effect. Thus, repeated loading may gradually increase the size of the plastic indentation and fatigue the material. The role of residual stresses released when the indenting load is removed is even more important in the fracture of brittle-plastic solids subjected to indentation. A brief reference to this theme appears on pp. 143 through 145.

In some cases, the chordal diameter of the indentation changes appreciably when the load is removed, particularly for solids with lower E/Y ratios. The theory for this is not well developed, but an order of magnitude for the diametral change has been estimated by Johnson [15]. If a uniform pressure, p, is applied over a circular zone of radius, a, in an elastic material, classic elasticity gives for the elastic displacement at the edge

$$[U_r]_{r=a} = \frac{(1 - 2\nu)(1 + \nu)}{2E} pa \tag{31}$$

Assuming $\nu = 0.3$ and $p = 3Y$, we have

$$\left[\frac{U_r}{a}\right]_{r=a} = -0.26 \frac{Y}{E} \tag{32}$$

Thus, even if plastic deformation occurs when p is applied, there is reversed elastic movement when p is removed approximately equal to that given by Eq 18. This implies a fractional *increase* in the diameter of the order of 0.26 (Y/E). For most solids for which Y/E lies between 10^{-2} and 10^{-3}, this dimensional change is not detectable.

Area of the Indentation

Most hardness indenters are either square pyramids (Vickers), elongated pyramids (Knoop), or triangular pyramids (Berkovitch). If the sides of the indentation are perfectly straight, there is no difficulty in calculating the area of the indentation from the length of the diagonals, though elastic recovery of the indentation after the load is removed may introduce some uncertainties, particularly in the shorter diagonal of the Knoop indentation.

A more contentious issue concerns the situation if the sides of the indentation are curved. For example, with annealed metals the Vickers indenter produces an indentation with concave boundaries (pincushion shape). In that case, the most reasonable procedure is to determine the true area of the indentation and use this instead of the area deduced from the length of the diagonals. The mean pressure over the pincushion indentation is the most meaningful measure of the yield property of the solid. With work-hardened materials there is upward flow of material around the indentation, and with the Vickers indenter, since the flow is less restricted on the faces than at the corners, an indentation is formed with convex boundaries (barrel shaped). Again, it is often customary to determine the true area of the indentation, taking into account the curved boundaries either by calculation or by projection of an image on a digitizing table [*41*]. This area is then used to calculate the "best" estimate of the hardness. This procedure may be open to criticism, since part of the curved portion of the indentation may be outside the load-bearing area. Such a situation could arise if the optical system focuses on the crest of the bulge, where the metal has already ceased to make contact with the indenter. It may also arise as a result of the detailed way in which the displaced metal flows away. There is one classic experiment in which the "ears" of the barrel-shaped indentation were gently lapped away, leaving a truly square indentation of much smaller area; the indenter was then replaced in the original indentation and the original load applied. There was very little increase (1.3%) in the indentation diagonal, and barreling did not occur [*44*].

Clearly, the operator must adopt some standard procedure, but he must be a little cautious in claiming that he has measured the "true" hardness.

Microindentation Hardness

Increase in Hardness at Small Loads

Large-scale hardness testing using pyramidal indenters is usually carried out at loads ranging from 10 to 300 N (about 1 to 30 kg weight). The diameters of the indentations usually lie between about 100 μm (0.1 mm) for the hardest materials and 1000 μm (1.0 mm) for the softer. With the Vickers indenters, the depth is about one seventh of the length of the diagonal, say, 15 to 150 μm. With a good optical microscope it should be possible to measure the diagonals to an accuracy of 1 μm or so, implying, on the average, an accuracy of better than 1% in the diagonal or 2% in the area. Over this load range the indentation hardness is almost independent of load unless the surface layers have been appreciably affected by work hardening or segregation.

Conventional microhardness equipment [*43,44*] operates over a load range of 0.1 to 5 N (10 to 500 g weight). The indentation diameters are about ten times smaller than those involved in macrohardness tests, say, 10 to 100 μm; the indentation depths will lie between 1 and 10 μm (10^3 to 10^4 nm). The accuracy of the measured indentation diameter is correspondingly poorer. Furthermore, there is a very real problem in identifying the true edge of the load-bearing part of the indentation. Thus, the accuracy at the lowest loads may be as poor as $\pm 10\%$. In some cases it is found that the hardness diminishes at very small loads, but this may be due to vibration, which becomes increasingly important as the load is reduced. More generally there is an increase in hardness at small loads. This could be due to the method of preparing the specimen, since the surface layers may be hardened by the polishing process. However, a similar trend is observed with carefully annealed specimens. Some workers have suggested that there is elastic recovery of the *diameter* of the indentation when the load is removed [*45*] and that this is proportionately more marked for small indentations, thus leading to an apparent increase in hardness at small loads [*40,44*]. However, because of geometric similarity, this explanation does not seem tenable. Again it has been suggested that, because of the limit of resolution of the optical system, the indentation appears smaller than its true size by approximately a constant absolute amount; the proportionate effect would then be most marked for the smallest indentations [*44*, p. 130]. Finally, a view (which I do not share) has been expressed that the whole effect vanishes if the correct statistics are applied to the experimental data. It is, of course, true that the smaller the indentation, the larger the experimental error. Nevertheless, the broad consensus, based on careful experiments by many independent workers, is that the hardness increases at small loads. More than 25 years ago Bückle [*43*] and Mott [*44*] attributed this, in somewhat different terms, to the limited number of dislocations in the small volumes being deformed.

Recently, indentation hardness measurements have been extended into a range of even smaller loads, from 0.05 N (5 g weight) down to a few micronewtons. One might refer to these as picohardness measurements [*46,47*]. Some of the relevant results will be discussed in this volume. They show that for indentation depths less than about 10^3 nm (1 μm) there is a steady rise in hardness, and, for a depth of 20 nm (0.02 μm) corresponding to a load of the order of 100 μN, the hardness of electropolished nickel is about four times the bulk macroscopic hardness. These values are based on depth measurements. They were carried out in air so that surface films might have been present. However, similar hardness increases were observed with gold and are therefore probably genuine [*47*]. Similar experiments have been carried out on atomically clean surfaces of nickel in ultra high vacuum (UHV) using electrical resistance measurements to deduce the area of contact [*48,49,51*]. They again show that for loads in the range 30 to 100 μN the hardness is of the order of 4.5 GPa, that is, about five times the bulk value. An interesting observation is that surface forces pull the surfaces together [*49*]. The attraction depends on the geometry and cleanliness of the surfaces, but a typical value for the attractive force is 5 to 10 μN. Evidently it would not be sensible to carry out picohardness measurements at applied loads of less than, say, 100 μN. We shall discuss the experimental aspects in the following section.

At one time it was considered that the hardness of a deformed volume so small that it contained no dislocations would approach the theoretical value of the perfect crystal. For nickel, for which the shear modulus, G, equals 79 GPa, we should expect a critical shear stress of the order of $\tau = G/30 \simeq 2.5$ GPa and a corresponding indentation hardness, H, of $3Y \simeq 6\tau = 15$ GPa. The highest observed value is about 4. However, the model is probably not valid since there is no sudden yielding, which would be expected if the indenter suddenly generated a large number of dislocations. The sudden yielding observed in the earlier pioneer experiments of Gane [*52*] was associated with the presence of a thin carbonaceous film. It is, indeed, probable that individual dislocations are generated rather easily and that multiplication takes place rapidly. However, in a single crystal the overall yielding required to accommodate the shape of the indenter will be complicated by the constraints imposed by the availability of local slip planes. Further, the limited size of the deformed zone probably entangles the dislocations and restricts their movement. It might, therefore, be simpler to regard the hardness of these very small volumes as resembling that of a fully work-hardened specimen for which the dislocation entanglement is so marked that the material approximates a solid containing no mobile dislocations [*1,48*]. This is a field in which critical thinking and experimentation are required.

Measurement of Microindentation Hardness

Apart from the operator himself, there are four elements involved in hardness measurement: the indenter, the loading device, the specimen, and the measurement of the indentation size.

1. *The indenter*—The main requirements are high modulus, absence of plastic deformation, low friction, smoothness of surface, and perfection of form. The first four of these are satisfied by using highly polished diamond as the indenter material. Perfection of form is a challenge to the diamond polisher. However, one general comment may be made concerning the geometry to be chosen. For microindentations the Vickers or Knoop indenter is perfectly satisfactory, since the chisel edge, which is unavoidable, can be kept small compared with the size of the indentation. For picoindentations, however, this is not so, and it is advisable to use the Berkovitch indenter in which the three faces can be brought to a point very much smaller in diameter than the size of the indentations used.

2. *The loading device*—Whatever device is used—dead loading, electrostatic or magnetic loading, or spring loading—the most important require-

ment is the minimization of steplike loading. At the lowest loads in microindentation and over the whole load range of picoindentation, it is important to reduce vibration to the lowest possible level. This includes vibrations produced by sound waves.

3. *The specimen*—In studying the performance of a particular piece of hardness equipment it is possible, with some effort and experience, to prepare a specimen with near-ideal characteristics, such as the absence of a work-hardened layer, extreme smoothness, and no contaminant film. If the equipment is to be used as a research tool to study, say, segregation, grain boundaries, or wear mechanisms, it is also possible to prepare *sections* with favorable characteristics. However, if the free surface itself is to be investigated, regions must be chosen which are uniform enough to contain the indentation. In studying films of one material on a substrate of a different material, the thickness of the film and the strength of its attachment to the substrate may be important.

4. *Indentation size*—The most difficult (and important) factor is the determination of the load-bearing area of the indentation. With microindentations, the imprint is generally studied after the load is removed. Even if elastic recovery is unimportant, the precise location of the edges of the indentation appears to depend on the technique of observation. For example, with optical systems the angle of illumination affects the apparent position of the edge. Transmission electron microscopy often gives a slightly different answer. One must not pursue precision which is not there.

Because of the difficulties outlined here, there is a good deal of interest in the use of depth measurements. This is unavoidable in picoindentation studies because the diameter of the indentation may be of the order of the wavelength of light. The depth measurement may be made while the load is acting, and any change in depth when the load is removed may provide some measure of the amount of elastic recovery. The main uncertainties in depth measurements concern the initial zero and piling up or sinking in of the material around the indentation. The initial zero may be established electrically if the indenter consists of a conducting diamond. But piling up and sinking in depend on the crystal orientation, on the work-hardening characteristics of the solid under investigation, and on the level of friction between the indenter and the specimen. All of these parameters may vary from one specimen to another and from one material to another. The depth may also be significantly affected by surface roughness. All of these considerations suggest that attempts at rigorous standardization may turn out to be nugatory.

There are other methods of determining the area of contact while the load is applied, but they provide estimates rather than precise values. These include electrical contact resistance, thermal resistance, and acoustic impedance. Such measurements are defensible only if no other methods are applicable, for example, in *in situ* deformation of solids in UHV.

Conclusions, Future Trends, and Challenges

The indentation hardness, H, of ductile materials is a measure of their plastic yield properties as modified by the constraints imposed by the elastic hinterland. If these constraints are small (low values of E/Y, where E is the elastic modulus and Y the uniaxial yield stress), the value of H is only a little greater than that of Y. The volume displaced plastically by the indenter is accommodated by elastic strains in the hinterland. As the constraint increases (larger values of E/Y), a point is reached at which the elastic hinterland can no longer accommodate the displaced volume, and there is an appreciable upward flow of material around the periphery of the indentation. At this stage, the deformation process can be best expressed in terms of classic slip-line field plasticity, and H reaches an upper value of the order of 3 Y. The hardness also depends on the shape of the indenter, on the friction between indenter and specimen, and on the work hardening produced by the indentation process itself. All these factors influence the value of H deduced from the load and the area of the indentation. Evidently, the detailed processes taking place at the indenter–specimen interface, in the plastically deformed material, in the free surface, and in the hinterland are necessarily complex. Thus, however accurately we may determine H, we cannot expect it to provide a simple quantitative correlation with a single yield property of the material. Rough correlations may, however, be made, and they are meaningful. Similarly, with plastic-brittle solids, the generation of cracks during or after indentation provides a measure of certain fracture properties; in some cases the correlation may be fairly precise and in others only approximate.

There is a general fundamental problem concerning the flow of material around the indentation. For both ductile and brittle solids this invariably involves the movement of dislocations. However, dislocations are not necessarily most concentrated immediately beneath the indenter, and there is some evidence, as yet unpublished, that the large pressure gradients expel the dislocations from this region.[4] The matter needs further study, and for this reason there is scope for the elaboration of a variety of techniques for determining local dislocation densities. This is part of a much more general problem that is associated with interpreting hardness measurements—our lack of understanding of the yield and failure properties of solids under high compressive stresses and under high stress gradients.

As far as the indentation hardness measurements themselves are concerned, some of the practical problems involved have already been discussed. These include the problem of applying the load, the avoidance of vibration, the effects of elastic recovery, and the precise area of the load-bearing indentation. All of these problems are accentuated in microhardness measurements. When the indentations become so small that the measurement of the

[4] L. M. Brown, Cavendish Laboratory, Cambridge, United Kingdom, personal communication, 1984.

size of the indentation is limited by the resolving power of the measuring instrument, it becomes easier to measure the *depth* of penetration of the indenter. These measurements can be made with extremely high accuracy: furthermore, they can be automated, digitized, and the calculated hardness displayed to many decimal places. But accuracy and reproducibility are not enough. The depth method introduces additional problems, namely, the geometric perfection of the indenter, the zero of the depth measurement, elastic yielding (or recovery if the load is removed), and, not least, the piling up or sinking in of material around the indentation. All of these factors introduce uncertainties, so that the hardness value deduced from such depth measurements, however precise, may not represent the true pressure over the indentation. However, with these precautions in mind we may use the hardness values as some measure of the yield properties of the material in and around the indentation. Consequently, microindentation measurements employing the depth-of-penetration technique can, if used with caution and good sense, become a valuable tool in the study of processes involving the surface and near-surface layers of solids [*53*,*54*].

Acknowledgments

My chief thanks are due to Prof. K. L. Johnson for wise guidance and for showing me chapters of his forthcoming book, to Dr. M. M. Chaudhri for helpful discussions, to Dr. J. B. Pethica for comments relating especially to microindentation hardness measurements, and to the referees for constructive criticism.

References

[*1*] O'Neill, H., *The Hardness of Metals and Its Measurement*, Chapman and Hall, London, 1934; *Hardness Measurement of Metals and Alloys*, Chapman and Hall, London, 1967.
[*2*] Lysaght, V. E., *Indentation Hardness Testing*, Rheinhold, New York, 1949.
[*3*] Tabor, D., *The Hardness of Metals*, Clarendon Press, Oxford, 1951.
[*4*] Weingraber, H. von, *Technische Härtemessung*, Carl Hauser Verlag, Munich, 1952.
[*5*] Westbrook, J. H. and Conrad, H., Eds., *The Science of Hardness and Its Research Applications*, American Society for Metals, Metals Park, OH, 1973.
[*6*] *Härteprüfing in Theorie und Praxis*, Verein Deutscher Ingenieur, Berichte 308, Stuttgart, 1978.
[*7*] Hill, R., *The Mathematical Theory of Plasticity*, Clarendon Press, Oxford, 1950.
[*8*] Hill, R., Lee, E. H., and Tupper, S. J., "Theory of Wedge-Indentation of Ductile Metals," *Proceedings of the Royal Society (London)*, Vol. A188, 1947, pp. 273-289.
[*9*] Samuels, L. E. and Mulhearn, T. O., "The Deformed Zone Associated with Indentation Hardness Impressions," *Journal of the Mechanics and Physics of Solids*, Vol. 5, 1957, pp. 125-134.
[*10*] Tabor, D., "A Simple Theory of Static and Dynamic Hardness," *Proceedings of the Royal Society (London)*, Vol. A192, 1948, pp. 247-274.
[*11*] Tabor, D., "The Hardness of Solids," *Proceedings of the Institute of Physics, F, Physics in Technology*, Vol. 1, 1970, pp. 145-179.
[*12*] Marsh, D. M., "Plastic Flow in Glass," *Proceedings of the Royal Society (London)*, Vol. A279, 1964, pp. 420-435.

[13] Bishop, R. F., Hill, R., and Mott, N. F., "The Theory of Indentation Hardness Tests," *Proceedings of the Physical Society (U.K.)*, Vol. 57, 1945, pp. 147-159.
[14] Johnson, K. L., "The Correlation of Indentation Experiments," *Journal of the Mechanics and Physics of Solids*, Vol. 18, 1970, pp. 115-126.
[15] Johnson, K. L., *Contact Mechanics*, Cambridge University Press, Cambridge, 1985.
[16] Chiang, S. S., Marshall, D. B., and Evans, A. G., "The Response of Solids to Elastic-Plastic Indentation. I. Stresses and Residual Stresses," *Journal of Applied Physics*, Vol. 53, 1982, pp. 298-311.
[17] Lee, C. H., Masaki, S., and Kobayashi, S., "Analysis of Ball Indentation," *International Journal of Mechanical Sciences*, Vol. 14, 1972, pp. 417-426.
[18] Atkins, A. G. and Tabor, D., "Plastic Indentation in Metals with Cones," *Journal of the Mechanics and Physics of Solids*, Vol. 13, 1963, pp. 149-164.
[19] Atkins, A. G., Silverio, A. A., and Tabor, D., "Indentation Hardness and the Creep of Solids," *Journal of the Institute of Metals*, Vol. 94, 1966, pp. 369-378.
[20] Johnson, K. L. *Contact Mechanics*, Cambridge University Press, Cambridge, 1985, Chapter VI.
[21] Griffith, A. A., "The Phenomena of Rupture and Flow in Solids," *Philosophical Transactions of the Royal Society (London)*, Vol. A221, 1920, pp. 163-198.
[22] Roesler, F. C., "Indentation Hardness of Glass as an Energy Scaling Law," *Proceedings of the Physical Society (U.K.)*, Vol. B69, 1956, pp. 53-60.
[23] Frank, F. C. and Lawn, B. R., "On the Theory of Hertzian Fracture," *Proceedings of the Royal Society (London)*, Vol. A299, 1967, pp. 291-306.
[24] Johnson, K. L., O'Connor, J. J., and Woodward, A. C., "The Effect of the Indenter Elasticity on the Hertzian Fracture of Brittle Materials," *Proceedings of the Royal Society (London)*, Vol. A334, 1973, pp. 95-117.
[25] Lawn, B. R. and Wilshaw, T. R., *Fracture of Brittle Solids*, Cambridge University Press, Cambridge, 1975.
[26] Hagan, J. T., "Shear Deformation under Pyramidal Indentations in Soda-Lime Glass," *Journal of Materials Science*, Vol. 15, 1980, pp. 1417-1424.
[27] Yoffe, E. H., "Elastic Stress Fields Caused by Indenting Brittle Materials," *Philosophical Magazine A*, Vol. 46, 1982, pp. 617-628.
[28] Matthews, J. R., "Indentation Hardness and Hot Pressing," *Acta Metallurgica*, Vol. 28, 1980, pp. 311-318.
[29] Arutiunian, N. Kh., "The Plane Contact Problem in the Theory of Creep," *Applied Mathematics and Mechanics (PMM)*, Vol. 23, 1959, pp. 1283-1313.
[30] Kutnetsov, A. I., "Penetration of Rigid Dies into a Half-Space with Power-Law Strain Hardening and Non-Linear Creep," *Applied Mathematics and Mechanics (PMM)*, Vol. 26, 1962, pp. 717-732.
[31] Johnson, K. L., "Inelastic Contact: Plastic Flow and Shakedown" in *Contact Mechanics and Wear of Rail/Wheel Systems*, J. Kalousek, R. V. Dukkipati, and G. M. L. Gladwell, Eds., University of Waterloo Press, British Columbia, 1982, pp. 79-100.
[32] Brookes, C. A., O'Neill, J. B., and Redfern, B. A. W., "Anisotropy in the Hardness of Single Crystals," *Proceedings of the Royal Society (London)*, Vol. A322, 1971, pp. 73-88.
[33] Tanaka, Y., Ido, M., Honda, S., and Koshima, M., "Asterisms of the Laue Spots Produced by Cone Indentation in the (100) Face of Copper," *Philosophical Magazine A*, Vol. 47, 1983, pp. 93-104.
[34] Westbrook, J. H., "Microhardness Testing at High Temperatures," *Proceedings of the American Society for Testing and Materials*, Vol. 57, 1957, pp. 873-897.
[35] Gerk, A. P., "The Effect of Work-Hardening upon the Hardness of Solids: Minimum Hardness," *Journal of Materials Science*, Vol. 12, 1977, pp. 735-738.
[36] Gilman, J. J., "Dislocation Sources in Crystals," *Journal of Applied Physics*, Vol. 30, 1959, pp. 1584-1594.
[37] Gerk, A. P. and Tabor, D., "Indentation Hardness and Semiconductor-Metal Transition of Germanium and Silicon," *Nature*, Vol. 271, 1978, pp. 732-733.
[38] Vereshchagin, L. F., Yakovlev, E. N., Vinogradov, B. V., Sakun, V. P., and Steanov, G. N., "Transition of Diamond into the Metallic State," *High Temperature High Pressure*, Vol. 6, 1974, pp. 505-508.
[39] Roberts, S. G., Pirouz, P., and Hirsch, P. B., "A Simple Technique for Measuring Doping

Effects on Dislocation Motion in Silicon," *Journal de Physique, Colloque C4*, Vol. 44, 1983, pp. C4-75–C4-83.

[*40*] Stilwell, N. A. and Tabor, D., "Elastic Recovery of Conical Indentations," *Proceedings of the Physical Society (U.K.)*, Vol. 78, 1961, pp. 169–179.

[*41*] Rowe, G. W., Smethurst, E., and Downing, A. H., "A Reassessment of Accuracy in Diamond Pyramid Hardness Testing," *Journal of Material Science, Letters*, Vol. 1, 1982, pp. 109–112.

[*42*] Bergsman, E. B., "Problems of Hardness Testing of Thin Sheet Metals," *Sheet Metal Industry*, Vol. 31, 1954, pp. 5–14.

[*43*] Bückle, H., "Progress in Micro-Indentation Hardness Testing," *Metallurgical Reviews (London, Institute of Metals)*, Vol. 4, 1959, pp. 49–100.

[*44*] Mott, B. W., *Microindentation Hardness Testing*, Butterworths, London, 1957.

[*45*] Tate, D. R., "A Comparison of Microhardness Indentation Tests," *Transactions of the American Society of Metals*, Vol. 35, 1945, pp. 374–389.

[*46*] Pethica, J. B., "Microhardness Tests with Penetration Depths Less than the Ion-Implanted Layer Thickness" in *Ion Implantation into Metals*, V. Ashworth, W. Grant, and R. Proctor, Eds., Pergamon Press, Oxford, 1982, pp. 147–156.

[*47*] Pethica, J. B., Hutchings, R., and Oliver, W. C., "Hardness Measurements at Penetration Depths as Small as 20 nm," *Philosophical Magazine A*, Vol. 48, 1983, pp. 593–606.

[*48*] Pethica, J. B. and Tabor, D., "Contact of Characterised Metal Surfaces at Very Low Loads: Deformation and Adhesion," *Surface Science*, Vol. 89, 1979, pp. 182–190.

[*49*] Pashley, M. D. and Tabor, D., "Adhesion and Deformation Properties of Clean and Characterised Metal Micro-Contacts," *Vacuum*, Vol. 31, 1981, pp. 619–623.

[*50*] Pollock, H. M., Shufflebottom, P., and Skinner, J., "Contact Adhesion between Solids in Vacuum: 1. Single Asperity Experiments," *Journal of Physics D. (Applied Physics)*, Vol. 10, 1977, pp. 127–138.

[*51*] Pollock, H. M., Maugis, D., and Barquins, M., "The Force of Adhesion between Solid Surfaces in Contact," *Applied Physics Letters*, Vol. 33, 1978, pp. 798–799.

[*52*] Gane, N., "The Direct Measurement of the Strength of Metals on a Sub-Micrometer Scale," *Proceedings of the Royal Society (London)*, Vol. A317, 1970, pp. 367–391.

[*53*] Gilman, J. J., "Hardness—A Strength Microprobe," *The Science of Hardness and Its Research Applications*, J. H. Westbrook and H. Conrad, Eds., American Society for Metals, Metals Park, OH, 1973, pp. 51–72.

[*54*] Bückle, H., "The Science of Hardness Testing and Its Research Applications," *The Science of Hardness and Its Research Applications*, J. H. Westbrook and H. Conrad, Eds., American Society for Metals, Metals Park, OH, 1973, pp. 453–494.

Philip M. Sargent[1]

Use of the Indentation Size Effect on Microhardness for Materials Characterization

REFERENCE: Sargent, P. M., **"Use of the Indentation Size Effect on Microhardness for Materials Characterization,"** *Microindentation Techniques in Materials Science and Engineering, ASTM STP 889*, P. J. Blau and B. R. Lawn, Eds., American Society for Testing and Materials, Philadelphia, 1986, pp. 160–174.

ABSTRACT: This paper describes several different physicochemical mechanisms that produce an indentation size effect on microhardness. The mechanisms are illustrated with reference to published data. A simple quantitative model for surface layer effects is presented, and its relevance to chemomechanical effects on ceramics is discussed.

A well-behaved empirical function describing some indentation size effects is introduced, and its advantages are discussed. It is used in a new method of graphically presenting microhardness data which is presented here for the first time. This "error-ellipsoid" diagram not only is visually compact and statistically rigorous but is based on a new physically meaningful parameter, the "10-μm hardness."

KEY WORDS: hardness, indentation, microhardness, microindentation, surface layers, surface plasticity, chemomechanical effects, indentation size effect (ISE), Meyer index, microindentation hardness testing

A great many interrelationships between cooperative and competing processes occur in a solid when it is indented. Figure 1 attempts to show all the causal relationships between these effects. This paper is concerned with only the subset of these processes that control the variations in measured microhardness values which occur when indentations are made using different loads. All of these processes are shown in the relevant part of Fig. 1 and will be discussed separately, and their independent existence will be demonstrated using examples from the published literature. The principal mechanisms causing these variations are those of microstructure size and the thickness of environmentally affected layers or superficial films. Since these have

[1] Research assistant, Department of Engineering, University of Cambridge, Cambridge CB2 1PZ, England.

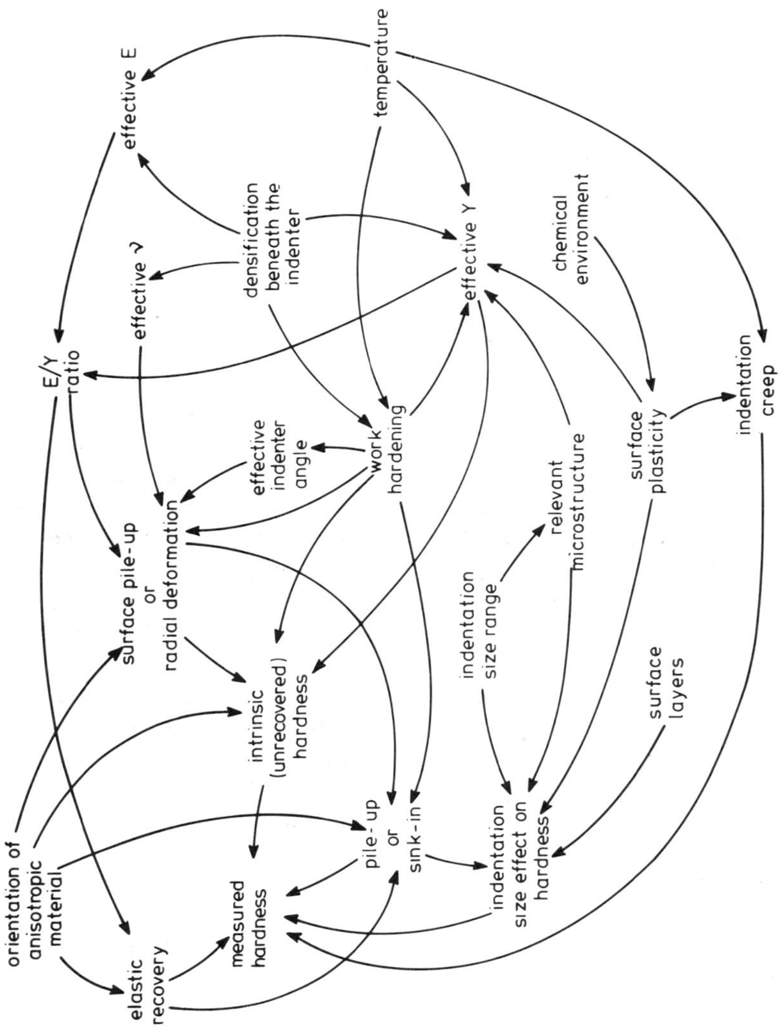

FIG. 1—*Some of the interrelationships between processes occurring on indentation.* E *is the Young's modulus,* Y *is a flow stress, and ν is Poisson's ratio* [1].

characteristic dimensions, the hardness variations are best termed indentation size effects (ISE) and not load effects.

Indentation Size Effects

For conical and pyramidal indenters, indentations of different sizes are, to a first approximation, geometrically similar, and so hardness should be independent of indentation size. This is generally true for large indentations but not for indentations less than about 10 μm deep. A number of mechanisms contribute to the variation of microhardness with indentation size, and since there is no good physical model for any of them on which a quantitative description of ISE can be based, this paper uses an empirical ISE index defined only in terms of the directly measured quantities of force on the indenter (the load) and the diameter of the indentation produced

$$n = \frac{\partial F/F}{\partial d/d} \tag{1}$$

where n is the ISE index and is a dimensionless, normalized parameter describing the shape of the curve of one experimental parameter plotted against the other. This index has been measured to vary between 1.5 for copper with a work-hardened surface layer [*1*] and 4.0 for lead telluride (PbTe) [*2*]. A value of 2.0 means that there is no size effect and that hardness is independent of indentation size, for microindentations, this is the exception rather than the rule. This ISE index describes the change in equilibrium stress around indentations of the same shape but of different sizes. It should be distinguished from the Meyer index, which is derived only for the case in which a ball indenter is used and the indentations are different in both size and shape.

Small indentations can most obviously measure different hardnesses from large indentations if there is a superficial layer of another substance, for example, oxide film in air, contaminant carbon film in oil-pumped vacuum systems, or preferential surface segregation of one component in an alloy. A similar effect occurs if the surface of the solid has a different mechanical behavior from that of the bulk, even when it has the same composition. This is especially important in insulators and semiconductors, in which the effects of surface electronic states and space charge penetrate much more deeply than they do in electronic conductors [*3-5*]. Dislocations in nonconducting materials commonly carry charged jogs, and so the density of charge carriers and the band bending near the surface affect the plasticity. This is the major cause of chemomechanical and optomechanical effects.

The second class of phenomena which can cause indentation size effects are those connected with the finite size of the "deformation carriers" (usually dislocations in crystalline materials, but also shear bands in some glasses) and of the microstructure and the indentations themselves. Mott [*6*] hypothesized that smaller indentations would require dislocation slip bands to be more closely spaced than would large indentations, and thus the flow stress for these finer bands would be higher. This is in accordance with most experimental observations that smaller indentations measure a higher hardness (which means that ISE indices are less than 2.0).

Figure 2 (*top*) illustrates how surface layers and microstructural features interact with an indentation. In the layer case, very small indentations measure only the hardness of the surface, and very large indentations are dominated by bulk interactions and measure only the hardness of the underlying material. For intermediate-sized indentations, the measured hardness depends on the size of the indentation in a smooth, monotonic manner (Fig. 2 *bottom*, *right*).

The lower part of the top of Fig. 2 shows diagrammatically how microstructural features influence mechanical behavior within a particular radius (shown as circles on the figure). Large indentations interact with a large number of features (whether they be grain boundaries, dislocation loops, precipitates, or regions of slightly different amorphous structure in a metallic glass). The statistical nature of the expected interaction is shown in Fig. 2 (*bottom*,

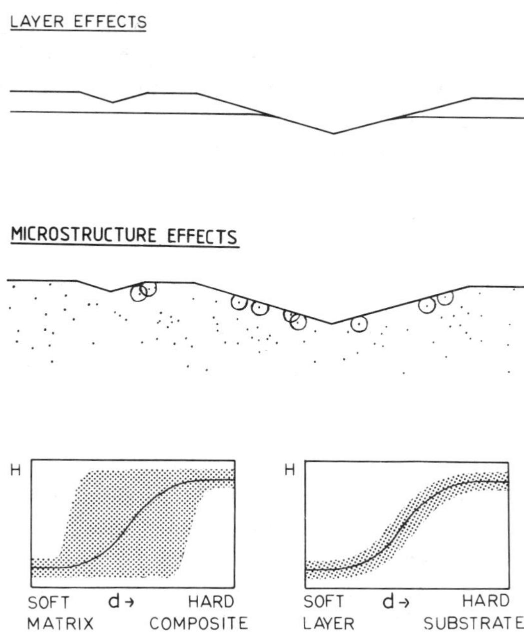

FIG. 2—*The expected effects of surface layers and microstructure on the relationship between indentation size and measured hardness:* (top) *Vickers indentations showing the sampling effect of layers and discrete microstructural elements;* (bottom) *what this implies for the measured hardness and the experimental scatter of individual measurements (the shaded areas).*

left), which shows a great deal of scatter (in the shaded area) in the transition between small indentations which measure the hardness of a pure matrix material and large indentations which "see" a harder composite material.

Layer effects need not be due to different superficially applied materials; different dominant plasticity mechanisms near the surfaces will also have such an effect. For example, the operation of image forces on dislocations, charge double layers in semiconductors and insulators, optomechanical effects, and dislocation etch-pit dragging in solvents will all affect behavior in the top micrometre or so of a material.

A Simple Single-Layer Model

A very simplified model of layer behavior has been devised and its properties investigated quantitatively, the results being expressed in terms of n, the ISE index [*1*]. The model bulk material has a hardness, H_2, and is covered with a layer of thickness, L, and hardness, H_1. This single-layer situation is a major simplification; no gradual change in hardness between the layers is assumed, nor are any criteria of strain compatibility across the interface taken into consideration.

A Vickers indentation is assumed (in this model) to measure only the hardness of a hemispherical volume centered at the original surface (before indentation), with a diameter equal to the diagonal length of the indentation. Within this hemisphere is the pyramidal space, in which the indenter penetrates, and two volumes: V_1 of bulk material and V_2 of surface material (the effects of elastic compression and surface pileup are ignored, and the layer is assumed to be the same thickness everywhere). The measured hardness is calculated by a simple law of mixtures

$$H = \frac{H_1 V_1 + H_2 V_2}{V_1 + V_2} \tag{2}$$

These simplifications have been made in order to reduce the number of parameters in the model; the form of the load/indentation size relationship and the quality of the data available are such that almost any function with more than two parameters can be made to fit. This model has only two parameters, H_1 and H_2. The ratio, t/d, of layer thickness to indentation diagonal varies over a range such that at one extreme the hemisphere is entirely contained within the layer and at the other the effect of the layer is negligible.

The model allows an estimate to be made of the variation of hardness with indentation size using the ISE index equation and the formula for calculating the hardness pressure, H, from Vickers indentations. All the hardnesses presented in this paper are calculated from the load divided by the projected area of the indentation and expressed in gigapascals. This is different by a geometric factor of 1.0785, and by a unit conversion factor of 1.0194×10^{-4}, from Vickers pyramid numbers (VPN), which are calculated from the surface area and expressed in kilograms force per square millimetre (1000 $kgf \cdot mm^{-2}$ is 9.81 GPa). The projected area method is used so that comparison can be made with Knoop and ball hardness measurements.

$$\text{VPN} = \frac{1854.4 \cdot \text{load}}{\text{diagonal}^2} \qquad \text{kgf} \cdot \text{mm}^{-2} \tag{3}$$

where the load is in kilograms force and the diagonal is in micrometres.

$$H = \frac{2000.0 \cdot \text{load}}{\text{diagonal}^2} \qquad \text{GPa} \tag{4}$$

where the load is in Newtons and the diagonal in micrometres (the factor 2.0 arises because the indentation area is half the square of the diagonal length).

The results of using the model are shown in Fig. 3. In the upper figure the

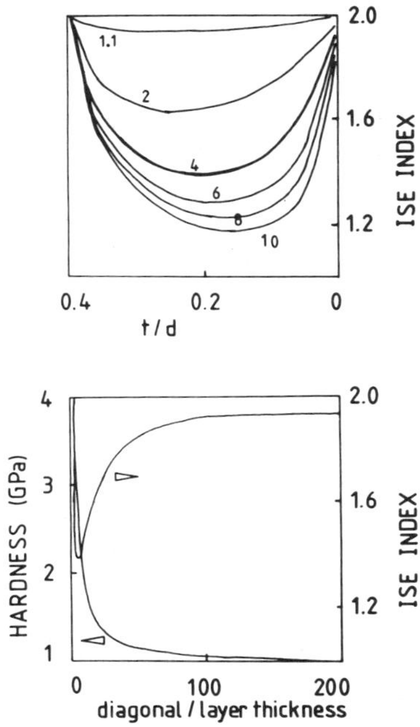

FIG. 3—*The results of the surface layer model with a substrate hardness of 1 GPA* [1]: (top) *calculated ISE indices plotted against the ratio of layer thickness and diagonal* (t/d) *for a substrate hardness of 1 GPa and superficial layer hardnesses of 1.1, 2, 4, 6, 8, and 10 GPa;* (bottom) *the same data as the 4-GPa line at the top, but plotted on reciprocal axes of diagonal/thickness. The calculated hardness for this combination of materials is also plotted; the scale is on the lefthand ordinate. The pseudoconstant regime, in which the ISE index approaches 2.0 asymtotically, is shown clearly in this graph.*

bulk hardness was set to 1 GPa, and the layer hardness was set to the value marked on each of the curves. The lower figure plots both the prediction of the ISE index, *n*, and the hardness prediction for the case in which the bulk is 1 GPa and the surface 4 GPa, but with the abscissa the reciprocal of the upper figure.

If experimental data for magnesium oxide (MgO) single crystals are compared with the predictions of the model, a rough fit may be obtained if a layer 0.05 to 1.2 μm thick, with a hardness of 14 GPa, is assumed to exist [*1*]. With such a crude model, the values cannot be more than an estimate, but space charge effects are thought to influence the top 1 μm [*2*].

The model shows that the ISE is more severe (that is, further from a value of 2.0) when the hardness of the layer and the bulk diverge and that, as the indentation size becomes markedly different from the layer thickness, the ISE index does approach 2.0 as expected. It is interesting that the lower figure shows that only a very thin layer has a strong effect on hardness and the ISE index and that as the indentations become very much larger than the thickness of the layer, there is still an effect on hardness, and *n* approaches 2.0 only asymtotically. The fact that *n* is roughly constant (about 1.96 for this model) over a wide range of indentation sizes in this region means that the ISE index equation fits well here. The value of the index is roughly similar to that found by Burnand for a very large number of inorganic single crystals. Burnand made an extensive review of published ISE indicies and found that they clustered about 1.8 to 1.9; he interpreted this as a fundamental property of the microhardness test [*7*]. On the basis of the model, it is proposed that this effect is due to measurements being made in the asymtotic regime for crystals whose surfaces are harder than the bulk because of surface-specific effects on plasticity. However, this can be only one influence; Fig. 1 shows that many other effects may be responsible.

Burnett and Page have since extended this model to describe the case of a soft amorphous layer on a hard crystalline substrate which has its own intrinsic indentation size effect from other causes [*8,9*]. Such a model describes relationships in which a peak occurs in the measured hardness at some particular indentation size, but it is not the only applicable model for this relatively common experimental result [*10*]. Figure 4 shows diagrammatically how a combination of size effects from grain size and from some other mechanism (for example, surface layers or work hardening) can lead to a hardness peak. The transition from single-crystal to polycrystal behavior occurs when indentations and grains are about the same size, so that for specimens with different grain size the peak will probably be displaced, and for ductile metals the hardness would tend to be higher for the finer-grained material because of the effect of grain size on yield stress and work hardening.

Figure 5 shows the experimental data from Raghuram et al [*11*] for single-crystal niobium in which the presence of a peak indicates the presence of two size effect mechanisms. However, for a single crystal no grain size effect ex-

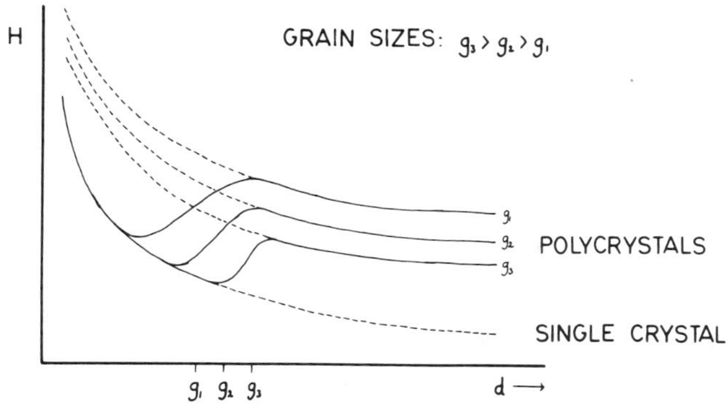

FIG. 4—*The expected effects of indentation sizes, similar to a material's grain size, in which the material is also subject to indentation size effects from other causes.*

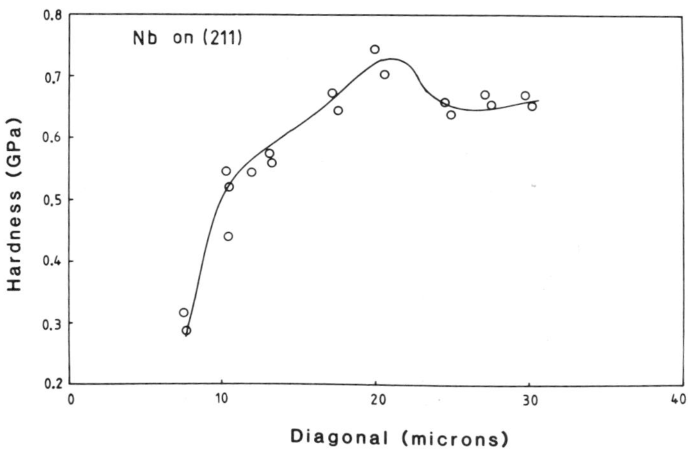

FIG. 5—*Data from Ref* 11 *replotted to show similarity with part of Fig. 4.*

ists, no chemomechanical effects or layers are expected, and surface or "Mott-model" work hardening would lead to a hardening and not a softening. In this case, the geometry of dislocation rosettes around indentations may have led to some dislocation-loop size effect with interactions with the production of crystallographically orientated pileups.

Examples of Indentation Size Effects

Figure 6 shows the more usual results from single crystals with an ISE index between 1.8 and 2.0 [*12*], for materials of the same isomechanical class [*13*]. Although these are very similar solids, they have been indented under

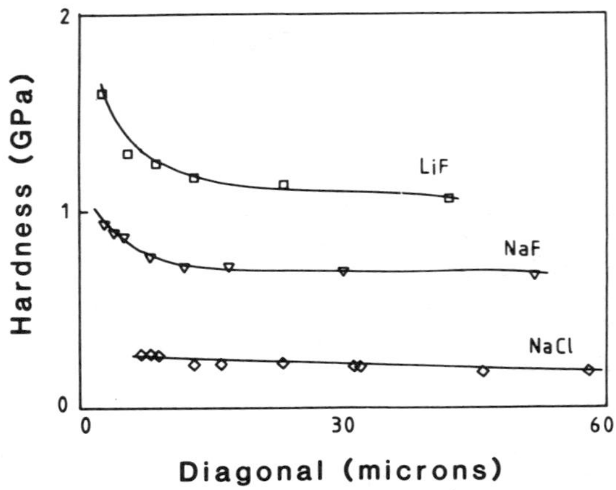

FIG. 6—*Data from Ref* 12 *replotted to show ISE variation within a group of materials of the same isomechanical class* [13].

different conditions in that room temperature is a different homologous temperature for each. Crystals with a lower melting point are expected (*a*) to be softer, (*b*) to be more strain-rate sensitive, and (*c*) to have activated secondary slip mechanisms and larger deformation volumes around the indentations so that ISE indices are closer to 2.0 [*1*]. For a comparison to be more useful, the tests should be made at the same fraction of the melting point over the same indentation size range (not the same load range), and hardness stresses should be considered with respect to the solid's shear modulus [*14*].

The effect of anisotropy in a single crystal of germanium is shown in Fig. 7, where indentations on the close-packed plane (111) show a different ISE from those on other surfaces [*15*]. Figure 7 shows softening for smaller indentations (that is, $n > 2.0$), but this should be contrasted with very extensive work showing values around 1.9 for silicon [*16*]. Measured hardnesses are also different for different orientations of a faceted indenter on a particular crystallographic surface [*17,18*], and ISE indices for these different orientations also differ. In lithium fluoride, the ISE index is higher for Vickers indentations with diagonals along ⟨100⟩ than for those along ⟨110⟩ [on (001) surfaces], whereas in magnesium oxide the opposite occurs [*1*]. This indicates that in those materials the indentation size effect is affected, if not controlled, by mechanisms other than those which control hardness anisotropy, because that has the same sense in both solids.

The interpretation of indentation size effects in semiconductors requires care because doping and temperature strongly affect the behavior. Figure 8 [*19*] shows that at 200°C indium antimonide behaves similarly to the alkali halides but that, as the temperature is reduced, the behavior changes to resemble that of germanium in Fig. 7. This shows the effect of a change in

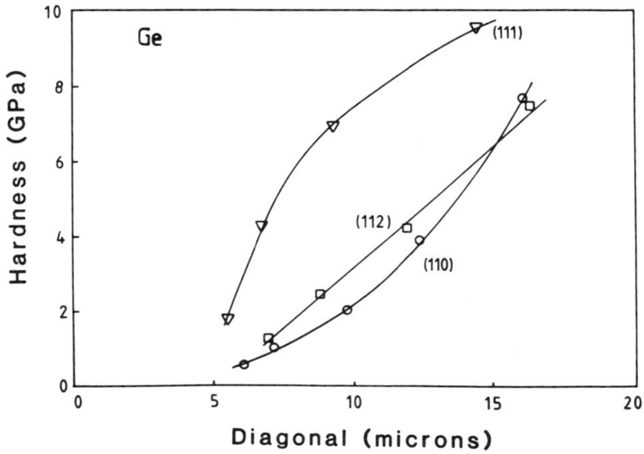

FIG. 7—*Data from Ref* 15 *showing both a change in ISE with crystallographic orientation and a surface softening effect.*

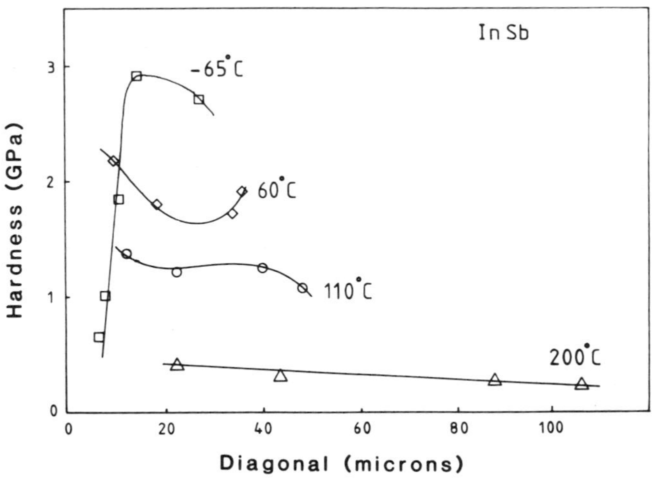

FIG. 8—*Data from Ref* 19 *showing the effect of different deformation mechanisms on ISE by testing one material over a range of temperatures with the mechanism dominance changing* [20].

deformation mechanism: at low temperatures the plasticity rates in ceramics and semiconductors are dominated by Peierls stress and kink nucleation, whereas at higher temperatures power law creep and diffusion creep dominate [*20,21*]. Figures 7 and 8 indicate that perhaps Peierls stress control implies an ISE index greater than 2.0, but the data for silicon (which are much more plentiful [*16*]) contradict this. More experiments under more closely controlled conditions are required in this area.

Gunasekera and Holloway's work in Fig. 9 shows that freshly cleaved glass is less sensitive to indentation size than is an old surface which has been exposed to damp air [*22*]. It is known that OH^- radicals attack silicon-oxygen bonds, but the dramatic surface softening illustrated in the figure, for quite large loads, emphasizes its importance.

Error Ellipsoids and 10-μm Hardness

In the past, the force-diagonal power law (Eq 5) has been used to describe microhardness data [*6*], where

$$\text{load} = a \cdot \text{diagonal}^n \tag{5}$$

and a is a materials parameter, with the strange dimensions of $N \cdot m^{-n}$. The parameter, n, is identical to that stated in Eq 1, and Eq 5 is a special case of the integral of Eq 1. From Eqs 1 and 4 it is possible to derive a 10-μm hardness, H_{10}, which is the hardness that would be measured if the load on a Vickers indenter were adjusted to give an indentation with a diagonal length of exactly 10 μm. This H_{10} value can be found very easily by interpolation within a set of indentations which span this size. The 10-μm hardness is a special case of a "standard diagonal" hardness, H_s

$$H_s = 19.7743 \cdot \left(\frac{\text{load}}{s}\right)^2 \cdot \left(\frac{\text{diagonal}}{s}\right)^{-n} \quad \text{GPa} \tag{6}$$

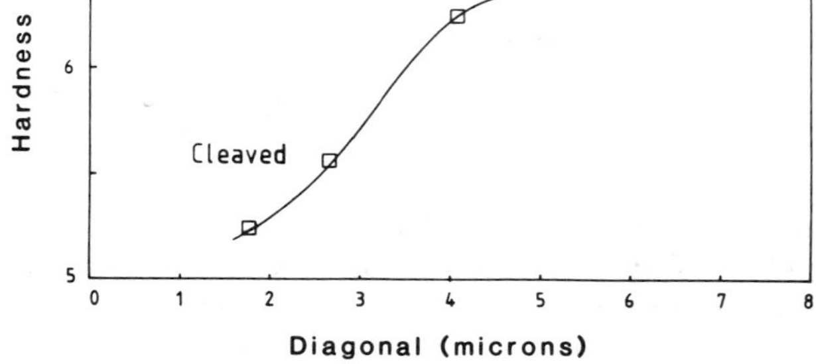

FIG. 9—*Data from Ref 22 showing the effect of an environmentally affected surface on the ISE.*

where the load is in gram force units (1 gf is 9.81 mN), and the diagonal and s are in micrometres. Picking a standard length, s, removes the awkward dimensionality because only the dimensionless ratio is taken to the power of the ISE index. Ten micrometres was chosen as a standard purely for convenience; most indentation size effects are manifested at this size, and a load range of 5 to 100 gf produces indentations of this size in a wide range of ceramics. For Knoop indentations, a standard length of 70 μm has also been used [*16*].

The empirical equation which can be fitted to data (load and diagonal measurements) is now of the form

$$\text{load} = 0.5 \cdot H_s \cdot s^2 \left(\frac{\text{diagonal}}{s}\right)^{n-2} \qquad \text{mN} \tag{7}$$

This should be fitted using a weighted, least-squares linear regression [*1*], in which the residuals are measured in the diagonal measurements and not, as is often done, in the load measurements [*1,23*]. Equation 7 leads directly to an equation relating indentation size and hardness

$$H = H_s \cdot \left(\frac{\text{diagonal}}{s}\right)^{n-2} \tag{8}$$

Both Eqs 7 and 8 describe the data as well as Eq 5, but with the advantage of a physically meaningful constant instead of a.

Consideration of Eqs 6 and 7 shows that a set of measurements of indentation size and load over a range of sizes can be directly transformed (using a weighted linear regression onto the log-index relationship) into just two well-defined parameters of simple dimensionality: n and H_{10}, with an experimental error associated with each [*1*]. Figure 10 shows data from more than 500 indentations on 13 materials plotted on a single ISE diagram, demonstrating the compactness of this method of presenting experimental data. Full details of these data and experimental methods have been published in Refs *1*, *17*, *18*, and *23*, but they are included here only as an example of this method of data reduction. The high-temperature, controlled-atmosphere, microindentation machine in which the measurements were made is of the BESSIE design [*23*], and a full-load indentation time of 15 s was used in all cases.

The small cross in the center of each ellipsoid in Fig. 10 is the weighted least-squares best fit value of n and H_{10}, and the ellipsoid is a 95% confidence limit, roughly equivalent to two standard errors in the mean [*1*]. The slanting angle for the ellipsoid for the copper data occurs because there is significant covariance between the errors in n and H_{10} because the 10-μm hardness was found by extrapolation of the experimental data: copper is so soft that all the indentations were larger than 10 μm. The 500°C high-load data are also slanted for the same reason, but the great disparity between this ellipsoid and

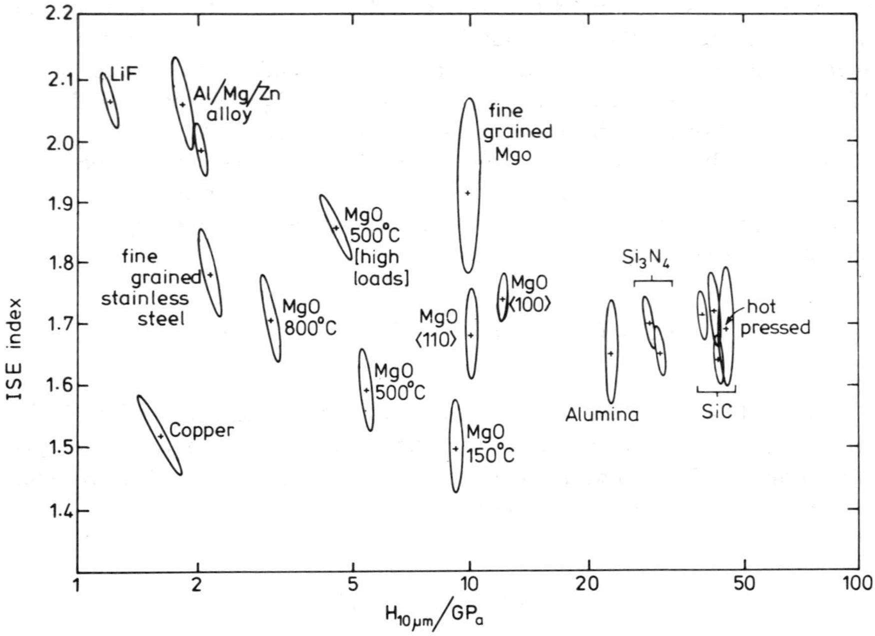

FIG. 10—*Ninety-five percent confidence ellipsoids summarizing data from indentations made using 10 to 100-gf loads; 1 gf is 9.81 mN [except for the ellipsoid labeled high loads, for which 100 to 1000 gf was used, and for copper (25 to 300 gf), lithium fluoride (5 to 200 gf), and aluminum-magnesium-zinc (25 to 200 gf). The abscissa is a logarithmic scale of the hardness for a "perfect, typical" 10-μm indentation made with a Vickers indenter* [1].

that for the standard load range at the same temperature occurs because the log-index relationship fits experimental data only "locally," over a relatively small range of indentation sizes. For this kind of plot to be useful, experiments must be chosen so that data are interpolated and not extrapolated to the standard indentation size.

Figure 10 shows that there is a significant difference in ISE between polycrystalline and single-crystal [⟨110⟩ orientation on (001)] magnesium oxide but that the hardness values at 10 μm are indistinguishable. By comparison, the different preparations of silicon carbide have different hardnesses but similar ISE indices.

Although it is possible, by inspecting the diagram, to determine whether or not two materials, or two treatments of the same material, are distinguishable, there is a statistical test which can be applied. The *F* test, which compares two normal distributions, can be used to show that the two preparations of silicon nitride are different to the 0.05 level, that is, with a probability of better than 95%.

This type of error-ellipsoid diagram is of completely general application and can be used to present any data which take the form of sets of two-dimensional linear regressions. Extension to three variables and three dimensions would easily be possible with suitable computer graphics.

Conclusions

The complexity of Fig. 1 is borne out in reality, as has been demonstrated by this discussion of indentation size effects. Therefore, there are too many distinct effects for physically based models to be generally useful at this stage, and so some empirically based formula is required to reduce the mass of experimental data to manageable levels and to aid communication between researchers in this field.

The log-index relationship used here (Eq 7) handles most cases (if only locally), relates directly to only experimentally measured quantities, gives error estimates, allows distinctions between materials to be drawn on a basis of indentation behavior in a statistically rigorous manner, and also permits a compact and clear visual representation of data. Thus, it forms an excellent basis for characterizing materials for consistency and quality control. Nevertheless, the replacement of this empirical approach with mechanistically based models must be the prime objective of future research in indentation science.

References

[*1*] Sargent, P. M., "Factors Affecting the Microhardness of Solids," Ph.D. thesis, University of Cambridge, England, September 1979. (The full description of the statistics used to calculate error ellipsoids is to be published in *Journal of Testing and Evaluation*, March 1986.)

[*2*] Crocker, A. J. and Wilson, M., *Journal of Materials Science*, Vol. 13, 1978, p. 833.

[*3*] Westwood, A. R. C. and Macmillan, N. H. in *The Science of Hardness Testing and Its Research Applications*, J. H. Westbrook and H. Conrad, Eds., American Society for Metals, Metals Park, OH, 1973, pp. 377–414.

[*4*] Haasen, P., Muller, H., and Zoth, G., *Journal de Physique C4* (Suppl. 9), September 1983, pp. 365–373.

[*5*] Cuthrell, R. E., *Journal of Applied Physics*, Vol. 49, 1978, pp. 432–436.

[*6*] Mott, B. W., *Microindentation Hardness Testing*, Butterworths, London, 1956.

[*7*] Burnand, R. P., "Preparation of Transition Metal Carbides and Indentation of Ionic Solids," Ph.D. thesis, University of Exeter, Exeter, England, 1974.

[*8*] Burnett, P. J. and Page, T. F., *Journal of Materials Science*, Vol. 19, 1984, pp. 845–860.

[*9*] Burnett, P. J., "Mechanical Properties of Ion Implanted Ceramic Surfaces," Ph.D. thesis, University of Cambridge, Cambridge, England, 1984.

[*10*] Bückle, H. in *The Science of Hardness Testing and Its Research Applications*, J. H. Westbrook and H. Conrad, Eds., American Society for Metals, Metals Park, OH, 1973, p. 453.

[*11*] Raghuram, A. C., Reed, R. E., and Armstrong, R. W., *Journal of Applied Physics*, Vol. 40, 1969, p. 4666.

[*12*] Upit, G. P. and Varchenya, S. A., *Physica Status Solidi*, Vol. 17, 1966, p. 83.

[*13*] Frost, H. J. and Ashby, M. F., *Deformation Mechanism Maps*, Pergamon Press, Oxford, 1982.

[*14*] Gilman, J. J., *Journal of Applied Physics*, Vol. 44, 1973, pp. 982–984.

[*15*] Bannerjee, R. K. and Feltham, P., *Journal of Materials Science*, Vol. 9, 1974, p. 1478.

[16] Roberts, S. G., "Modification of the Surface Mechanical Properties of Ceramic Materials by Ion Implantation," Ph.D. thesis, University of Cambridge, Cambridge, England, 1982.
[17] Sawyer, G. R., Sargent, P. M., and Page, T. F., *Journal of Materials Science*, Vol. 15, 1980, p. 1001.
[18] Sargent, P. M. and Page, T. F., *Scripta Metallurgica*, Vol. 25, 1981, p. 245.
[19] Bannerjee, R. K. and Feltham, P., *Journal of Materials Science*, Vol. 11, 1976, p. 1171.
[20] Sargent, P. M. and Ashby, M. F., *Scripta Metallurgica*, Vol. 18, 1984, pp. 219-224.
[21] Sargent, P. M. and Ashby, M. F., *Scripta Metallurgica*, Vol. 17, 1983, pp. 951-957.
[22] Gunasekera, S. P. and Holloway, D. G., *Physics and Chemistry of Glasses*, Vol. 14, 1973, p. 50.
[23] Naylor, M. G. S., "The Effects of Temperature on Hardness and Wear Processes in Engineering Ceramics," Ph.D. thesis, University of Cambridge, Cambridge, England, 1982.

Franz H. Vitovec[1]

Stress and Load Dependence of Microindentation Hardness

REFERENCE: Vitovec, F. H., **"Stress and Load Dependence of Microindentation Hardness,"** *Microindentation Techniques in Materials Science and Engineering, ASTM STP 889,* P. J. Blau and B. R. Lawn, Eds., American Society for Testing and Materials, Philadelphia, 1986, pp. 175–185.

ABSTRACT: The effect of the test load on the sensitivity to stress of the diamond pyramid hardness (DPH) was investigated on a martensitic spring steel, a Type 304 stainless steel, and an aluminum bronze with duplex microstructure. For this purpose, a loading frame and load cell of 1200-N capacity were fitted to the microhardness tester, which provided for application of uniaxial tensile loads to a flat test specimen. The test materials were heat treated differently and had different surface preparations, ranging from polished before vacuum annealing to electropolished following the annealing treatment.

Within the load range of 15 to 500 gf, the hardness increased in most cases to a peak from which it leveled off to a more or less constant value as the test load increased. Applied uniaxial tensile stresses did not change this trend except when plasticity initiated. In general, the hardness decreased linearly with increasing applied tensile stress. The rate of decrease was independent of the load for the hardness measurement. Yielding affected the hardness sensitively, particularly at small loads, thus changing the trend of the effect of applied stress on hardness as well as the hardness-load relation. Surface films or residual stresses do substantially affect the hardness-load relationship and the trend of the stress effect.

KEY WORDS: hardness load dependence, hardness stress effect, yield stress, aluminum bronze, stainless steel, spring steel, microindentation hardness testing

It is commonly observed that within a load range of 1000 to 1 gf the diamond pyramid indentation hardness (DPH) at first increases as the test load is decreased, reaches a maximum, and decreases again at very low loads. Bückle [*1*] explained this trend in terms of a transition from the hardness of a single crystal at small loads to that of a quasi-homogeneous polycrystalline structure at larger indentations. The basic trend of the hardness-load relationship is influenced by a number of experimental factors [*2–4*] such as the geometry of the indenter [*5*], faults in the aperture of the microscope [*6*],

[1]Professor, Department of Mechanical Engineering, University of Alberta, Edmonton, Alberta T6G 2G8, Canada.

friction between indenter and specimen [*7*], rate of descent of the indenter [*8*], vibrations [*9*], method of specimen preparation [*10*], and factors relating to the mechanism of the indentation such as elastic recovery, formation of a bulge, formation of a new surface, and the effects of residual stresses.

The aim of this study is to investigate the effect of stress on hardness at low loads and to determine how the hardness-load relationship is affected by stress.

Various studies were performed on the effect of stress on macroindentation hardness. Fink and Van Horn [*11*] found a maximum decrease of 5% of Rockwell E-scale values on the tensile side of elastically bent brass and an aluminum-copper alloy. Kostron [*12*] employed the Ludwik-cone indenter with a 150-kgf test load and Brinell balls with 62.5 and 250-kgf test loads. Kokubo [*13*] studied the effect of stress on the Vickers pyramid hardness using a 5-kg load, and Sines and Carlson [*14*] used the Rockwell B-scale hardness. Oppel [*15*] proposed that internal stresses may be determined by Knoop hardness, which, because of the rhombic shape of the indenter, may indicate the direction of the uniaxial stress that causes the hardness change. Bückle [*6*] reported that the microindentation hardness varies linearly with the applied stress as in the macrohardness range. Blain [*16*] defined hardness as the load on an indenter which initiates plastic deformation. He showed that such defined hardness is much more sensitive to stress in the test material than conventional hardness. This would indicate that the effect of stress should increase as the test load is decreased.

Experimental Procedures

Testing Facility

For the experimental study, a Leitz Miniload hardness tester was used with loads of 15, 25, 50, 100, 200, 300, and 500 gf. A small load frame was fitted to the table of the hardness tester to apply tensile loads to the specimen. This is shown in the photograph of Fig. 1. The load is applied by turning the screw, and is indicated by the deflection of the double cantilever load cell and measured by the dial gage. The elongation of the test specimen can be measured by the micrometer of the crosshead of the hardness tester. The load frame was first calibrated by dead weights to a force of 1200 N. In this range, the deflection of the load cell changed linearly with the load. The flat test specimens had heads with a pinhole. To minimize eccentricities in load application, a calibration specimen was used which had strain gages applied to both sides. Simultaneous measurement of the two strains on the calibration specimen was used to align the grips and reduce the probability of eccentric load application.

The flat test specimens had a nominal cross section of 1.25 mm thick by 2.5 mm wide and a straight test section of 12.5 mm; the fillet radii were 2 mm.

FIG. 1—*Microindentation hardness tester with load frame.*

Before the test program was embarked on, the alignment of the hardness tester was checked. A Vickers diamond pyramid indenter was selected which corresponded with the requirements of ASTM Test for Microhardness of Materials (E 384-84). The calibration of the hardness tester was verified with commercial test blocks and cross-checked with a Shimadzu microhardness tester.

Test Materials

Test materials with different microstructures (fine grained, single phase, and dual phase) were selected. These were a quenched and tempered flat spring steel, a commercial Type 304 stainless steel, and an aluminum bronze which was induction melted from oxygen-free high-conductivity (OFHC)-copper and 9.5 weight percent aluminum. The spring steel was used in the as-received condition, mechanically polished, and etched with 4% Nital. The austenitic stainless steel was heat treated in two different ways: (*a*) a mechanically polished specimen was vacuum annealed at 1050°C, water quenched, and hardness tested without further preparation; (*b*) after heat treatment as in Point *a*, a specimen was annealed at 370°C (700°F) for 2 h to reduce the quenching stresses, and then electropolished. One aluminum bronze specimen was tested in the as-cast condition after it was polished and etched. Another specimen was metallographically prepared and vacuum annealed at 650°C (1200°F).

The spring steel had a uniformly martensitic microstructure. Sufficient material was removed by polishing to eliminate any surface decarburized layer. The austenitic stainless steel had a uniformly equiaxed microstructure with an average grain size of 20 μm. The aluminum bronze had a α-α/δ dual-phase microstructure. The average grain size of the α phase was 38 μm. All the hardness indentations were in the center of the α phase.

The mechanical properties determined from microhardness test specimens are listed in Table 1.

Test Procedure

The damper of the testing machine produced a rate of descent of the penetrater of 120 μm/s. Because the changes of hardness caused by applied stress are small compared with the scatter at low loads, the hardness data were determined as the average value of eight indentations. Statistical evaluation of the data showed an increase of the standard deviation and the standard error with decreasing test load. Specifically, the standard deviation for the spring steel increased from 1.35 at 500-gf load to 7.21 at 15-gf load, that for stainless steel increased from 3.21 to 5.72, and that of the aluminum bronze from 2.86 to 6.85. The hardness tests were performed immediately after the surface preparation of the test specimen.

TABLE 1—*Mechanical properties of test materials.*

	Spring Steel	Stainless Steel		Aluminum Bronze	
		Quenched	Annealed	As Cast	Annealed
Yield strength, MPa	1140	230	230	210	230
Tensile strength, MPa	1490	505	625	370	470
Strain to fracture, %	6.5	45	56	14	20

Results

Hardness as a function of the test load for the spring steel is shown in the diagram of Fig. 2. The curves for different tensile stresses applied to the specimen show a significant initial increase of hardness with increasing test load. This increase may be due in part to some surface decarburization which may have been present although it could not be observed in the microstructure. It may also be due in part to the rapid increase in constraint as the depth of indentation increases from about 2 μm at 15-gf load. The curves are shifted to lower hardness values by tensile stresses.

Hardness-load curves for the stainless steel are shown in Fig. 3. The upper set of curves is for the steel in the water-quenched (WQ) condition; the lower is for the steel in the stress-relieved condition. Both sets of curves show an initial decrease of hardness as a tensile stress is applied.

It was generally observed that plastic deformation caused a substantial increase of hardness, particularly at low loads; thus hardness responded sensitively to the beginning of yielding. Yielding also showed up as relaxation of the load cell due to the rigid screwdrive grip displacement and the rigid load frame.

One observes that as the tensile stress is increased to the beginning of yielding (215 MPa for the WQ steel and 200 MPa for the stress-relieved steel) the hardness at low loads increases while that at larger loads continues to decrease. The stress at which this is observed is slightly below the conventionally determined yield strength. Only as the stress is further increased does cross-

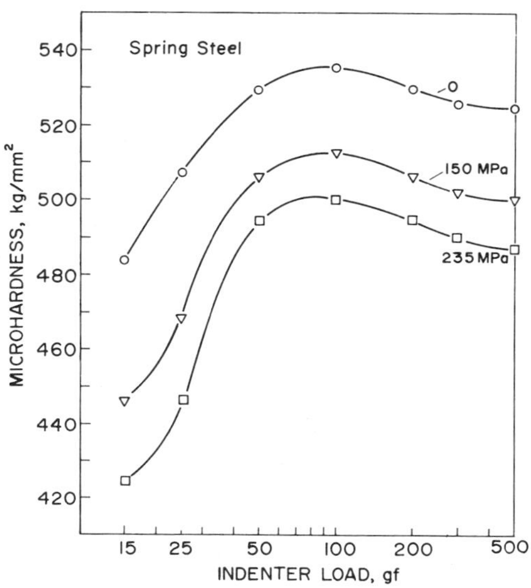

FIG. 2—*Hardness versus load for spring steel at different tensile stresses.*

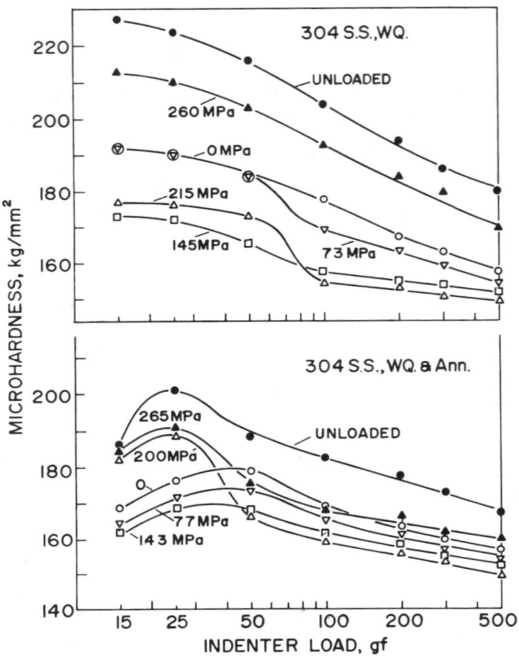

FIG. 3—*Hardness versus load for stainless steel at different tensile stresses.*

plastic deformation occur up to approximately 1.5% strain, and the hardness of the entire load range increases. Nevertheless, superimposed on the strain-hardening effect caused by the plastic deformation is the hardness-reducing effect caused by the applied tensile stresses. This is indicated by the further increase in hardness as the tensile stress is removed by unloading the specimen. The curves for the water-quenched steel decrease monotonically with increasing test load while those of the stress-relieved and electropolished steel show an initial increase of hardness. This difference may be traced to the presence of an oxide film and to residual stresses in the water-quenched specimen.

The curves for the aluminum bronze specimens, shown in Fig. 4, exhibit a trend similar to that of the stainless steel. Initially, the hardness is decreased as the tensile stress is applied, and as yielding begins the hardness at low loads increases while that at high loads decreases further. Cross-plastic deformation causes a shift of the curve to higher hardness, and removal of the tensile load further increases the hardness.

The difference between the two sets of curves is somewhat similar to that of the stainless steel in that the annealed bronze shows a more pronounced initial increase of hardness with increasing test load.

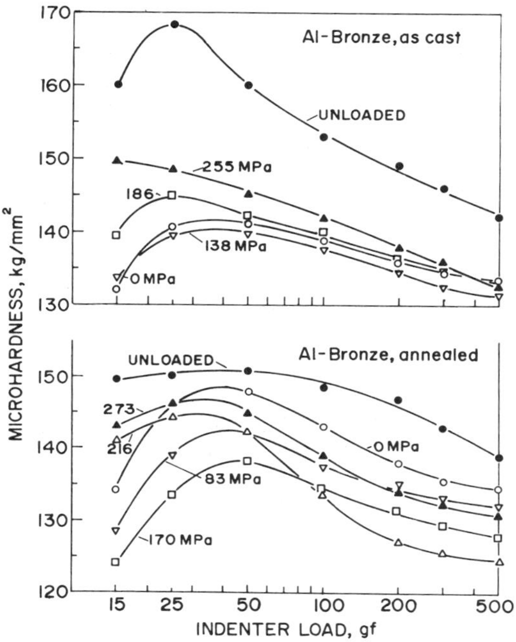

FIG. 4—*Hardness versus load for aluminum bronze at different tensile stresses.*

Discussion

Details of the effect of stress and local plasticity on hardness are well displayed in diagrams of hardness versus applied tensile stress, as shown in Figs. 5 to 7.

The decrease in the hardness in spring steel (Fig. 5) caused by tensile stress is practically linear. The lines for the hardness at different loads are almost parallel, and the decrease at the lower loads is only slightly larger than at the higher loads.

The curves for the stainless steel are shown in Fig. 6. The hardness at the low loads of the water-quenched steel at first remains unaffected by stress, as may be expected in the presence of a separate surface film. The hardness values decrease more slowly with increasing stress, compared with the trend shown by the spring steel. When the applied stress is increased beyond 145 MPa, the hardness at low test loads, that is, 15 and 25 gf, increases while that at larger loads continues to decrease until the applied stress reaches 215 MPa. From 215 MPa on, cross-plasticity occurs, and the hardness increases at all test loads. The rise of hardness at low loads could be caused by a slight eccentricity of the tensile load, which may cause yielding to initiate at the surface and propagate toward the core of the specimen. Regardless of the mecha-

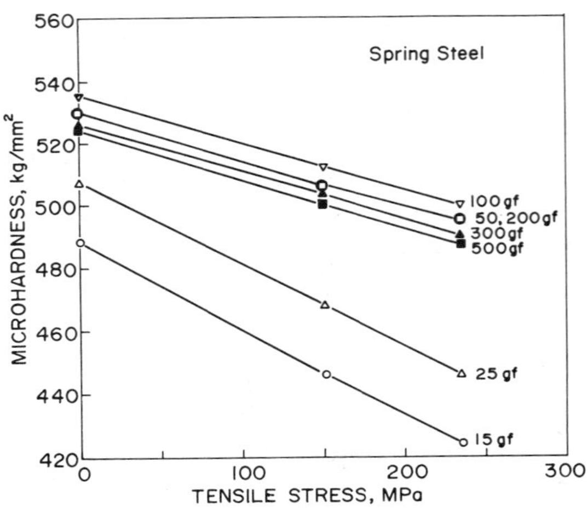

FIG. 5—*Hardness of the spring steel versus the applied tensile stress for different indenter loads.*

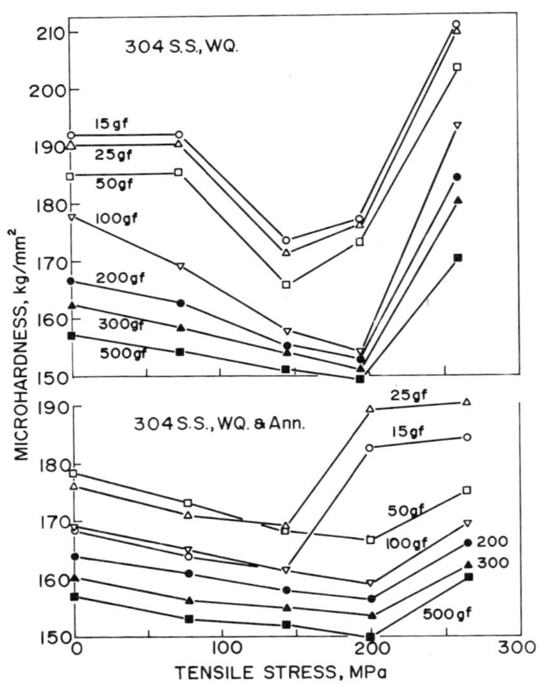

FIG. 6—*Hardness of the stainless steel versus the applied tensile stress for different indenter loads.*

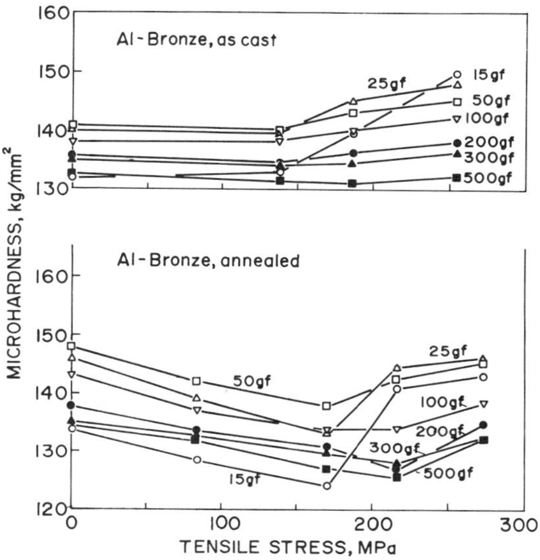

FIG. 7—*Hardness of the aluminum bronze versus the applied tensile stress for different indenter loads.*

nism, this behavior shows that low-load hardness responds sensitively to local plasticity and, as a result, may affect the hardness–load relation. This characteristic is even more pronounced in the stress-relieved and electropolished specimens. However, apparently because of the absence of a surface film, the hardness at all load levels decreases initially with increasing tensile stress.

The aluminum bronze also shows the effects of surface preparation, stress relief, and yielding at the surface, as shown in Fig. 7. The material in the as-cast condition shows only the effect of yielding on hardness but no effect of tensile stresses. Heat treating at 650°C restores the sensitivity to tensile stresses, and as in the stress-relieved stainless steel, the hardness at all load levels decreases at more or less the same rate as the tensile stress increases. At 216 MPa, the hardness of the annealed bronze at 15 and 25-gf load increases more than at other loads. Beyond 216 MPa cross-yielding occurs, and at 273 MPa about 1.6% permanent strain has accumulated, which causes an increase of hardness at all loads.

The change of hardness caused by an applied tensile stress gives an indication of the effects of residual stresses. These can change the hardness load relationship since they are commonly associated with a stress gradient.

The slope of hardness decreases caused by tensile stresses is 0.265 kg/mm^2/MPa (2.59 MPa/MPa) for the spring steel, and 0.05 kg/mm^2/MPa (0.49 MPa/MPa) for the stainless steel and the aluminum bronze. The difference in values could be related to the differences in constraint caused by the body-centered martensite and the face-centered structure of the stainless steel and the α phase of the aluminum bronze. However, calculation of the ratio of hardness and yield strength, that is, the constraint factor according to Shaw and De Salvo [*17*], shows the reverse—a factor of four for the spring steel and of six for the other two alloys. Theoretically, the constraint factor should be three or smaller. A theoretical analysis of the hardness variation at low loads in terms of constraint effects requires further study.

Conclusions

For materials which are homogeneous up to the surface, are free from residual stresses, and have no detrimental surface film, the hardness is decreased by tensile stresses regardless of the test loads used. For such polycrystalline metallic materials, the hardness also increases initially with increasing test load and decreases toward a constant value after passing through a maximum. Tensile stresses which are uniformly distributed across the test section do not affect the hardness-load relationship. Surface films and residual stresses can change this trend, as well as the sensitivity to applied tensile stresses. When plastic yielding initiates, due to applied tensile stresses, the hardness may increase at first at small test loads and then progressively at larger loads as the tensile stress is increased. Plastic deformation increases the hardness but does not change the effect of the elastic stress, as shown by a further increase of hardness on unloading after tensile straining.

Acknowledgments

The author expresses his appreciation to the Natural Science and Engineering Research Council of Canada for financial support of this work.

References

[*1*] Bückle, H. in *The Science of Hardness Testing and Its Research Applications,* J. H. Westbrook and H. Conrad, Eds., American Society for Metals, Metals Park, OH, 1972, Chapter 7, pp. 453-491.

[*2*] Bückle, H., *Metallurgical Reviews,* Vol. 4, 1959, pp. 49-100.

[*3*] Blau, P. J., *Metallography,* Vol. 16, 1983, pp. 1-18.

[*4*] Braunovic, M. in *The Science of Hardness Testing and Its Research Applications,* H. J. Westbrook and H. Conrad, Eds., American Society for Metals, Metals Park, OH, 1972, Chapter 6, pp. 329-376.

[*5*] Campbell, R. F., Henderson, Q., and Donleavy, M. R., *Transactions of the American Society for Metals,* Vol. 40, 1948, pp. 954-982.

[*6*] Buckle, H., *Zeitschrift für Metallkunde,* Vol. 45, 1954, pp. 623-632.

[*7*] Bischof, W. and Wenderott, B., *Archiv für das Eisenhüttenwesen,* Vol. 15, 1941-1942, pp. 497-505.

[*8*] Campbell, R. F., *Materials Research and Standards,* October 1967, pp. 443-449.

[*9*] Kennedy, R. G. and Marrotte, N. W., *Materials Research and Standards,* November 1969, pp. 18-22.

[*10*] Vitovec, F. H. and Binder, H. F., "The Effect of Surface Preparation and Condition on Microhardness," Technical Report 55-372, Wright Air Development Center, Wright-Patterson Air Force Base, OH, October 1955.
[*11*] Fink, W. L. and Van Horn, K. R., *Journal of the Institute of Metals*, Vol. 44, 1930, pp. 241-248.
[*12*] Kostron, H., "Härte und Werkstoffspannungen," *Mitteilungen des Technischen Versuchsamtes—Austria*, Vol. 21, 1932, pp. 17-31.
[*13*] Kokubo, S., "On the Change in Hardness of a Plate Caused by Bending," *Science Reports*, Tohoku Imperial University, Sendai, Japan, Vol. 21, 1932, pp. 256-267.
[*14*] Sines, G. and Carlson, R., *ASTM Bulletin*, No. 180, February 1952, pp. 35-37.
[*15*] Oppel, G. U., *Proceedings*, Society for Experimental Stress Analysis, Vol. 21, 1964, pp. 135-140.
[*16*] Blain, P. A., *Metal Progress*, January 1957, pp. 99-100.
[*17*] Shaw, M. C. and De Salvo, G. J., *Metals Engineering Quarterly*, May 1972, pp. 1-9.

David R. Kelley,[1] Chris E. Johnson,[1] and David S. Lashmore[1]

Fabrication and Certification of Electroformed Microhardness Standards

REFERENCE: Kelley, D. R., Johnson, C. E., and Lashmore, D. S., **"Fabrication and Certification of Electroformed Microhardness Standards,"** *Microindentation Techniques in Materials Science and Engineering, ASTM STP 889,* P. J. Blau and B. R. Lawn, Eds., American Society for Testing and Materials, Philadelphia, 1986, pp. 186–195.

ABSTRACT: A commonly quoted characteristic of a material is its hardness. Accurate measurements of this characteristic require the use of precise standards to verify that the testing instrument, procedures, and comparisons with the results of another laboratory are accurate; however, it is recognized that currently available standards vary considerably in hardness from point to point across the surface. Electroplating technology was utilized to fabricate new microhardness standards. This technology provides a means for obtaining uniform hardness by close control of process variables which determine grain structure and composition.

Two microhardness standards are now in production—one with a hardness of ~125 kg/mm^2 and the other at ~600 kg/mm^2. The hardness values are certified at loads of 0.245, 0.490, and 0.981 N (25, 50, and 100 gf) with both Vickers and Knoop indenters. These electroplated materials have standard deviations in hardness, particularly at low loads, that are significantly better than the available standards. The fabrication of the new standards, their certification procedures, and testing instrument characteristics are discussed.

KEY WORDS: hardness, Knoop hardness, Vickers hardness, hardness testing metrology, nondestructive testing, testing instruments, microindentation hardness testing

Microhardness standards serve as an important means of quality control, not only for electrodeposited coatings, but also for other applications, and they can be used to ensure that the testing instruments are operating properly. Presently, the only standards available are produced by the makers of the testing instruments, and these not only lack a uniform standard for certi-

[1]Physical science technician, physical scientist, and group leader and metallurgist, respectively, Electrodeposition Group, Metallurgy Division, Center for Materials Science, National Bureau of Standards, Washington, DC 20234.

fication of hardness but also exhibit significant variation of hardness across their surfaces.

At present, all commercial standards are produced from alloys prepared from a metallurgical cast process. The main drawback with this process is the difficulty in controlling cooling of the alloy melt. Failure to cool evenly produces a variable grain structure and composition. Electrodeposition methods of fabricating microhardness standards which do not have these drawbacks have been developed at the National Bureau of Standards (NBS). By very closely controlling current density, temperature, and electrolyte agitation, it is possible to produce extremely uniform electrodeposited material, which results in less variability in hardness from point to point across the surface. To date, two standards have been produced, one from a "bright copper" electrolyte and another from a "bright nickel" electrolyte. Electroplated bright copper was chosen because its nominal hardness of 125 kg/mm^2 is not only widely used but closely represents the hardness of some commonly used noble metals. Electrodeposited bright nickel was chosen because its nominal hardness of 600 kg/mm^2 closely represents the hardness of some commonly used ferrous metals.

Electroplating, using proprietary brighteners, produces grain refinement and grain distribution superior to that achieved with no brighteners, resulting in significantly more uniform hardness, as well as bright, smooth surfaces even before any polishing. Proprietary brighteners contain a leveling agent, usually an organic, absorbed on the surface of the high current density areas or peaks on the plating surface. This organic may inhibit plating on these high current density areas but allows plating to continue in the valleys, thereby providing a leveling effect. Another characteristic of the bright plating electrolytes is very good micro-throwing power, that is, the ability to deposit metal in grooves and cracks where these surface imperfections are of a microscopic nature.

Experimental Procedures

Fabrication Procedure

The electrolytes used in the electroplating of microhardness standards are commercially available copper and nickel solutions. The anodes were 0.00490% phosphorized copper and low-sulfur, low-cobalt nickel bars placed in anode bags. The power supply was a 15-A, 100-V constant-current source. The electrolyte agitation was provided by pumping filtered air through a sparger [perforated tube of polyvinyl chloride (PVC)]. The substrate used in this process was a 22.5 by 45-cm sheet of polished AISI 1010 steel, mounted in a Teflon box. Deposits of 500 μm of copper on a copper substrate and on a steel substrate showed no significant difference in hardness, therefore, the less expensive steel substrate was chosen. Mounting the substrate in a box with the open side facing the anodes in the electrolyte provides for a much more uniform current distribution. A uniform current density is necessary for uniform grain size, which is essential for this project. Electroplating is complete when 1 mm of copper or nickel is deposited.

After the copper or nickel has been deposited, a 2.5-cm strip is removed from all four sides of the plate to ensure uniformity of thickness across the plate. The plate is then cut into 1.35-cm-square specimens and placed in a stainless steel ring, 2.5 cm in outside diameter and 1.0 cm in height. The stainless steel ring was polished to ensure coplanar surfaces. This ring is then filled with an epoxy medium, used as a mold to enhance uniform polishing. The mold is polished on an automatic system. The polishing procedure removes approximately 125 μm of material from the original coating of 1 mm, thus leaving a substantial coating thickness to prevent the steel substrate from having any effect on the hardness measurement. The resultant coating thickness allows for a ratio of a minimum of 100:1 for coating thickness to depth of indentation when the hardness is determined at the maximum certified load of 0.981 N (100 gf).

Certification Procedure

The certification of these test blocks was made in accordance with the ASTM Test for Microhardness of Electroplated Coatings (B 578-80) and Test for Microhardness of Materials (E 384-84). The test instrument load-time response was measured using a compression load cell, calibrated with NBS-certified weights, to ensure conformity to ASTM Test E 384, Part B. The load cell was used to determine the actual load being applied to the test block during the time of indentation, as well as to determine the dwell time for full-load application. The optical measuring system of the test instrument was calibrated by an NBS-certified stage micrometer. The hardness indentations were made in all four corners and in the center of the test blocks and were measured using a $\times$100 objective lens having a numerical aperture of 0.90. The hardness values are certified at loads of 0.245, 0.490, and 0.981 N (25, 50, and 100 gf). Vickers and Knoop indentations are made on separate specimens.

The first step in the certification of the microhardness standards was to identify and investigate sources of error which influence the accuracy of the certified values. The identifiable sources of error are (1) indentation measurement due to the optics or operator variability, (2) test instrument loading, and (3) test block variation.

It is generally acknowledged that the greatest source of error is the optical measurement. The first step that was taken to reduce this error was to use an optical system with a $\times$100 objective with a numerical aperture of 0.90, thus achieving a total magnification of 1000 with better resolution than is normally found on most test instruments, in which the objective power is typically $\times$40 or ×50 and the numerical aperture is 0.65. The tips or ends of the microindentations were more readily defined, resulting from the use of the higher magnification and resolving power. In using this optical system, each operator determined the calibration factor using a certified stage micrometer. This calibration factor was verified each time a test block was measured. It was also found that better reproducibility of a measurement was obtained when using a filar micrometer eyepiece, which incorporated a double line just below the single cursor line. The tips of the microindentation are centered between the double lines and referenced to the single line just above the tip of the microindentations. The incorporation of all the previously mentioned steps has greatly reduced the "scatter" of the measurements.

The second source of error that was investigated was test instrument loading. A number of test instruments were evaluated, and it was discovered that impact loading, ringing (oscillations) resulting from impact loading, and undamped vibration were associated with the various test instruments, especially at loads of 0.245 N (25 gf) and less.

A miniature precision load cell was used to determine the actual load being applied during the time of indentation. The load cell was calibrated with certified weights, and an indentation was then made directly on the load cell. The load cell featured a peak/hold option which recorded the peak load applied during the test. The load-time responses of the test instruments were recorded on a digital oscilloscope. One instrument indicated a 15 to 20% impact load above the preset load value, for example, a 10 gf load was actually measured to be 12 gf. Also associated with this impact load is a drop in load before a steady-state preset load value is reached. A second instrument gave an impact load 20% greater than the preset load, which resulted in a long ring down time before a steady load value was reached. The load-time response of a third test instrument showed smooth uniform load application with an impact load of 20% greater than the preset value but with little ring down. None of these three instruments incorporated any damping mechanism to prevent impact loading or ringing. A fourth instrument was evaluated, and the associated load-time response showed no impact loading. This instrument was hydraulically damped. An example of a load-time response from an instrument which produces an impact "overshoot," is shown in Fig. 1*a*, along with a load-time characteristic from an instrument which applies the load more gradually, in Fig. 1*b*. The microhardness standards are presently being indented on a hydraulically damped instrument in which the actual load is within 0.5% or less of the applied load. Although impact loading may be large only at low applied loads, this error becomes significant when producing NBS-certified standards.

The third source of error investigated was variation in the hardness of the test block resulting from the grain size or composition variation. The variation in hardness across the test block is illustrated in Fig. 2, which represents hardness impressions on NBS standards compared with two different com-

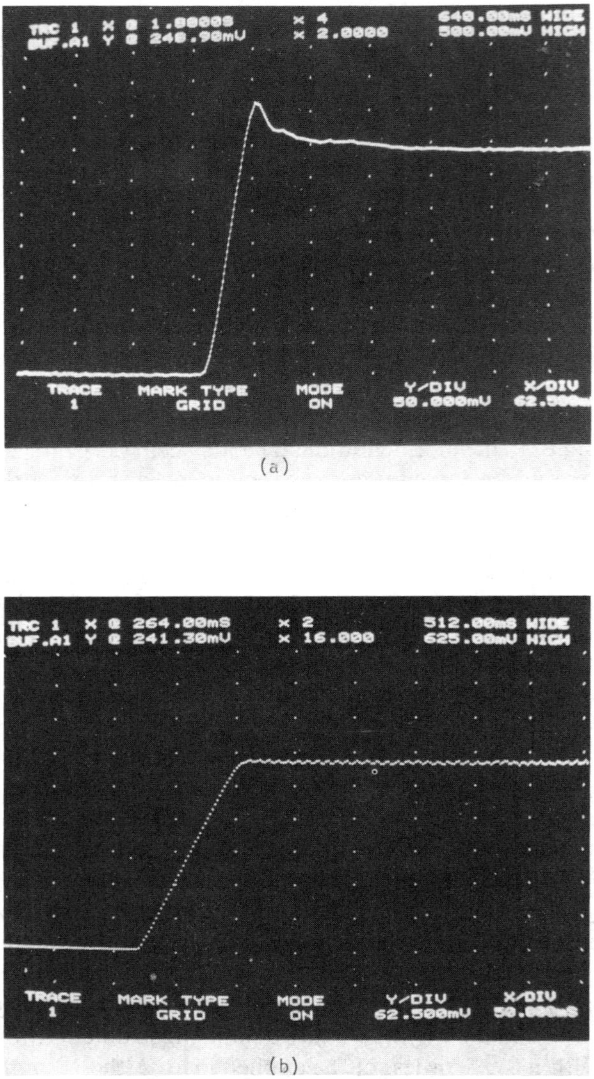

FIG. 1—*Plots of load-time response for microhardness test instruments showing* (a) *20% impact loading above the preset value of 10 gf resulting from an undamped test instrument and* (b) *no-impact loading above the preset value of 10 gf from a hydraulically damped instrument: 1.0 gf = 33.33 mV;* x *axis = time, 62.5 ms/cm;* y *axis = load, 50 mV/cm.*

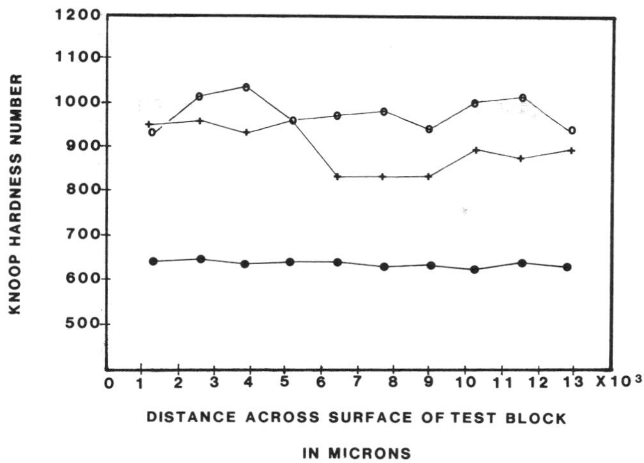

FIG. 2—*Comparison of the NBS nickel microhardness standard with two commercially available steel test blocks illustrating the variation in hardness across the test surfaces when using a 25-gf load: (•), NBS nickel standard; (○) commercial steel Test Block A, (+) commercial steel Test Block B.*

mercial steel test blocks. An impression was made every 1250 μm across the surface, and the hardness was plotted as a function of distance. The load used was 25 gf.

Results and Discussion

The results of a comparison of measurements of the same indentation using an optical system with an objective with a numerical aperture of 0.65 and a total magnification of ×400 (×40 objective and ×10 eyepiece) and an optical system using an objective with a numerical aperture of 0.90 and a total magnification of ×1000 (×100 objective and ×10 eyepiece) are tabulated in Table 1. One indentation at each of three values of load was made at five positions on the test blocks. The average hardness value and standard deviation in hardness units are reported for each set of five indentations for each load value. Each of these indentations was measured using both optical systems. The lower values of hardness and smaller standard deviations were obtained with the use of the optical system that incorporated the higher numerical aperture objectives and higher magnifications. This objective numerical aperture effect was also demonstrated by Tarasov and Thibault [*1*] and Brown and Iveson [*2*]. Higher resolution and magnification increased the ability to determine the ends of the indentations, resulting in greater diagonal lengths and less variation.

The use of a precision load cell to determine the load applied during the time of indentation resulted in the discovery of impact loading, oscillations

TABLE 1—*Optical measurement: comparisons of ×400 and ×1000 objectives.*

Nickel				Copper			
Load Value, g	Position	Instrument A, ×400	Instrument B, ×1000	Load Value, g	Position	Instrument A, ×400	Instrument B, ×1000
			KNOOP HARDNESS				
25	1	721.7	627.9	25	1	133.6	124.7
	2	702.6	644.1		2	128.5	124.2
	3	690.3	633.7		3	135.1	126.6
	4	696.4	658.0		4	138.3	126.6
	5	721.7	666.6		5	131.5	126.6
	Average	706.5	655.8		Average	133.3	126.0
	SD	14.5	10.0		SD	4.3	1.6
50	1	730.8	622.7	50	1	131.3	128.5
	2	735.5	637.7		2	130.6	127.8
	3	661.2	641.5		3	130.2	127.1
	4	699.1	637.7		4	129.2	126.1
	5	703.5	626.4		5	137.2	129.9
	Average	706.0	633.2		Average	131.7	127.8
	SD	29.8	8.2		SD	3.2	1.4
100	1	690.3	627.9	100	1	129.3	128.3
	2	660.8	625.3		2	132.5	128.3
	3	715.3	633.3		3	135.4	129.0
	4	666.6	635.9		4	131.8	132.3
	5	663.7	633.3		5	131.3	127.1
	Average	679.3	631.9		Average	132.0	129.0
	SD	33.3	4.6		SD	2.2	2.0
			VICKERS HARDNESS				
25	1	724.3	578.7	25	1	114.1	115.3
	2	626.8	612.4		2	115.8	112.4
	3	689.4	585.2		3	115.3	111.9
	4	641.6	612.4		4	121.9	114.7
	5	752.3	619.5		5	129.0	112.4
	Average	697.7	601.6		Average	119.2	113.3
	SD	54.6	18.4		SD	6.2	1.5
50	1	683.1	612.8	50	1	123.4	116.5
	2	695.0	633.2		2	118.6	115.3
	3	683.1	617.8		3	120.8	115.7
	4	649.2	617.8		4	126.2	119.1
	5	689.0	643.8		5	128.1	115.7
	Average	679.8	625.0		Average	123.4	116.4
	SD	17.8	13.0		SD	3.9	1.5
100	1	672.9	645.4	100	1	124.4	115.8
	2	626.8	653.1		2	124.4	118.2
	3	664.9	630.4		3	121.2	115.6
	4	664.9	660.9		4	128.7	117.0
	5	689.4	645.4		5	126.0	115.8
	Average	663.7	647.0		Average	124.9	116.4
	SD	23.0	11.3		SD	2.7	1.1

TABLE 2—*Operator comparison of Knoop hardness across test block by optical measurement using a ×100 objective.*

		Load Values					
		25 g		50 g		100 g	
Position	Indentation	Operator A	Operator B	Operator A	Operator B	Operator A	Operator B
1	1	605.4	605.4	617.7	627.9	612.4	615.9
	2	612.4	612.4	617.7	612.7	619.5	626.7
	3	612.4	598.5	617.7	612.7	612.4	615.9
	4	591.8	598.5	622.8	612.7	615.9	615.9
	5	598.5	591.8	607.8	612.7	612.4	619.5
	Average	604.1	601.3	616.8	615.8	614.5	618.8
	SD	9.0	7.8	5.5	6.8	3.2	4.7
2	1	598.5	591.8	617.7	622.8	608.9	605.4
	2	612.4	585.2	612.7	622.8	612.4	612.4
	3	598.5	585.2	617.7	612.7	608.9	608.9
	4	605.4	585.2	607.8	607.8	612.4	601.9
	5	598.5	591.8	612.7	612.7	608.9	605.4
	Average	602.7	587.8	613.7	615.8	610.3	606.8
	SD	6.2	3.6	4.2	6.7	1.9	4.0
3	1	598.5	605.4	612.7	607.8	612.4	615.9
	2	598.5	585.2	617.7	617.7	605.4	605.4
	3	591.8	598.5	617.7	607.8	612.4	619.5
	4	598.5	598.5	612.7	607.8	608.9	605.4
	5	591.8	598.5	612.7	602.9	608.9	612.4
	Average	598.8	597.2	614.7	608.8	609.6	611.7
	SD	3.7	7.4	2.7	5.4	2.9	6.3
4	1	591.8	605.4	602.9	622.8	615.9	605.4
	2	591.8	591.8	612.7	607.8	608.9	612.4
	3	591.8	598.5	612.7	612.7	612.4	608.9
	4	598.5	605.4	598.1	612.7	612.4	623.1
	5	591.8	598.5	612.7	622.8	612.4	608.9
	Average	593.1	599.9	607.8	615.8	612.4	611.7
	SD	3.0	5.7	6.9	6.7	2.5	6.8
5	1	598.5	598.5	617.7	622.8	619.5	612.4
	2	598.5	598.5	598.1	607.8	612.4	615.9
	3	578.6	585.2	617.7	622.8	615.9	612.4
	4	585.2	578.6	612.7	617.7	612.4	612.4
	5	591.8	612.4	622.8	617.7	608.9	608.9
	Average	590.5	594.6	613.8	617.8	613.8	612.4
	SD	8.7	13.1	9.5	6.1	4.0	2.5
All indents							
	Average	597.2	596.2	613.4	614.8	612.1	612.3
	SD	8.1	8.9	6.4	6.6	3.4	6.1

TABLE 3—*A comparison of NBS electroplated and commercially available standards, Knoop hardness (25-g load).*[a]

Indentation	NBS Copper Standard	Commercial Brass Standard	NBS Nickel Standard	Commercial Steel Standard A	Commercial Steel Standard B
1	140.0	111.4	644.1	945.1	954.9
2	134.6	168.8	649.6	1017.0	964.9
3	136.2	120.2	638.6	1039.0	935.4
4	138.9	122.8	644.1	964.9	964.9
5	136.7	126.6	644.1	975.0	838.2
6	140.0	134.1	633.3	985.3	838.2
7	136.2	130.5	638.6	945.1	838.2
8	140.0	130.0	627.9	1006.0	898.2
9	135.6	123.8	644.1	1017.0	880.4
10	135.1	117.1	633.3	945.1	898.2
Average	137.3	128.5	639.7	983.9	903.6
SD	2.2	15.7	6.7	34.4	49.4

[a]The average coating thickness for both the NBS standards exceeded 500 μm.

resulting from impact loading, and undamped vibrations. With regard to the effect of vibrations on microhardness testing, the results of this work are in agreement with those of prior work by Campbell [*3*].

Once the errors associated with the optical measuring system and test instrument loading were addressed and reduced, operator variability was investigated. An NBS bright nickel standard was used for this portion of the study. Five indentations at three values of load were made at five positions on the test block. Two operators (A and B) measured the same indentations with very similar results in average hardness values and standard deviation for the 25 indentations made at each load. These results are shown in Table 2 for each set of indentations at different positions for the three load values. The difference in the average hardness value determined by each operator for the 25 indentations at each load was a maximum of 1.4 hardness units. The improved optical measuring system greatly reduced the variability between operators.

The comparative results of multiple indentations on NBS standards and commercial available test blocks are shown in Table 3. The lower standard deviation obtained for the NBS standards is indicative of small variations in hardness from point to point across the test surface, whereas, the standard deviations for the commercial test blocks were much larger because of variation in grain sizes or material composition (different phases).

Conclusions

Errors resulting from optical measurement can be reduced by using high numerical aperture objectives.

Many commercial microhardness test instruments were found to introduce errors from impact loading and internal vibrations when used at low loads of 25 gf and less.

Many commercial standards have nonuniform microstructures, resulting in point to point variation in the hardness value.

Electrodeposited materials have a much more uniform microstructure.

Microhardness standards prepared by electrodeposition technology are significantly better than commercially available test blocks when used for low load testing of 200 gf and less.

References

[*1*] Tarasov, L. P. and Thibault, N. W., *Transactions of the American Society of Metals,* Vol. 38, 1946, pp. 331-353.

[*2*] Brown, A. G. R. and Iveson, E., *Journal of the Iron and Steel Institute,* Vol. 169, 1951, pp. 377-388.

[*3*] Campbell, R. F., Henderson, Q., and Donleavy, M. R., *Transactions of the American Society of Metals,* Vol. 40, 1948, pp. 954-982.

Robert M. Westrich[1]

Use of the Scanning Electron Microscope in Microhardness Testing of High-Hardness Materials

REFERENCE: Westrich, R. M., **"Use of the Scanning Electron Microscope in Microhardness Testing of High-Hardness Materials,"** *Microindentation Techniques in Materials Science and Engineering, ASTM STP 889,* P. J. Blau and B. R. Lawn, Eds., American Society for Testing and Materials, Philadelphia, 1986, pp. 196–205.

ABSTRACT: Thin coatings of titanium nitride (TiN), zirconium nitride (ZrN), and hafnium nitride (HfN) are difficult to measure for hardness by any technique other than microhardness measurement. The question of apparent hardness increase with decreasing load has been the subject of much discussion, and many explanations have been offered, such as those of Gane and Cox and of Bulter. Impressions measured at high and low loads on a single hardness standard typically show higher hardness readings with lower loads. Correction factors have been suggested, however, there is still some question concerning the absolute hardness at low loads.

The purpose of this work is to show a technique, using the scanning electron microscope (SEM) and a diffraction grating for a measurement standard, that produces more accurate microhardness readings. Further, data suggest that very thin coatings, on the order of 2 to 5 μm, can be measured for hardness. A hardness standardization and comparison technique was also developed to relate microhardness readings from different hardness scales more accurately. Computer techniques were utilized to fit the hardness data for hardness scale comparisons.

KEY WORDS: hardness, microhardness. microindentation, Vickers hardness, wear, scanning electron microscopy, microindentation hardness testing

The reason for this work is the current interest in high hardness sputter and evaporated coated tools. In the past, most work reported with microhardness readings included penetrator type and load. This effectively provided the ability to compare hardness results on a relative basis. However, relating microhardness readings from carburized steels to their bulk hardness as measured by a Rockwell hardness testing machine, produced quite different results. Work by Gane and Cox [*1*] indicated that as the load decreases in microhard-

[1]Senior engineer, Borg-Warner Corp., Des Plaines, IL 60018.

ness testing, the apparent hardness increases, and, therefore, comparisons can be made only at similar loads. Further, the Leitz [2] Miniload hardness tester manual refers to the increase of hardness numbers below 100 g. Leitz suggests adding a correction factor to the diagonal measurement to compensate for this inaccuracy. Figure 1 presents data taken from the Leitz Miniload guide showing that Vickers microhardness values can be corrected to agree with macrohardness.

Hardness Conversion Numbers

Mohrnheim [3], by using a regression analysis technique, showed that a semilogarithmic relationship existed in the conversion of Knoop to conventional macro Rockwell C-scale hardness (HRC) numbers. The accuracy of the conversion was shown to be within ±2 HRC 97.5% of the time with this relation. Using a VAX 11/780 computer and a NAG library program, which uses an algorithm for finding an unconstrained minimum of a sum of squares of nonlinear functions in a number of variables, accurate conversions of less than 0.8 HRC were produced.

Coefficients for various nonlinear functions, such as Vickers diagonal measurements (DPH)/HRC, micrometres/HRC, and others, are listed in the Appendix for insertion in a polynominal equation of the form

$$f(x) = a_0 + (a_1x) + (a_2x^2) \tag{1}$$

This equation and coefficients are used as part of a larger algorithm to routinely convert measurements to hardness numbers of different scales using a VAX 11/780.

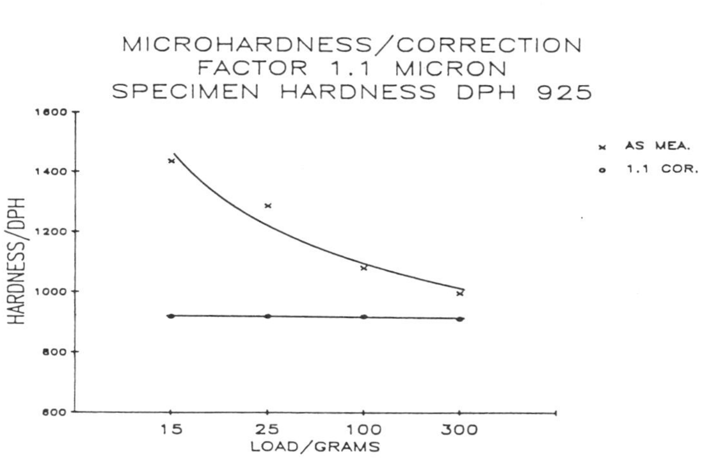

FIG. 1—*Factor applied to hardness measurement at decreasing loads.*

A small Hewlett-Packard calculator can easily handle the conversion, for which a sample program listing is also included in the Appendix.

Standardization of HRC Readings

In order to obtain hardness data for relating Rockwell C-scale to Vickers diagonal measurements, diagonal lengths were measured at various loads on HRC hardness standards. The hardness block values used are those shown in Figs. 2 and 3. The same least-squares fit program used in Eq 1, to find any intermediate hardness number, was utilized to obtain coefficients. Hardness numbers measured by this technique have been used with great success in comparisons with Rockwell machine-generated numbers. The technique does take into consideration some of the errors in microhardness measurement alluded to by Leitz [*2*] and others.

Scanning Electron Microscope Technique

In measuring hardness when no standards are available, acceptance of DPH or Knoop hardness readings as an absolute number is desirable. Microhardness readings without standards are difficult to accept as absolute readings, as has been shown earlier. Further, there are inaccuracies resulting from the resolution limit of the optical microscope and the point-to-point measuring accuracy. A better measuring tool would be the scanning electron microscope (SEM), with its high resolution capability. A problem in using the SEM is the accuracy of magnification; therefore, a technique was needed to obtain accurate magnification and, hence, a calibrated linear measurement of the impressions. A measuring device similar to an optical stage micrometer

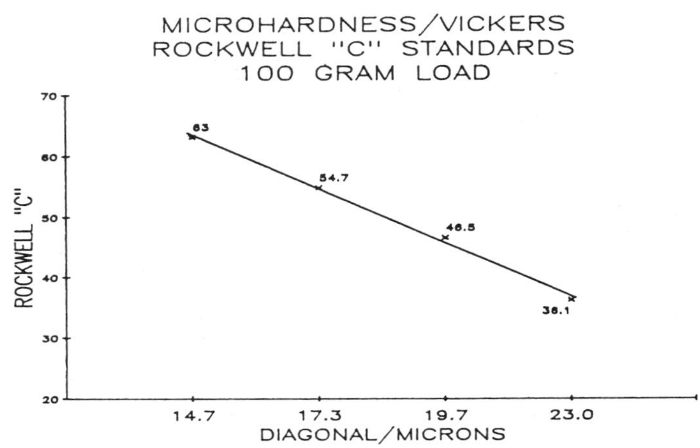

FIG. 2—*Relationship of the Rockwell C-scale to diagonal measurement at 100 g.*

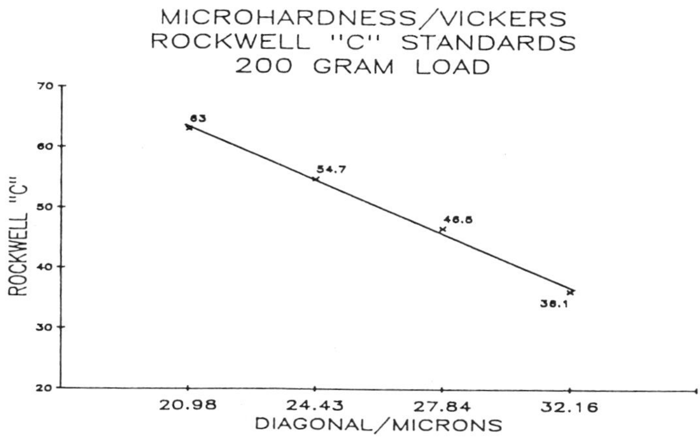

FIG. 3—*Relationship of the Rockwell C-scale to diagonal measurement at 200 g.*

was sought for measuring magnification on the SEM. A Bausch & Lomb 600-lines-per-millimetre diffraction grating was selected because of its high resolution and precision line spacing. Its specifications are shown in the Appendix. Microhardness impressions were made on a Shimadzu microhardness tester and measured both optically and on the SEM for comparison. Figure 4 shows the grating attached to the side of a coated tool for observation on the SEM. Figure 5 shows the results of the optical and SEM measurements made on the same impressions. The standard used is a Wilson test block at 63.0 HRC.

It can be seen that the optical measurements of the diagonals of hardness indentations do increase with decreasing load; however, the SEM measure-

FIG. 4—*Grating attached to a coated tool.*

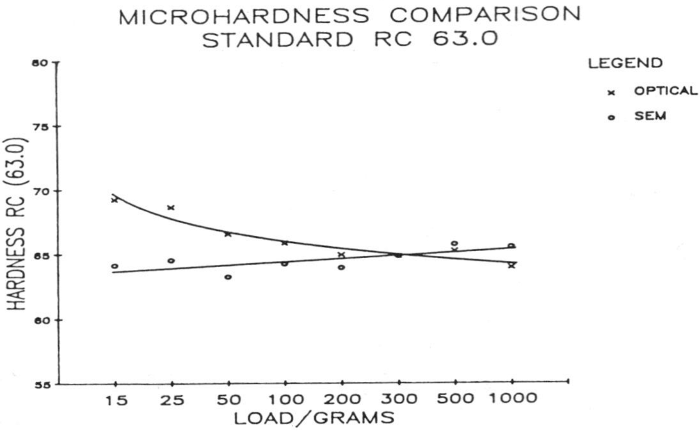

FIG. 5—*Optical and SEM measurements on the same hardness impressions.*

ments of the same impressions tend to decrease in hardness only slightly. The SEM measurements are made directly from photographs that incorporate both the impressions and the diffraction grating, as shown in Fig. 6.

Microhardness readings on titanium nitride (TiN) and hafnium nitride (HfN) coatings in Figs. 7 and 8 show that the SEM measurements are fairly linear, whereas the optical measurements continually increase. Further, as the hardness levels off, as shown in Fig. 8, the indication is that adequate coating thickness is present to support the impression and also that the true hardness of the coating has been reached.

Previously, it was shown that a 1.1-μm constant could be added to the diagonal measurement for correction purposes. Figures 9 and 10 show the TiN and HfN data with this 1.1-μm constant added to the optical readings. The correlation suggests that optical data can be corrected to be made more accurate.

The SEM/grating technique has been shown to produce accurate readings. It is, however, time-consuming to transfer the specimen from the microhardness tester to the SEM for measurement. It would be a distinct advantage to be able to make impressions with the specimen inside the SEM chamber and then proceed to measure the impressions.

Conclusions

Low-load hardness testing is usually conducted on very small impressions, and hence the limits of spatial resolution of the optical microscope are exceeded. Therefore, the SEM provides greater resolution to accomplish more accurate measurements. If SEM measurements are made using an accurate

FIG. 6—*SEM photomicrograph showing impressions and diffraction grating.*

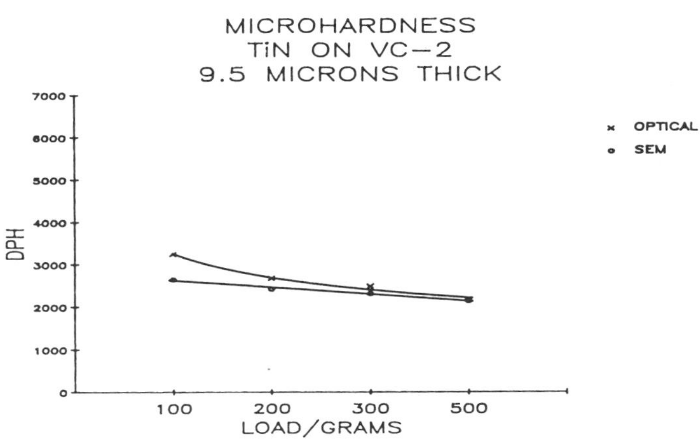

FIG. 7—*Comparison of optical and SEM-measured impressions on TiN.*

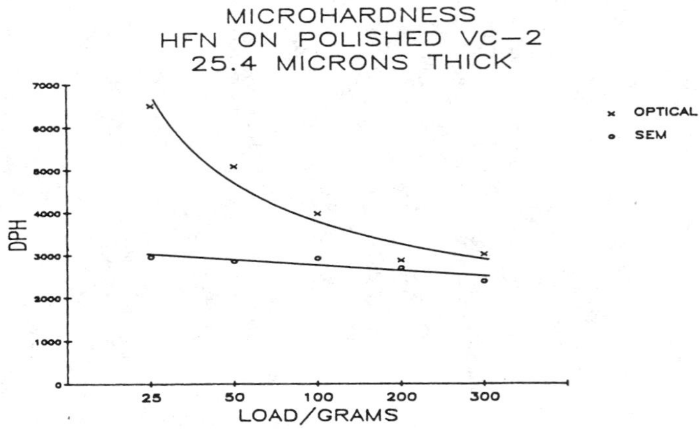

FIG. 8—*Comparison of optical and SEM-measured impressions on HfN.*

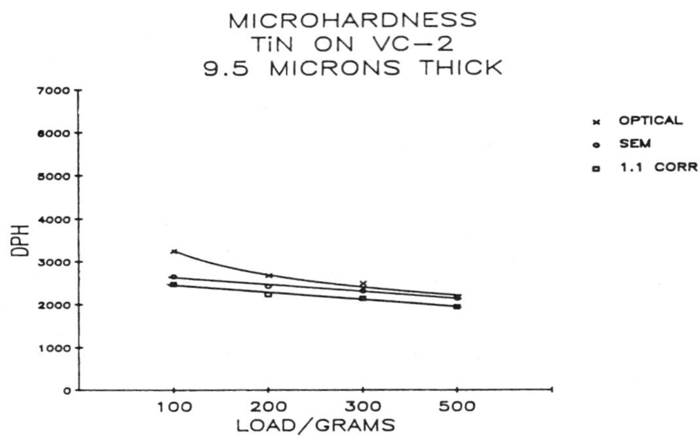

FIG. 9—*Data with corrected optical readings for TiN.*

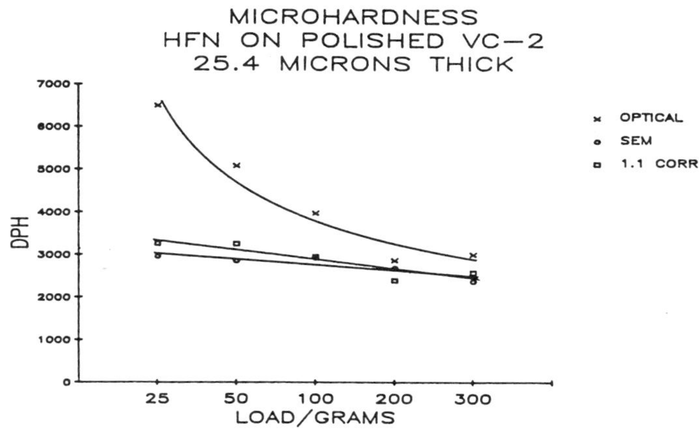

FIG. 10—*Data with corrected optical readings for HfN.*

magnification calibration standard, such as a diffraction rating, materials show minor hardness/load dependence at low loads.

Two techniques have been shown to allow accurate microhardness measurements. The first uses hardness standards with a curve-fitting routine to relate diagonal measurements to HRC values. The second uses the SEM to obtain accurate or absolute diagonal measurements so that hardness can be related to the actual volume displacement of metals. Both techniques have been used successfully depending on the hardness and loads required.

In relating Vickers diagonal measurements to Rockwell C-scale numbers, measurements on the optical microscope are limited to 15 μm, with reliance on SEM measurements for smaller impressions. Generally, hardness readings below 65 HRC with loads of 100 g or more are measured optically. Curve-fitting routines for optically measured readings are accurate only within the hardness standard range. No attempts should be made to extrapolate data beyond the standards or the limits set.

The hardness values of the coatings measured are in the range of 2000 to 4000 HV, and therefore, to date, the only reported readings are in Vickers or Knoop hardness numbers. In the future, with production of hardness standards that can be measured by a Rockwell hardness testing machine, it is possible that these very hard Vickers and Knoop numbers can be related to HRC by curve-fitting routines.

Acknowledgments

I wish to thank Borg-Warner Corp. for supporting this work and the permission to publish it. Also, J. Horwath and W. Sproul for their support and assistance.

APPENDIX

General Conversion of One Reading to Another, Hewlett-Packard Calculator

Equation: $f(x) = a_0 + (a_1x) + (a_2x^2)$

Key Strokes	Storage Registers
Enter	STO 1 = a_0
STO 4	STO 2 = a_1
RCL 1	STO 3 = a_2
RCL 2	
RCL 4	
×	
+	
RCL 3	
RCL 4	
x^2	
×	
+	
RTN	

Table of Coefficients

	a_0	a_1	a_2
DPH to HRC			
Limits, 20 to 65 HRC ASTM E 140-83 Table 1	−0.1546E + 02	0.1775E + 00	−0.9760E − 04
Limits, 61 to 69 HRC ASTM E 140-83 Table 1	0.7041E − 03	0.1264E + 00	−0.5750E — 04
HRC to DPH			
Limits, 20 to 65 HRC ASTM E 140-83 Table 1	0.4942E + 03	−0.1645E + 02	0.3351E + 00
Micrometres to HRC			
200-g load Limits, 20 to 70 HRC	0.1135E + 03	−0.2434E + 01	0.9507E − 03
100-g load Limits, 20 to 70 HRC	0.1078E + 03	−0.2919E + 01	−0.8536E − 02

	Grating
Bausch & Lomb[a]	Cat No. 35-99 No. E-012 1/2-in. plate glass substrate with evaporated aluminum Odd generation 90°V 600 lines/mm

[a]820 Linden, Rochester, NY 14625.

References

[1] Gane, N. and Cox, J. M., "Microhardness of Metals at Very Low Loads," *Philosophical Magazine*, November 1970, pp. 88–89.

[2] *Miniload Hardness Tester Manual*, E. Leitz, Inc., Rockleigh, NJ, 1958.

[3] Mohrnheim, A. F., "Microhardness Testing and Hardness Numbers," in *Interpretive Techniques for Microstructural Analysis*, Plenum, New York, 1977, pp. 117–126.

Engineering Applications

Peter J. Blau[1]

Applications of Microindentation Methods in Tribology Research

REFERENCE: Blau, P. J., **"Applications of Microindentation Methods in Tribology Research,"** *Microindentation Techniques in Materials Science and Engineering, ASTM STP 889*, P. J. Blau and B. R. Lawn, Eds., American Society for Testing and Materials, Philadelphia, 1986, pp. 209–226.

ABSTRACT: Microindentation techniques and tribology research have traditionally been closely linked. Three aspects of their historical relationship are discussed: (1) the use of microindenatations and scratches as indicators of small amounts of wear, (2) the characterization of candidate metal and alloy microstructures for wear applications, and (3) studies of microstructures that have been altered by various wear processes. Proper application of microindentation techniques can provide information beyond simply obtaining microindentation hardness numbers. For example, one can estimate mild wear losses, crystallographic texturing due to sliding, work hardening, and fine structure sizes below eroded surfaces, strength of transfer patches on severely worn surfaces, and the variation in contact conditions across a single wear scar. Many of these areas need to be explored in more extensive quantitative ways before the full power of microindentation methods in tribology can be utilized effectively. The author reviews several published studies as well as his recent work to illustrate future research opportunities.

KEY WORDS: friction, microhardness, microindentation hardness, metals, microstructure, tribology, wear, microindentation hardness testing

Microindentation-based techniques of many varieties have long been used in tribological investigations (that is, studies of friction, lubrication, and wear). The objective of this paper is to acquaint the reader with three aspects of the application of microindentation techniques to metallurgical work in tribology: (1) the use of microindentations and scratches for wear measurement, (2) hardness characterizations of pretest (or preservice) contact surfaces, and (3) studies of worn surfaces and cross sections of worn surfaces to learn about the fundamental behavior of materials in response to wear surface processes.

[1]Materials research engineer, Metallurgy Division, National Bureau of Standards, Gaithersburg, MD 20899.

Microindentation testing techniques have many features in common with the multitude of wear processes. To name just a few, they deform and displace the surface material by vertical impact, sliding, or scratching; they result in localized microstructural damage; and they are sensitive to the pretest surface preparation conditions. Microindentation techniques are significantly dissimilar to wear processes in other respects. For example, they usually do not involve repeated stressing as do most wear processes; they usually do not involve the influences of a chemical environment (including lubricant films); they usually do not involve thermoelastic factors; and they usually fail to characterize the strain-rate-dependent properties that are often important to wear behavior. For the preceding reasons, microindentation tests are important tools in many aspects of tribological work, but they have definite limitations which users must recognize. These limitations tend to make them more qualitative than quantitative in some instances. This paper provides tribology-related examples of microindentation testing techniques from previously published work and from the author's investigations.

Microindentations for Wear Measurement

By assuming that certain geometrical characteristics of indenters are retained by the impressions they produce, one can use microindentations as indicators of wear. For example, if one assumes that elastic and anisotropic shape recoveries of impressions are not significant, one can calculate the depths of penetration for Knoop and Vickers indenters as shown in Fig. 1. Measurements of the diagonals of impressions before and after relatively mild wear can then be used to estimate the amount of wear loss.

Glaeser has used a differential diagonal length measurement to calculate the wear on the inside of a bronze bushing [*1*]. A Knoop indenter on a shaft was inserted into the bushing and tapped with a hammer to produce a prewear indentation. Its size was replicated in plastic and compared before and after wear had occurred.

Begelinger and de Gee have used Knoop indentations to measure both the rate and the pattern of wear on magnetic recording tape heads [*2*]. Using a series of sequential measurements on a pattern of impressions, they traced the progressive development of wear damage across the contact face of a dummy tape head. Correlations were obtained between the tape roughness and the average wear depth.

Ives has used a slightly different method to estimate wear from indentations.[2] He used various loads to produce a range of indentation depths and then observed the smallest remaining impression not obliterated by wear. This provided a boundary value for the surface wear loss.

[2]L. K. Ives, National Bureau of Standards, personal communication, 1983.

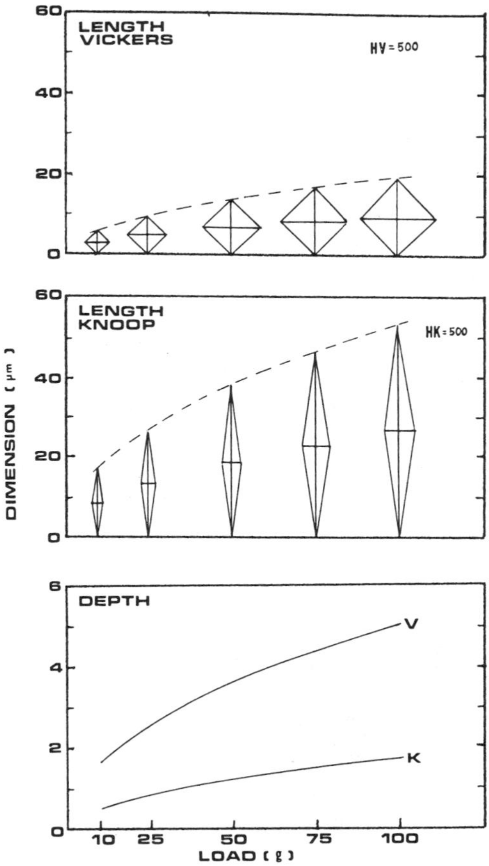

FIG. 1—*Comparison of the relative sizes and depths of ideal Knoop and Vickers impressions in a material with a microindentation hardness number of 500 kg/mm²*.

In these cited studies, it was assumed that vertical elastic shape recovery could be neglected; that is, the length to depth ratio of the Knoop impressions was 30.51.

Scratches on wear surfaces have also been used to indicate small amounts of wear. The smaller the amount of material loss, the more difficult it is to achieve accuracy and precision in wear measurements. To demonstrate this problem, an experiment in abrasive wear was conducted on polished oxygen-free high-conductivity (OFHC) copper. The copper was first mechanically polished to a 1-μm diamond lap finish and then scratched using both 100 and 200-g loads on a commercial scratch tester (see Fig. 2). Three conditions of lubrication were used on the tester:

(*a*) no lubricant,
(*b*) hexanes, and
(*c*) heavy paraffin oil (viscosity: 335 to 365 Saybolt).

The average scratch widths were measured optically, and the depths were measured with a stylus profiling instrument.

The surface was then abraded across the scratch direction with 15 strokes (length, 0.1 m) on fresh, dry 400-grit silicon carbide (SiC) paper with a load of 1.25 kg. The abrasion grooving was lightly polished away with 1-μm diamond paste, and the widths and depths were remeasured. Figure 3 shows the scratches before and after abrasion.

Figure 4 shows that using initial scratch widths on the polished surfaces to calculate reference depths correlated well with the geometric prediction based on the known indenter shape and tip radius (the solid line). The data for both loads and all three lubricating conditions fell near the predicted line. Post-abrasion and repolishing measurements of scratch widths and depths did *not* agree well with the prediction from the indenter geometry. This was probably because of overhangs produced at scratch edges during abrasion. The initial

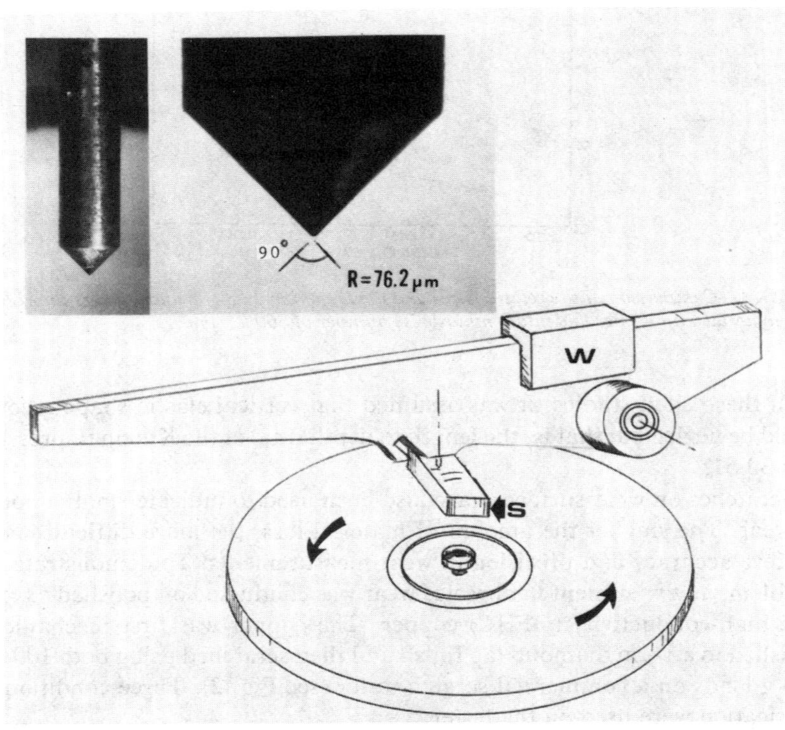

FIG. 2—*Diagram of a commercial scratch tester. The inset shows the indenter shape.*

and postabrasion scratch depths (that is, surface wear loss) agreed to within about 0.2 μm when the scratches made at 100 and 200 g loads were compared. The final scratch widths were not useful in accurately measuring the wear loss, as Fig. 4 illustrates. For example, if the final apparent widths of the scratches made with the 200-g load were converted to a depth value (about 2.5 μm, as seen from the solid line at that width), a difference in the wear loss of a factor of four, compared with the stylus tracing measurements, would be obtained. If overhangs caused most of the errors, then using larger scratches or less ductile metals would reduce these errors.

The previous examples show how microindentations and scratches can be used as sensitive indicators of small amounts of wear. The following sections describe some of the more traditional roles of microindentations—namely, to obtain hardness information about metal surfaces before and after wear.

Microindentation Hardness for Unworn, Prewear Surface Characterization

Wear takes many forms, and numerous wear classification schemes have appeared in the literature of tribology. Some of these schemes group wear by contact geometry and arrangement of contacting surfaces (for example, two-body versus three-body abrasion), some by the proposed mechanism of material removal, and others by what the investigator observes on worn surfaces. Whatever the wear mode may be called, methods for treating wear problems have usually involved some considerations of hardness. As shall be demonstrated, hardness/wear relationships are not always straightforward and clearly understood. Part of the problem may lie with the selection of the proper hardness test for data comparison with wear results. For example, it would initially seem more appropriate to use a scratch hardness test method to correlate hardness with abrasive wear. On the other hand, a high-strain-rate, vertical microindentation test may correlate better with impact wear behavior. Simulation of the wear environment may be a key factor in obtaining clearer correlations between wear and hardness.

A much-cited paper by Kruschov and Babichev [*3*] contains a detailed discussion of abrasive wear/hardness correlations for many metals and ceramics. Different classes of materials had different empirical correlations. Sometimes abrasive wear resistance can actually *increase* when hardness *decreases*. Zum Gahr [*4*] reported such an effect when wear testing a martensitic steel which contained retained austenite. The steel was given various refrigeration treatments to alter the amount of austenite. Relative abrasion resistance *decreased* by about 10% while the relative Vickers hardness *increased* by about 30% as a result of the various refrigeration treatments. The explanation for this behavior was based on the abrasion-induced transformation of greater amounts of retained austenite to martensite during wear in the softer specimens.

For metal-on-metal sliding wear, development in the hardness correlation

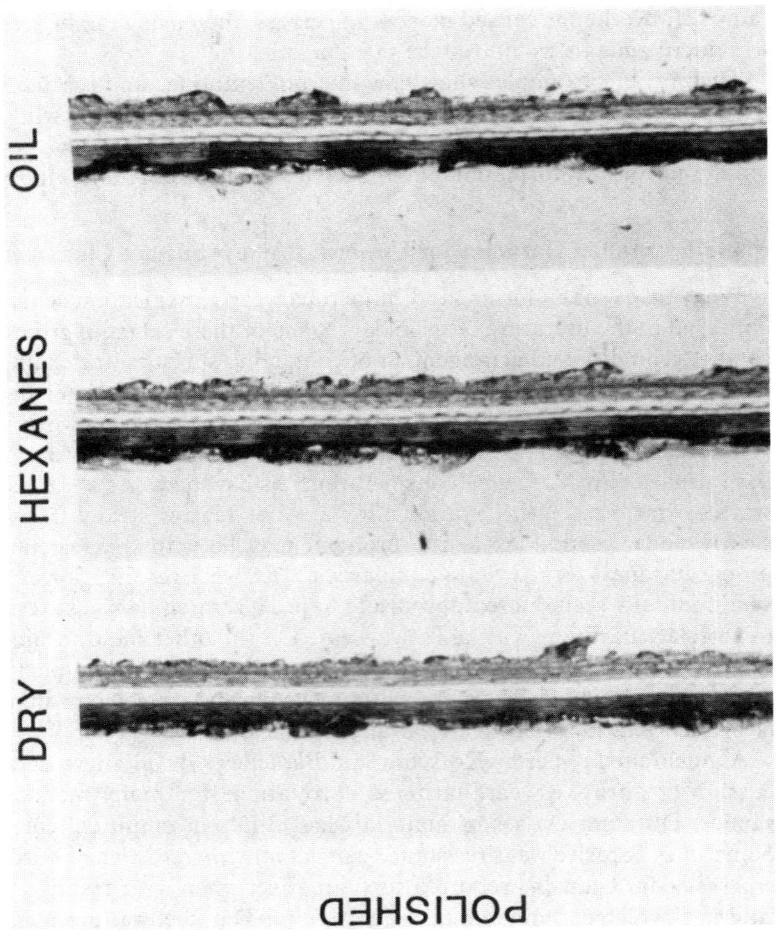
OIL
HEXANES
DRY
POLISHED

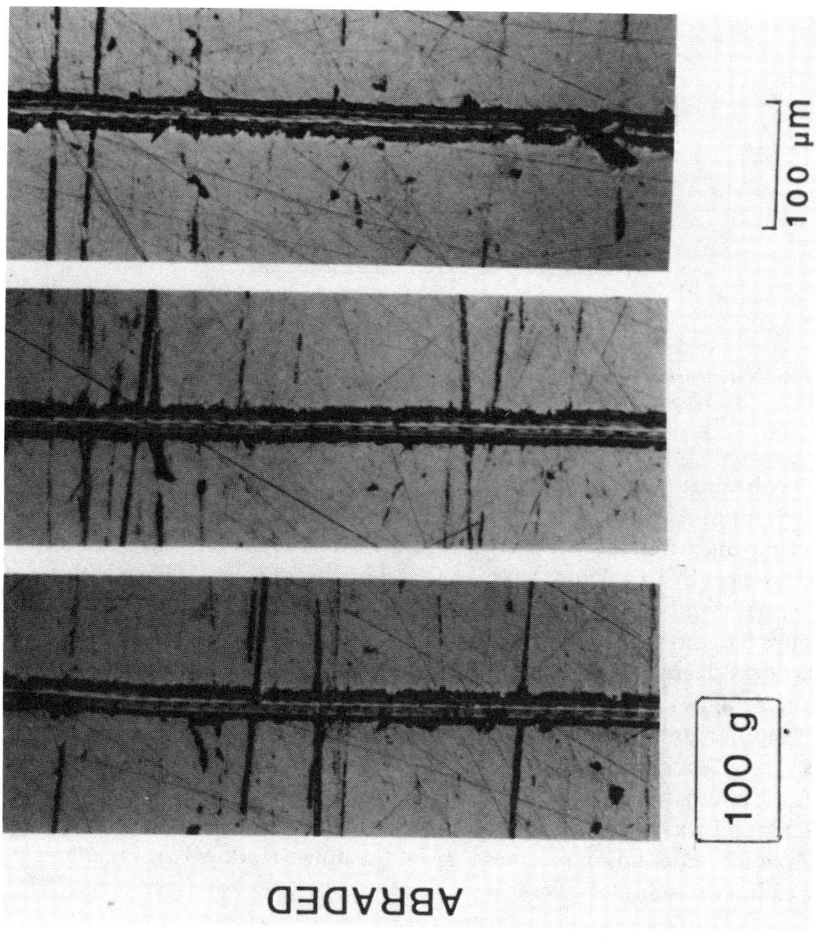

FIG. 3—(Top) *Indenter scratches on the polished copper surface. Lubricants had little effect on the widths or readability of the edges:* (bottom) *the same set of scratches after abrasion and light repolishing.*

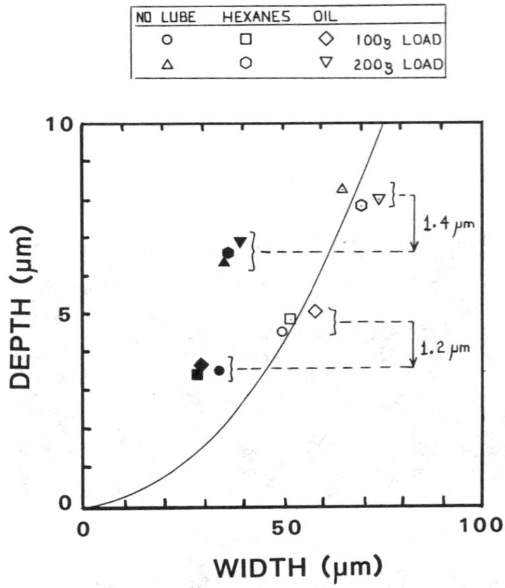

FIG. 4—*Relationship between optically measured scratch widths and stylus-measured depths (*open symbols, *before abrasion;* solid symbols, *after abrasion).*

area has been slower. Part of the problem is that various types of wear may occur simultaneously on the same contacting surface and that the contribution of hardness-related properties to the various types of wear may differ. Recent papers by Engel and Adams [*5*] and by this author [*6*] have addressed the competition between simultaneously operating wear modes.

Figure 5 demonstrates that sliding wear resistance can sometimes improve as hardness decreases. Polished blocks of a series of copper-aluminum (Cu-Al) binary alloys and dual-phase steels were worn against 52100 steel rings under similar unlubricated sliding conditions (load = 10 N; velocity = 20 cm/s, argon environment, 1 h). As the data show, even though the chemistries and microstructures of these two alloy systems differed greatly, a very similar trend was observed in the block wear volume, *W*, versus the Knoop (100-g) microindentation hardness, *H*, of the unworn alloys [*8*]. Linear least-squares fits to each plot gave

$$W = -0.399 + 6.65 \times 10^{-3} H \quad (1)$$

correlation coefficient = 0.991

for copper–aluminum and

$$W = -1.015 + 6.48 \times 10^{-3} H \quad (2)$$

correlation coefficient = 0.948

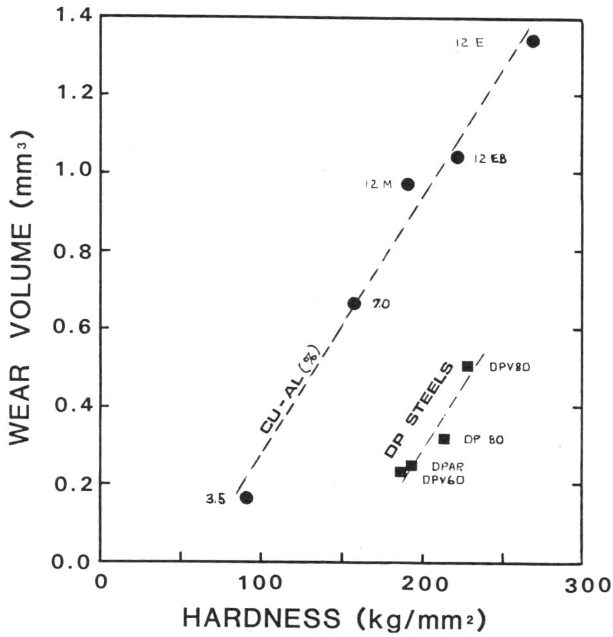

FIG. 5—*Knoop microindentation hardness data versus wear volume for several dual-phase steels and Cu-Al binary alloys. The Cu-12 Al alloy at the top was given three different heat treatments to change its hardness.*

for dual-phase steels, where W is in cubic millimetres and H is in kilograms per square millimetre. *Decreasing* hardness *improved* the wear resistance in this example, which supports the argument that wear and hardness values can result from different combinations of basic materials properties.

Examination of Eqs 1 and 2 reveals several additional characteristics of the microindentation hardness/wear correlations for these two different series of alloys. First, one observes high correlation coefficients for both alloy series, suggesting that the indicated trends were real. Next, there is close agreement between the values of the constants in the final terms of Eqs 1 and 2, suggesting that in two alloy systems with quite different compositions, thermomechanical treatments, and microstructures, a commonality in operative wear modes and attendant response to microindentation testing exists.

Finally, in comparing the wear volumes calculated from Eqs 1 and 2 for Cu-Al alloys and dual-phase (DP) steels of equal microindentation hardness, one finds that as the microindentation hardness number increases, the steel has less of a wear advantage over the copper alloys. For example, at $H = 180$ kg/mm^2 the ratio of Cu-Al to DP steel wear volume is 5.3, but at $H = 240$ kg/mm^2 the ratio is only 2.2. Therefore, even though the current experimental data suggest relationships contrary to the commonly held notion of "raise the hardness to lower the wear," the results are consistent enough to warrant further fundamental tribological explorations into the origins of this kind of behavior.

In evaluating the wear of composite materials or polyphase alloys, it is often important to consider the micromechanical properties of the individual constituents. For example, Vander Voort [*8*] and Exner [*9*] used microindentation-induced cracking behavior in the evaluation of the fracture toughness in cermets and cemented carbide tooling materials, respectively.

The present author used the microindentation hardnesses of the separate phases in an Al-Si-Cu alloy to help characterize its frictional run-in behavior and wear in dry sliding [*10,11*]. In these studies, the ductile Al-Cu continuous phase surrounding a more brittle silicon phase had a microindentation hardness which was about one tenth that of the silicon phase. Figure 6 shows both the heavily etched and the unetched surfaces of flat wear blocks before testing in the ball-on-flat unidirectional sliding geomeotry (the fixed balls were AISI 52100 steel). Stroke-by-stroke tests were run unlubricated in air. As illustrated in Fig. 7, friction run-in behavior and wear surface conditions are strongly affected by the tendency of the softer phase to smear over the harder one, fill in areas between hard protrusions during sliding, and serve as an embedment medium for brittle phase debris.

One can therefore conclude that while prewear microindentation hardness characterization of the mating surfaces is important, it may not be sufficient in itself for investigating the underlying mechanisms of the wear process in metals and alloys. For fundamental studies, more detailed microindentation hardness analysis of worn components is frequently required. Such analyses are exemplified and discussed in the next section.

Post-Test Characterization of Wear-Induced Microstructural Damage

As discussed in the preceding section, prewear microindentation hardness testing is a standard part of many wear and friction studies. Post-test characterizations are much less common. Although it has long been recognized, particularly by the materials science community, that wear results in changes in the contact surface and near-surface mechanical properties, difficulties in microindentation hardness testing of rough, worn surfaces have limited the number of studies in this area. Sometimes, metallographic cross sections or taper sections have been used rather than indenting the as-worn surface.

The relationship between worn-surface microindentation hardness and sliding wear damage can depend on how it is measured. Figure 8 illustrates such a case, comparing the results of two investigations in which Cu-Al single-phase alloys were slid against harder counterfaces. Wert et al [*12*] used cylindrical sapphire sliders and Blau [*13*] used AISI 52100 cylindrical rotating rings against alloy flats. The loads were 29.4 and 10 N, and the sliding velocities were 0.32 and 20 cm/s, respectively. Both studies used tests run in

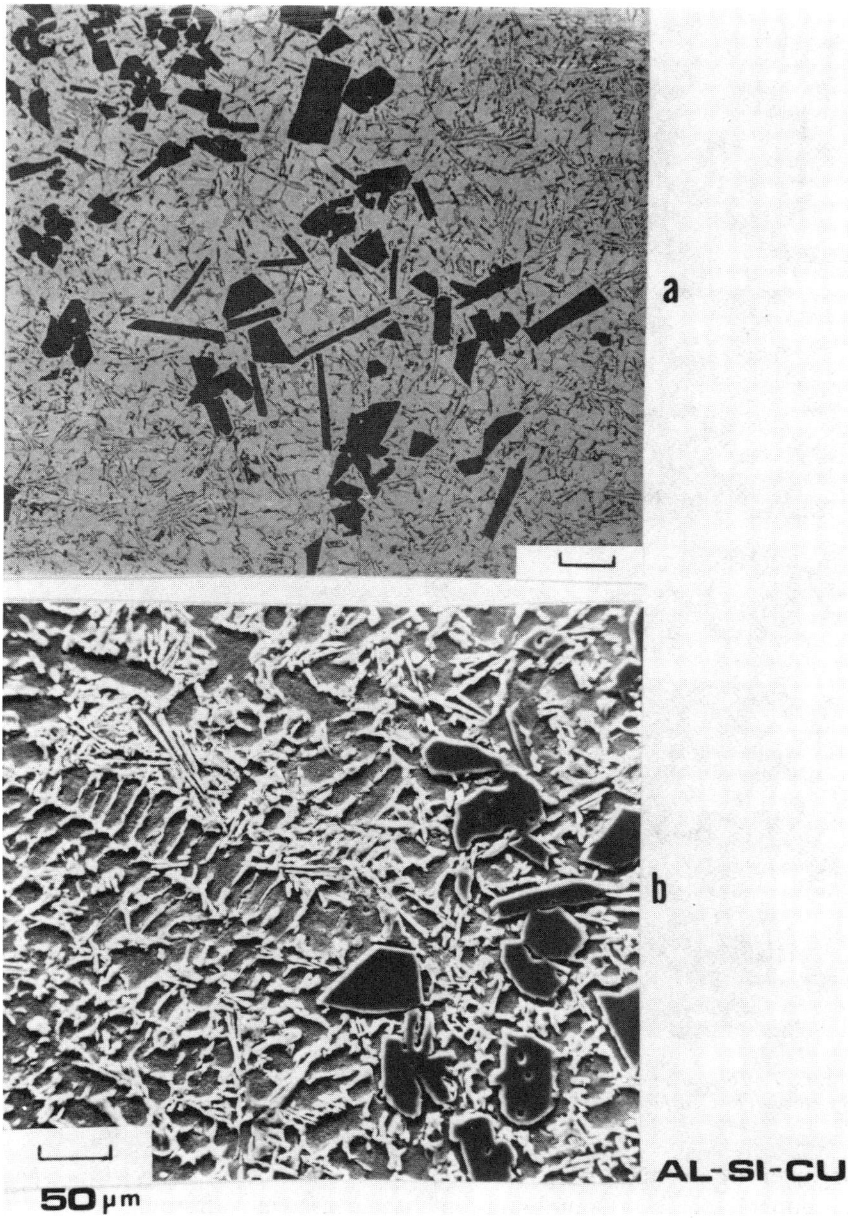

FIG. 6—*Microstructures of polished and heavily etched surfaces of Alloy C390-T5 (Al-Si-Cu) showing blocky silicon in a eutectic matrix of silicon and aluminum-copper phases.*

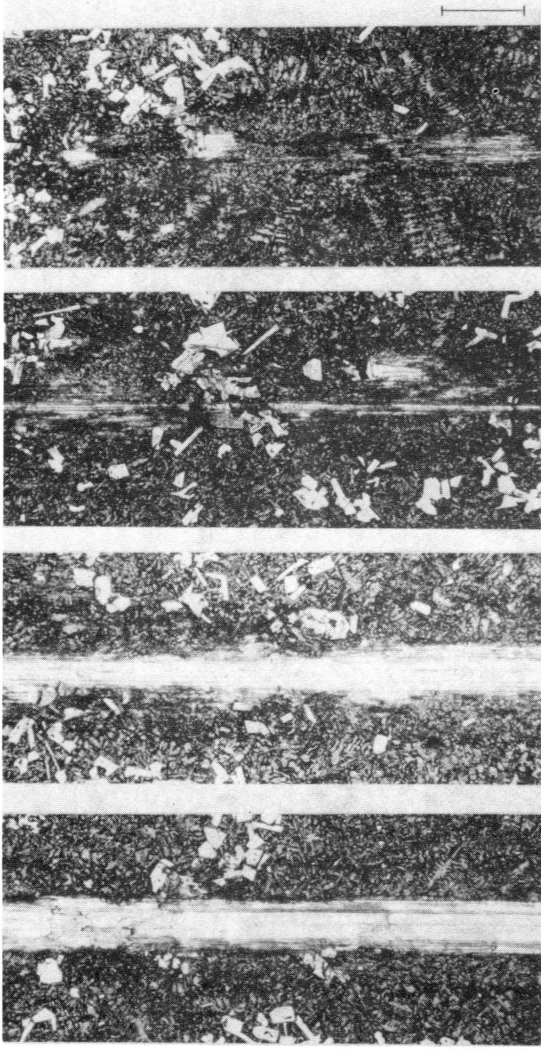

LOAD N	μ_{15} Ave.	St Dev μ_{15}	N_μ
2.5	.54	.13	13
4.9	.61	.10	11
7.4	.53	.10	10
9.8	.43	.11	7

FIG. 7—*Comparison of the wear tracks and friction run-in behavior for four sliding loads on the etched block surface.*

argon environments. As Fig. 8 shows, the wear rate of the alloys with respect to copper increased as aluminum content increased in both studies. However, the relative microindentation hardness (Knoop indenters used in both studies) of strained sliding surfaces versus bulk values, H_s/H_0, showed opposite trends for the two studies.

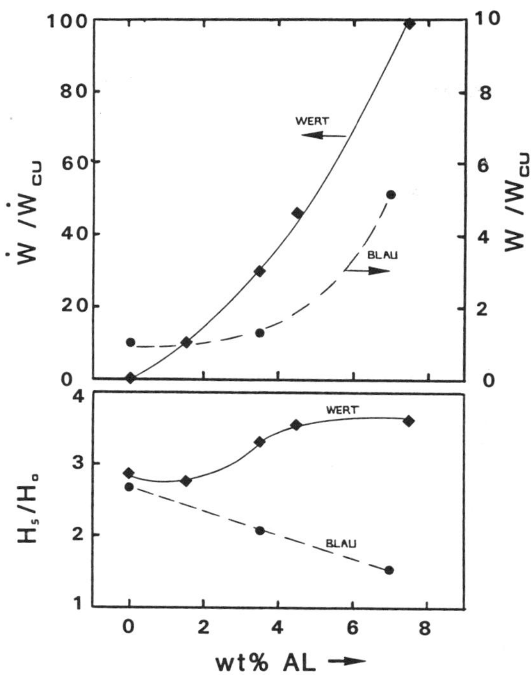

FIG. 8—*Comparison of two studies of the wear of Cu-Al alloys showing how different strain hardening results can be obtained using different microindentation hardness testing conditions.*

The reason for this seeming contradiction may be a result of how the relative hardness data were obtained. Wert et al measured "worn" surface, H_s, hardness values on tracks produced on their specimen materials after three strokes of a diamond hemisphere under 9.81 N load. Blau measured H_s on tapered metallographic sections below the actual wear surfaces of the test blocks. It is likely that the data of Wert et al were obtained for incompletely worn-in surfaces which were subjected to sliding contact conditions different from those used for the primary wear test specimens. Therefore, the two studies did not produce similar trends.

Another study of the microindentation hardness versus depth gradients in copper alloys subjected to unlubricated sliding dealt with the relationship between wear processes and surface work hardening behavior [*13*]. Figure 9 depicts such behavior. In that study, it was proposed that hardness may not always increase continuously from the undeformed bulk up to the free wear surface but may reach a limiting saturation value in the highly deformed layer several micrometres below the sliding surface. It was concluded that the presence of a subsurface hardness "plateau" depends on both the contact stress

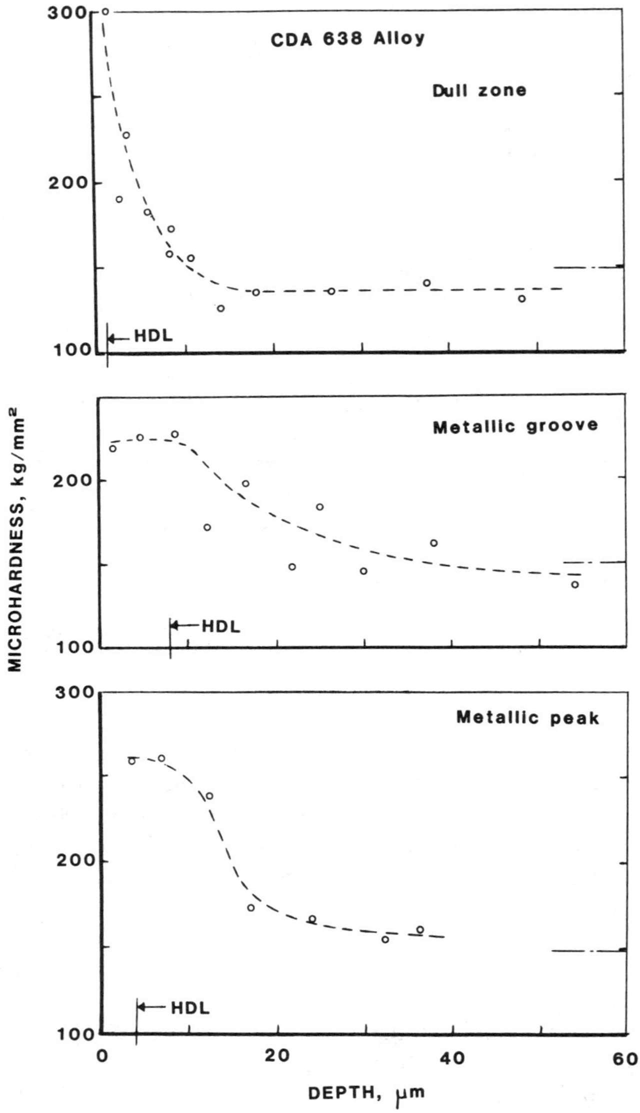

FIG. 9—*Microindentations hardness (Knoop indenter, 10 g) versus the depth below several locations on the same worn surface of copper Alloy CDA 638.*

conditions and the shear stress/shear strain characteristics of the materials involved. In such studies indentation sizes must be small compared with the depth of deformation below the wear surface.

In a paper on the wear of steels, Vingsbo discusses the modeling of subsurface microindentation hardness profiles in terms of nine separate contributions [*14*]. These are dislocation mobility, strain hardening, recovery, recrystallization, austenitization, strain-induced martensitic transformations, thermal martensitic transformations, precipitation hardening, and tempering.

The length-to-width (that is, aspect) ratio of Knoop impressions has been used to indicate sliding-induced texturing in copper [*15*]. On a polished cross section of a worn surface, the author located a large single grain and placed a reference rosette of indentations well away from the wear surface. In the same grain, he placed a row of parallel Knoop microindentations at various depths below the wear surface and noted that the aspect ratio changed markedly as the near-surface deformed zone was sampled. By examining the behavior of aspect ratios in the reference rosette, one could roughly estimate the extent of crystallographic rotation in the grain toward the sliding direction.

Richardson discusses the maximum hardness of strained metal alloy surfaces subjected to abrasive wear [*16*]. He considered the role of flow stress, microstructural transformations, temperature, and strain rate effects. Comparisons of maximum Vickers microindentation hardness were made between shot-peened, trepanned, and soil-abraded surfaces (see Table 1). The trepanned surface hardness values were generally the highest. The values for maximum hardness correlated well with abrasion resistance values.

Other work relating the scratch characteristics of alloys to their abrasive wear behavior is described by Kosel in another paper in this volume [*17*]. He

TABLE 1—*Selected data for the hardening of strained surfaces (after Richardson* [16], *Vickers scale).*

Metal or Alloy	Unstrained Hardness, kg/mm^2 [a]	Maximum Hardness of Worn Surfaces, kg/mm^2 [a]	Factor of Hardness Increase
Annealed aluminum	30.1	87	2.9
Cold-rolled copper	43.3	136	3.1
Annealed tantalum	85	415	4.9
Hot-rolled iron	105	356	3.4
Austenitic manganese steel	220	885	4.0
Bainitic steel, 0.37% carbon, by weight	438	754	1.7
Cold-worked tungsten	471	866	1.8

[a] For megapascal equivalents, multiply values by 9.81.

used irregular, nontraditional indenter shapes to reach a closer approximation of the abrasion application he was evaluating and studied the production of scratching damage in a scanning electron microscope.

Microindentations have also been used in determining the depth and nature of material damage in erosive wear. A recent paper by Wey, Moteff, and Ives [*18*] used a two-diagonal Knoop method [projected area hardness (PAH)] [*19*][3] to obtain the hardness versus depth profiles below eroded surfaces of copper. Because the PAH correlated so well with the dislocation cell sizes in complementary experiments on tension and compression behavior of the same material, the microstructures below the eroded surfaces could be inferred from the PAH versus depth profiles obtained on cross sections.

Sheldon [*20*] also appreciated the value of comparing erosion behavior with the microindentation hardness values of the as-eroded surfaces in investigating the erosive wear process. He compared annealed with eroded surface Vickers microindentation hardness numbers and found a much better correlation of wear data with the work-hardened surface hardness data. It was concluded that the latter hardness data were the most appropriate for use in modeling the erosion process.

A final example of the uses of microindentations in tribology relates not to the acquisition of hardness numbers but rather to the use of the indenter as a qualitative measure of the strength of transferred deposits on sliding wear surfaces. By selecting a relatively large patch of transferred material on a worn surface and attempting to place Knoop microindentations on it with increasing loads, beginning with, say, 0.2 to 0.5 N (a 2 to 5-g load), the critical cracking load can be estimated. Of course, many such tests should be done to obtain a fair evaluation, because the underlying support for the indenter on such patches is likely to vary considerably from one patch to another (see Fig. 10). Therefore, even though they are largely qualitative, such experiments may nevertheless be useful in assessing the relative strength and tenacity of transferred deposits on sliding surfaces.

Summary

Microindentation and scratch hardness impressions can provide useful tools for research and development in tribology as wear indicators, wear surface characterization parameters, and micromechanical properties probes in fundamental friction and wear research. By choosing the test methods properly, preparing metallographic specimens so as to minimize hardness ambiguities, and recognizing the limitations of the chosen methods, significant insight has been and will continue to be gained in clarifying the role of microstructure in the wear of materials.

[3] PAH (kg/mm^2) = 2000 [$P/(D \cdot d)$] where P = load (g), D = longer Knoop diagonal (μm), and d = shorter Knoop diagonal (μm).

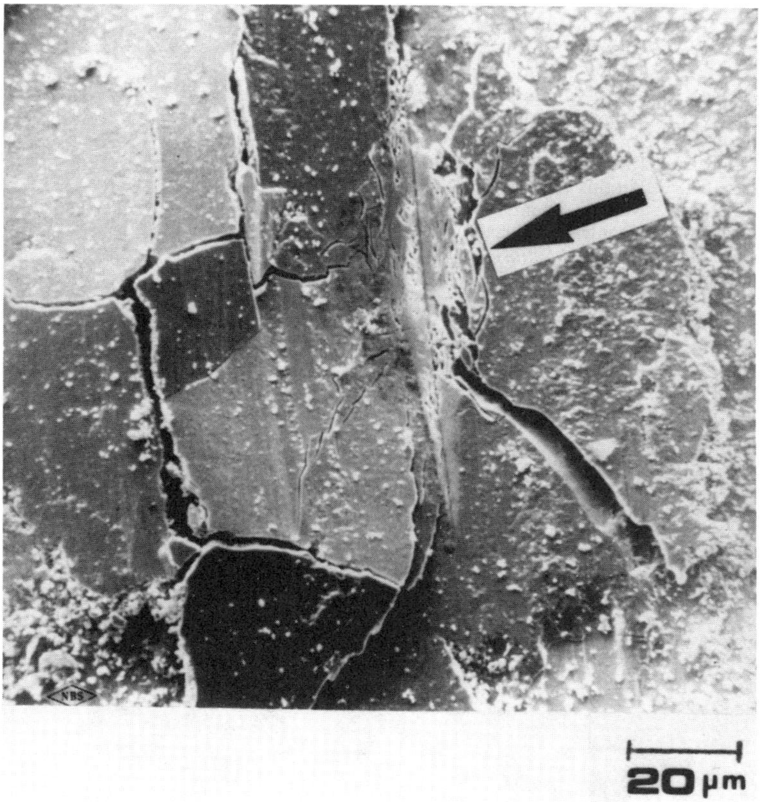

FIG. 10—*Microindentation on a patch of transferred material on a copper Alloy CDA 638 wear scar.*

References

[1] Glaeser, W. A., *Wear*, Vol. 40, 1976, pp. 135-137.

[2] Begelinger, A. and de Gee, A. W. J., *Wear*, Vol. 43, 1977, pp. 259-261.

[3] Khruschov, M. M. and Babichev, M. A., *Friction and Wear in Machinery*, American Society of Mechanical Engineers, Vol. 12, 1958, pp. 5-23.

[4] Zum Gahr, K.-H., *Proceedings, Wear of Materials 1979*, American Society of Mechanical Engineers, New York, 1979, pp. 266-274.

[5] Engel, P. A. and Adams, C. E., *Proceedings, Wear of Materials 1979*, American Society of Mechanical Engineers, New York, 1979, pp. 181-191.

[6] Blau, P. J., *Proceedings, Wear of Materials 1983*, American Society of Mechanical Engineers, New York, 1983, pp. 526-533.

[7] Ruff, A. W. and Blau, P. J., *Microscopic Aspects of Wear Processes in Metals: II*, internal report, National Bureau of Standards, Washington, DC, 1983.

[8] Vander Voort, G. F. and Hinton, R. W., *International Conference on Productivity Improvement Through New Tools and Applications*, Chicago, 11-12 June 1979.

[9] Exner, H. E., *Transactions of the Metallurgical Society of A.I.M.E.*, Vol. 245, 1969, pp. 677-683.

[*10*] Blau, P. J., *Wear*, Vol. 81, 1982, pp. 187–192.
[*11*] Blau, P. J. and Whitenton, E. P., *Tribology International*, August 1982, pp. 208–217.
[*12*] Wert, J. J., Singerman, S. A., and Caldwell, S. G., *Wear*, Vol. 92, 1983, pp. 213–229.
[*13*] Blau, P. J., *Journal of Materials Science*, Vol. 19, 1984, pp. 1957–1968.
[*14*] Vingsbo, O. and Hogmark, S. in *Fundamentals of Friction and Wear of Materials*, D. A. Rigney, Ed., American Society for Metals, Metals Park, OH, 1981, pp. 373–408.
[*15*] Blau, P. J., *Scripta Metallurgica,* Vol. 13, 1979, pp. 95–98.
[*16*] Richardson, R. C. D., *Wear*, Vol. 10, 1967, pp. 353–382.
[*17*] Kosel, T. H., in this volume, pp. 227–242.
[*18*] Wey, Y. G., Moteff, J., and Ives, L. K., *Proceedings, Wear of Materials 1983*, American Society of Mechanical Engineers, New York, 1983, pp. 354–359.
[*19*] Blau, P. J., *Scripta Metallurgica*, Vol. 14, 1980, pp. 719–724.
[*20*] Sheldon, G. L., *Journal of Engineering Materials and Technology*, April 1977, pp. 133–137.

Thomas H. Kosel[1]

Review of Scratch Test Studies of Abrasion Mechanisms

REFERENCE: Kosel, T. H., **"Review of Scratch Test Studies of Abrasion Mechanisms,"** *Microindentation Techniques in Materials Science and Engineering, ASTM STP 889*, P. J. Blau and B. R. Lawn, Eds., American Society for Testing and Materials, Philadelphia, 1986, pp. 227–242.

ABSTRACT: The use of scratch tests to simulate the material removal mechanisms which occur during abrasion is reviewed. Although useful studies of the effect of the rake angle on material removal have been carried out using diamond tools, closer simulation of the mechanisms of material removal can be obtained using actual irregular individual abrasive particles as scratch tools.

Previous studies are reviewed in which scratch tests have been performed with both conventional scratch test instruments and a specially designed system used for *in situ* scratch tests in the scanning electron microscope (SEM). Multiple-pass scratch tests over the same scratch path have been shown to create surface features and wear debris particles which are very similar to those produced by low-stress abrasion. Alumina (Al_2O_3) particles have been shown to produce continuous micromachining chips from the hard, brittle carbide phase of Stellite alloys, establishing direct cutting as the important mechanism of material removal for this type of abrasive. An *in situ* study of material removal from white cast irons by quartz particles has provided conclusive evidence that carbide removal does not occur by direct cutting but rather always involves microfracture.

Previously unpublished work which has compared scratch tests with crushed quartz and alumina particles is included. Also described is a new scratch test system which controls the depth of cut rather than the scratch load in order to simulate high-stress abrasion, in which abrasive particles are constrained to a fixed depth of cut. Preliminary new results show substantially different carbide fracture behavior under fixed-depth conditions.

KEY WORDS: abrasion, hardness, scratch testing, metals, white cast irons, wear, microindentation hardness testing

It is useful to divide the materials used in abrasion applications into two categories, which we will refer to as Class I and Class II materials. Class I can be taken to include both truly single-phase materials and materials which

[1]Associate professor, Department of Metallurgical Engineering and Materials Science, University of Notre Dame, Notre Dame, IN 46556.

contain second-phase particles that are small compared to the size of the grooves made by the abrasive particles, since the second-phase particles do not then interact with the abrasive as discrete particles. Abrasion of Class I materials is controlled primarily by the material hardness, abrasive angularity, and hardness of the abrasive in relation to the abraded material. Scratch test studies of the effect of the rake angle (defined as the angle between the rake face of the particle and the surface normal) have been made in Class I materials; these will be reviewed in the first section of this paper.

Class I materials as a group are not as abrasion-resistant as Class II materials, which are dual-phase materials that contain large, hard second-phase particles such as carbides. This class includes white cast irons and Stellite alloys, both of which contain chromium-rich M_7C_3 carbides in the 10 to 100-μm size range with hardnesses on the order of 1300 to 2000 HV. Since the carbides are hard compared with most environmental abrasives and are large compared with the size of the contact areas of the abrasive particles, they greatly enhance the abrasion resistance of the material. The material removal mechanisms for Class II materials are more numerous than those for Class I; they include cutting the matrix and second-phase particles and cracking and pulling out of the second-phase particles. Published studies of such mechanisms by means of scratch testing will be reviewed in the second section of this paper, together with previously unpublished work comparing scratch tests made with angular quartz and alumina (Al_20_3) particles.

It is necessary to distinguish between two types of abrasive loading conditions, which we will designate as fixed-load and fixed-depth conditions. The load per particle is essentially constant in the former, whereas the depth of the groove is constant in the latter. The two are equivalent for a homogeneous material (Class I), but when an abrasive particle encounters a hard second-phase particle in a Class II material under fixed-depth conditions, the magnitude of the normal force on the particle must increase as the abrasive particle tries to rise over the second-phase particle because of its greater hardness. Fixed-depth conditions occur for Class II materials when the abrasive particles are bonded together into a single mass, such as a grinding wheel or a sheet of abrasive paper. When an individual abrasive particle tries to deflect over a second-phase particle, the entire abrasive body must also deflect, leading to a rapid increase in load with the entire external load resting ultimately on a single contact. Under such conditions, the abrasive or the second phase may either fracture or deform plastically. Essentially fixed-load conditions occur for a Class II alloy when it is abraded in the commonly employed dry sand/rubber wheel abrasion test (ASTM Practice for Conducting Dry Sand/Rubber Wheel Abrasion Test, G 65-81), in which loose abrasive particles are fed between a rotating rubber-rimmed wheel and the specimen. When an abrasive particle encounters a second-phase particle, it can rise and pass over it by pressing slightly more deeply into the rubber wheel with a relatively small increase in normal load. The study of material removal mechanisms in Class

II alloys under fixed-depth conditions has recently been simulated using a special scratch test apparatus, and preliminary results of this work will be described in the third section of this paper.

Scratch Tests with Class I Materials

The effect of the rake angle (or attack angle, defined as the angle between the surface and the rake face of the tool) on material removal during abrasion has been studied effectively by means of scratch tests [*1,2,4,5*]. The existence of a critical attack angle, α_c, was demonstrated by Mulhearn and Samuels [*4*] in experiments with a plain carbon steel (0.35% carbon) using regularly shaped diamond and tungsten carbide tools. The attack angle was varied, and the resulting scratches were observed. Above the well-defined critical angle $\alpha_c = 90°$, a micromachining chip was cut by the tool used to simulate an abrasive particle. In contrast, below α_c the interaction of the tool and the specimen produced a groove by a ploughing action, which pushed material to the sides of the groove without removing it. The authors also investigated the attack angles present in silicon carbide (SiC) abrasive papers and suggested that a fairly small fraction (12 to 23%) of the particles were oriented so as to produce direct material removal by micromachining.

Mulhearn and Samuels [*4*] developed a theory for abrasion which took into account the shape of the abrasive particles. This theory was extended by Sedriks and Mulhearn [*5*], who measured the cross-sectional area, A, of grooves produced in lead by ground pyramidal tungsten carbide tools. The tangential force opposing the motion of the tool was also measured, and relationships for A as a function of α during both cutting and ploughing were proposed and shown to be in good agreement with the measured values. The intersection between the theoretical A-α curves for cutting and ploughing was found to correspond reasonably well to the experimental value of α_c, thus providing a means of predicting α_c if certain other parameters are measured using scratch tests. The results were in agreement with the assumption that for $\alpha > \alpha_c$, a chip is cut and the external load on the tool is supported by the frictional force exerted by the flow of the chip upwards along the rake face, whereas for ploughing the force is supported by the inclined contacting area. The theories just described were confined to non-work-hardening metals but were later extended [*1*] to include work-hardening metals.

Murray, Mutton, and Watson [*2*] used single-point scratch tests to demonstrate that α_c depends on the microstructural condition produced by different heat treatments in the same steel. This result was used to support the suggestion that the nonlinear dependence of the wear resistance of a heat-treated steel on its hardness may be due to a change in α_c with hardness.

Aghan and Samuels [*3*] studied the mechanisms of material removal during fine polishing of pure annealed copper with diamond and $Al_2 0_3$ particles. The specimens were abruptly removed from the polishing cloth so that it was possible to observe micromachining chips which were still attached to the ends and sides of the scratch grooves, and debris particles from the copper were also recovered and observed in the scanning electron microscope (SEM). Although scratch tests were not utilized in this work, some of the observations of Aghan and Samuels regarding earlier scratch test work are relevant. They noted that previous studies had been confined to tests involving orthogonal cutting with tools of idealized shapes and that these studies had not produced evidence of certain features observed in their own work.

Scratch Tests with Class II Materials

The use of scratch tests to simulate material removal mechanisms during abrasion of Class II materials has been discussed in a recent series of papers. Shetty et al [*6–8*] employed single and multiple-pass scratch tests performed with different tools in a conventional scratch testing apparatus to examine material removal mechanisms in Stellite alloys. Later, Prasad and Kosel [*9, 10*] performed *in situ* scratch tests in the SEM in order to investigate material removal mechanisms in quartz abrasion of white cast irons. The fixed-load work of Shetty et al will be reviewed first, followed by that of Prasad and Kosel.

Scratch Tests Performed Outside the SEM

Most of the work described in the previous section on Class I materials employed scratch tools of idealized shape and specimens which were metallographically polished prior to scratch testing. Since abrasion normally involves irregularly shaped abrasive particles which remove material from an irregular surface that has been shaped and work hardened by the abrasive, Shetty et al [*6*] reasoned that abrasion material removal mechanisms might differ significantly from those observed in scratch tests with idealized particles and surfaces. An additional factor not always properly simulated by conventional scratch testing is that of the relative hardness of the tool compared with that of the material. This factor is especially important in considering the important case of abrasion of Class II alloys by quartz abrasive, since the abrasive hardness (about 1000 HV) is usually greater than that of the matrix but less than that of the carbides in abrasion-resistant alloys.

In view of these considerations, Shetty et al [*6*] compared the material removal mechanisms in Stellite powder metallurgy (PM) alloys using both Vickers diamond indenters and individual particles of the semirounded Ottawa quartz abrasive which is commonly used in the dry sand/rubber wheel abrasion test (RWAT). The abrasive particles were cemented to the tip of a conical holder for use as scratch tools, and the scratch tests were performed using a Bergsman microhardness tester (Fig. 1), which mounts on a standard inverted metallurgical microscope with the scratch tool in place of one of the

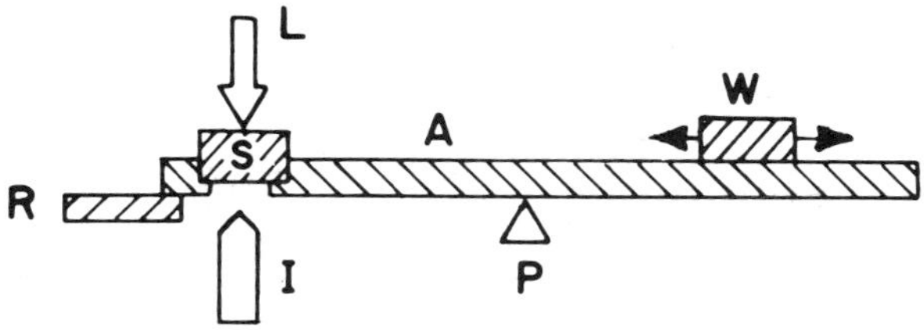

FIG. 1—*Schematic drawing of Bergsman apparatus* [6-8]. *Specimen S mounts on Arm A. After the arm is balanced on Pivot P by adjustable Counterweight W, a fixed load, L, is added. The specimen stage, to which P and Rest R are fixed, is lowered until Identer I lifts the arm from Rest R. The translation of the stage then produces a scratch.*

microscope objective lenses and the specimen in the usual specimen position. Scratches were made by selecting the area to be scratched, rotating the scratch tool into position, lowering the specimen until it contacted the tool with a known deadweight load, and translating the specimen. In order to simulate the irregular work-hardened surface of an abraded specimen, as many as 100 sequential scratches were made in the same scratch groove with the same tool. Some tests were also done using specimens which had been preworn in the RWAT.

Important differences were found between single and multiple-pass scratch tests. Although both single-pass quartz and diamond scratches produced a slip-band cracking phenomenon which was attributed [*6*] to transformation of the matrix from face-centered cubic (FCC) to hexagonal close-packed (HCP) [*11*], the phenomenon was not observed in the multiple-scratch tests with either quartz or diamond tools. The material removal mechanisms during steady-state abrasion are thus best studied in a way that simulates the roughening and work hardening of the surface during abrasion.

This was also found in multiple-pass quartz scratches, in which surface features were produced that were very similar to those generated by RWAT abrasion of the same alloy, as can be seen in Fig. 2. In particular, there is a tendency for the carbides to protrude from the surface because they are attacked much less efficiently than the matrix by this relatively soft abrasive. The fine wear debris particles on the two surfaces are also of similar size and shape. Some of these are quartz fragments formed by damage done to the quartz by the carbides.

Although the single-pass tests generated some phenomena which were not reproduced in the multiple-pass tests or in RWAT abrasion, one phenomenon was discovered first by the single-pass tests, and evidence of its occurrence during RWAT testing was found only later. This phenomenon occurred only in the finest-carbide alloy investigated, in which cracks were generated along the carbide–matrix interface and between carbides, forming detached particles that were much larger than the carbides and that contained both matrix material and several carbides. Subsequent examination [*12*] of the

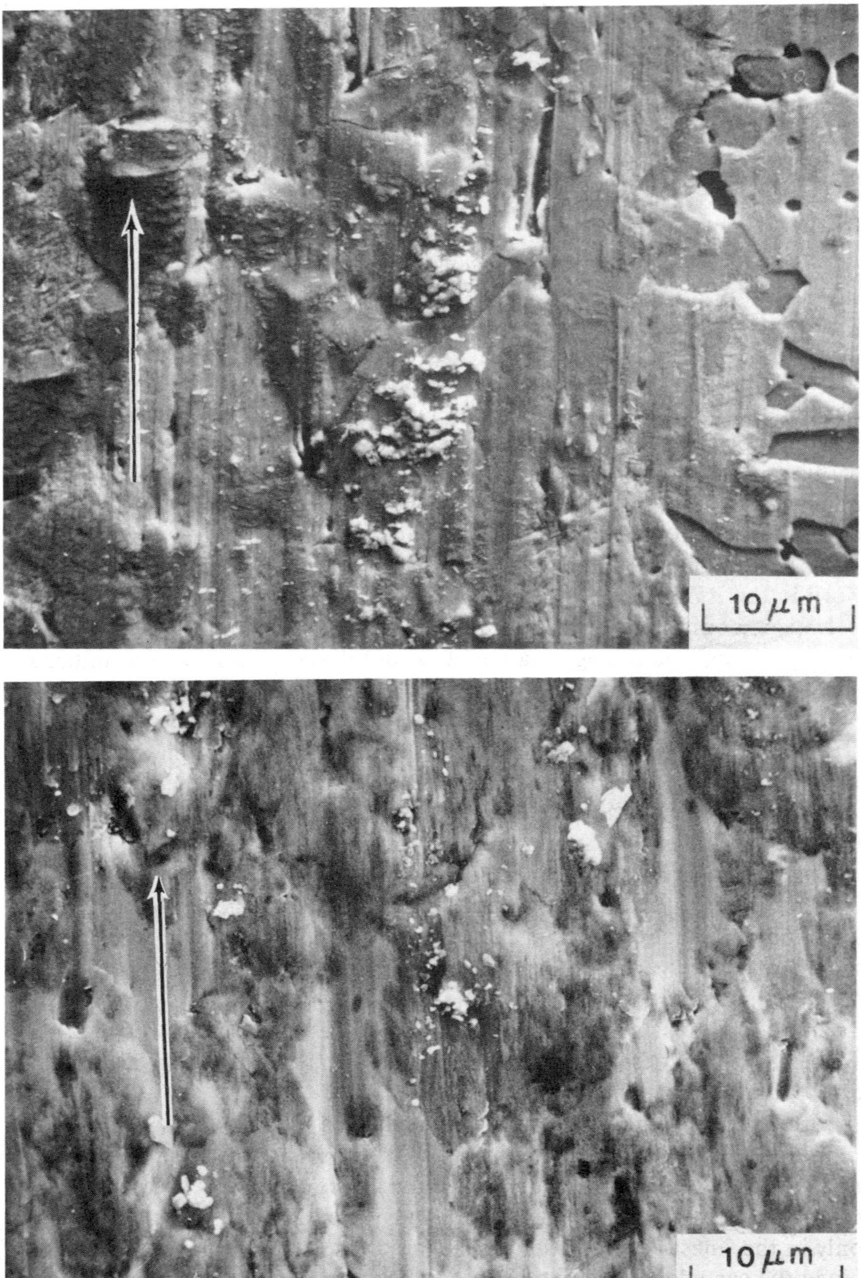

FIG. 2—*Comparison of* (top) *multiple-pass rounded quartz scratch (50 passes, 2-N load) with* (bottom) *quartz RWAT-worn surface. Both are Stellite 6 powder metal (PM) with fine carbides. The scratch in* a *was made outside the SEM* [6].

RWAT surfaces showed evidence of large pits left behind by the removal of such particles; the pits had not been observed previously because they become rounded by subsequent abrasive particles.

As mentioned earlier, the hard carbides cause damage to the quartz scratch tools. This is a problem in performing multiple-pass quartz scratch tests, since the particle tip is gradually flattened during the test. However, as seen in Fig. 2, such tests can still provide a reasonable simulation of the worn surface.

Comparison of the results obtained with diamond and quartz tools showed that a much better simulation of actual abrasion was obtained with quartz particles as tools, since diamond tools can cut through the carbides whereas quartz tools cannot. Diamond scratches also frequently produced large cracks in the carbides, whereas carbide cracking with quartz tools was less frequent and less severe. Multiple-pass diamond scratches produced features quite unlike those produced by abrasion, as may be seen in Fig. 3, which shows extruded lips of material at the edges of the scratch groove and thin sheets of matrix material within it. Thus, as expected, it was concluded that simulations of quartz abrasion were best done using actual irregular quartz particles such as those used in the abrasion tests.

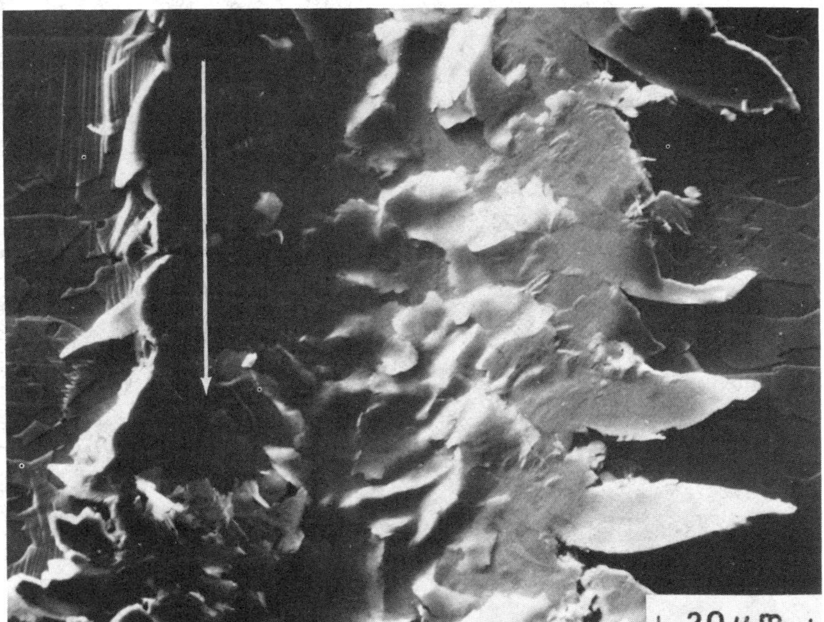

FIG. 3—*Multiple-pass diamond scratch (100 passes, 3-N load) on Stellite 19 PM alloy. The scratch was made outside the SEM* [6].

Since a number of differences had been observed in comparisons of the abrasion behavior of Class II alloys using quartz and Al_2O_3 abrasives [*13–15*], Shetty et al also compared scratches made using Al_2O_3 abrasive particles with those obtained with a Vickers diamond [*7*]. This allowed the observed differences to be attributed primarily to the shape difference between the two types of tools, since both were of sufficient hardness to produce direct carbide cutting. Figure 4 shows a micromachining chip composed of alternating sections of the matrix and M_7C_3 carbide phases. Other evidence of carbide plasticity was also observed. Such plasticity is probably favored by the compressive component of the complicated stress field ahead of the tool, which suppresses crack propagation.

Al_2O_3 particles used as scratch tools were found to be subject to substantial tool wear, which influenced the type of wear debris particles formed. SEM examination of the Al_2O_3 particles used as tools for single-pass scratch tests showed chiplike matrix debris particles and relatively little tool wear, whereas those used for 100-pass tests had a flattened area on the tip about 80 μm wide and showed very fine irregular debris particles and fine micromachining chips in addition to the larger chips formed during the first passes. A corresponding

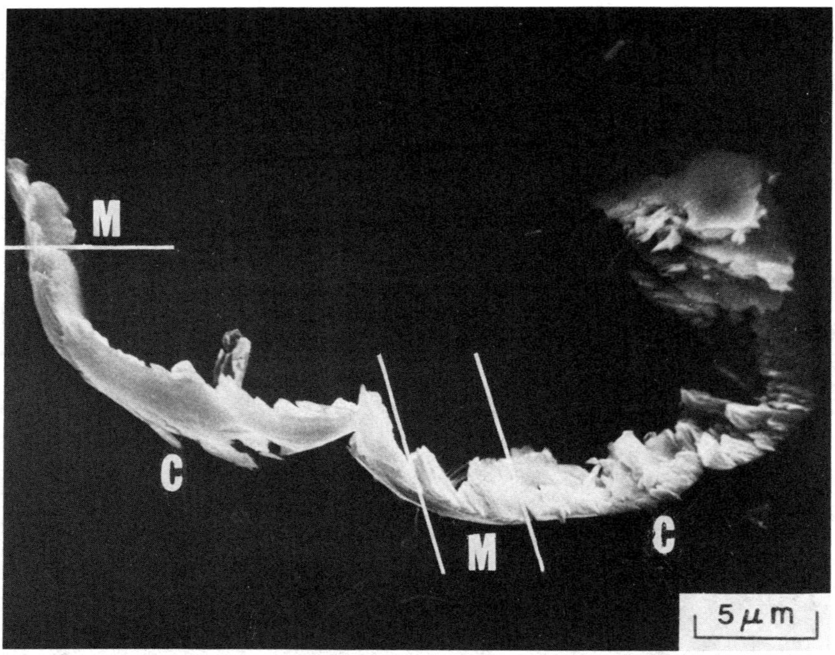

FIG. 4—*Micromachining chip containing sections of both carbide (C) and matrix (M) phases; the white lines separate the carbide and matrix areas. The chip was formed by making a scratch on Stellite 19 with an Al_2O_3 abrasive particle, outside the SEM* [7].

increase of the widths of the scratch paths on fresh areas of the specimen was observed, with considerable reduction in the depth of the scratches. The flattening of the Al_20_3 particle tip during multiple-pass scratch testing must be greater than that experienced by Al_20_3 particles during RWAT testing, since the loads and distances traveled during the scratch test are both greater. However, because wear debris extracted from spent Al_20_3 abrasive used in RWAT tests of the same alloys contained micromachining chips and fine irregular debris particles similar to those found on the Al_20_3 particle tips, the differences introduced by tool wear do not appear to be too great to invalidate the use of scratch tests. It is nonetheless apparent that tool wear should be kept to a minimum by minimizing the number of passes made with the same particle.

Al_20_3 single-pass scratches on preworn specimens produced little damage and were difficult to locate. This is probably because the carbides protrude slightly from the surface after Al_20_3 RWAT abrasion, so that the Al_20_3 particle comes in contact almost entirely with carbides, which protect the matrix.

A third, still unpublished study by Shetty et al [*8*] compared the material removal mechanisms found using individual crushed quartz particles with those resulting from Al_20_3 particles. Since both crushed abrasives have similar angular shapes, the differences could be attributed to the different relative hardnesses of the particles. The earlier work [*6*] used rounded quartz particles.

The crushed quartz particles produced micromachining chips from the matrix phase in single-pass scratches and in the first passes of multiple-pass scratches. The lack of chip formation in later stages was due to tool wear, as in the case of Al_20_3, but the amount of tool wear in the case of crushed quartz was more severe, resulting in flattened areas wider than 200 μm on the particle tips after 100-pass scratches. In addition, quartz fragments were often observed on the scratched specimen in or near the scratch, and observations of the used quartz particles showed extruded tongues of quartz, which had the appearance of machining chips extending from the edge of the flattened area. This extensive damage results from the fact that the carbides are harder than the quartz tool and therefore do more damage to the tool than the tool does to the carbides. The lack of damage to the carbides was also apparent in the development of carbide protrusion from the scratch groove during single and multiple-pass scratches.

The single-pass scratch tests with crushed quartz on surfaces preworn with rounded quartz abrasive (resulting in carbide protrusion) produced no observable damage, except for the trail of quartz debris particles left behind. An example is included in Fig. 5. Since the same area was not observed before and after scratch testing, it was difficult to be certain whether the damage done to the carbides was too small to observe or actually nonexistent. This question is important in the practical problem of abrasion of Class II alloys by quartz or other naturally occurring, relatively soft minerals. It was suggested

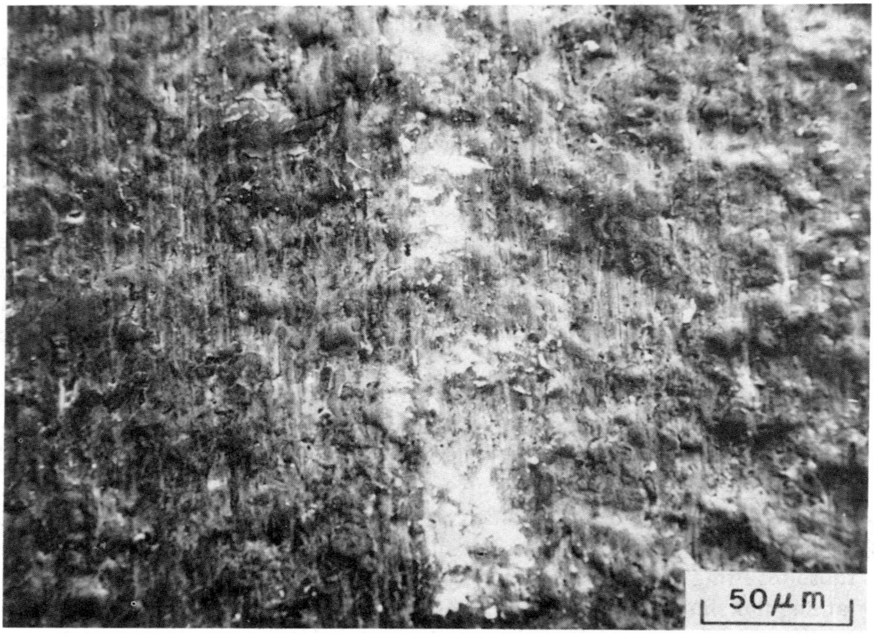

FIG. 5—*Single-pass crushed quartz scratch on a specimen of Stellite 6 pm which had been preworn with quartz abrasive in the RWAT. The white particles were identified as quartz fragments using energy-dispersive X-ray spectroscopy. The scratch was made outside the SEM* [8].

[*8*] that *in situ* SEM scratch experiments could resolve this question by successive scratch tests.

In Situ *SEM Scratch Tests*

Bates et al [*16*] and Bates [*17*] designed and used an *in situ* scratch test apparatus to perform single-pass scratch tests in the SEM, using hemispherically tipped diamond tools on quenched and tempered 1075 steel and a white cast iron containing large chromium-rich M_7C_3 carbides. Their observations led them to propose that there are three important ranges of load in which spherical tools harder than the target material interact differently with the specimen. In Range I (very low loads) they observed plastic deformation but no visible wear debris. In Range II the higher loads resulted in greater penetration and in flow of material toward the outer edges of the groove, and at higher loads (Range III) a semicontinuous chip made up of platelets (lamellae) was formed ahead of the tool. The authors suggested that in Range III a shear crack is initiated ahead of the tool and approximately parallel to the tool face and that this travels into the material. As the resulting platelet flows upward along the rake face it is compressed against the next platelet, resulting in bonding and the formation of a chip consisting of many lamellae. The sequential formation of platelets was directly observed and recorded on videotape.

Prasad and Kosel [*9*] used *in situ* SEM scratch tests with individual abrasive particles to study the material removal mechanisms in a Class II material with rounded quartz abrasive particles. The apparatus used is illustrated schematically in Fig. 6, and was designed [*18*] with a load arm which was instrumented with strain gages to measure the normal and frictional forces, after Bates [*16,17*]. The load arm was fixed inside the SEM and the load was applied by moving the specimen upward until contact was made and the desired load was reached (as determined by the strain gages). The scratch was then made by translating the specimen in relation to the scratch tool. A hypereutectic white cast iron was used for the studies because Fulcher et al [*15*] had found that the very large (200-μm) chromium-rich M_7C_3 primary carbides developed large pits due to apparent gross fracture during RWAT abrasion with rounded quartz, resulting in a loss of abrasion resistance. The primary question addressed by Prasad and Kosel [*9,10*] was that of the operative mechanisms of carbide removal during quartz abrasion of a Class II alloy.

Since the work of Shetty et al [*8*] had not resolved the question of whether or not direct removal of carbide occurred during a quartz scratch on a preworn surface having protruding carbides, a number of sequential passes were made [*9*] over a large primary carbide in a specimen preworn in the RWAT with quartz abrasive. The results showed no evidence of detectable material removal from the carbide. After 25 passes at a load of 5 N, debris particles were observed on the carbide, but they were found to be quartz fragments.

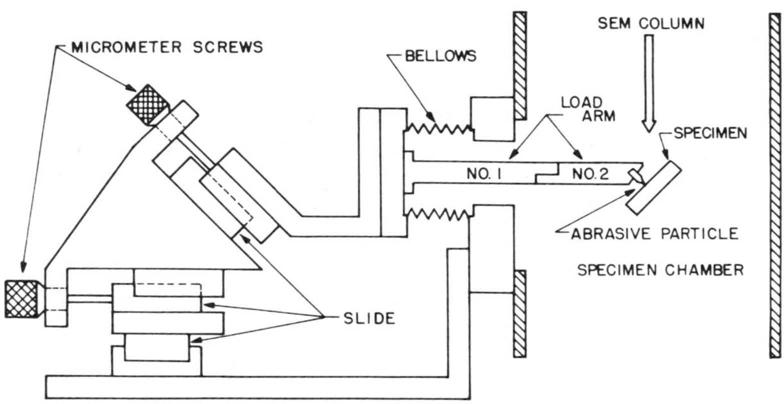

FIG. 6—*Schematic drawing of the* in situ *scratch test apparatus* [18]. *The removable load arm inside the column vacuum is separated from external positioners by bellows. The SEM stage motions are used for application of load and for translation to produce a scratch.*

Removal of these fragments revealed that there was no damage to the underlying carbide, and no carbide debris was found on the quartz particle. Thus, it was conclusively shown that no significant damage is done to previously uncracked carbides by rounded quartz particles, even at scratch loads which are much higher than the 0.08-N load per particle [*19*] in the rubber wheel abrasion test.

Large primary carbides in the white cast irons often contain matrix-filled cavities. It was shown [*9*] that a short period of RWAT abrasion resulted in the removal of the matrix material, resulting in the formation of a small pit. During sequential *in situ* scratch tests of carbides containing such pits, the pits were found to be enlarged by successive microfracture events at their edges. This was concluded to be the operative mechanism for the formation of the large pits observed by Fulcher et al [*15*] in primary carbides. This observation was greatly facilitated by the ease of sequential observations of the same carbides with the *in situ* technique.

In situ scratches made on specimens which were previously deeply etched to remove the matrix and leave the carbides artificially protruding showed that microchipping of the initially sharp carbide edges occurred, suggesting that removal of the matrix between the carbides would initiate the carbide rounding action. The microgrooves which are observed in the leading edges of carbides cannot be caused directly by the soft quartz abrasive particles and are apparently caused by small carbide fragments generated by fracture events in other carbides. The thin eutectic carbides are often cracked by the abrasive particles, providing a source of carbide fragments which leads to microgrooving.

In a further study of carbide removal from the white cast irons during quartz abrasion [*10*], it was shown that some of the initially rounded Ottawa quartz abrasive particles are fractured during the RWAT test and that these particles are responsible for the initiation of most of the pits in the large primary carbides in the hypereutectic alloy. This is because fracture results in a decrease in the radius of curvature of the abrasive particle tip and therefore creates a substantial increase in the Hertzian elastic contact stress induced in the carbides. Application of elastic Hertzian analysis is justified by the demonstration that the carbide does not deform plastically. Gross cracking of the large carbides was demonstrated to be readily produced by *in situ* scratch tests using fractured quartz particles, whereas it could not be induced using fresh rounded particles.

The *in situ* scratch test work described here has demonstrated that the carbides in Class II alloys are removed only by mechanisms originating in fracture events and that, since the carbide removal rate is believed to control the overall material removal rate, the abrasion resistance of these practical abrasion-resistant materials with quartz abrasives is probably controlled by the fracture resistance of the carbides. The *in situ* scratch test technique was invaluable in obtaining the necessary information on material removal mechanisms.

Fixed-Depth Scratch Tests

Prasad and Kosel [*20*] have recently developed the fixed-depth scratch test apparatus shown schematically in Fig. 7. The scratch tool, an irregular abrasive particle, is mounted at the end of the central arm, while the majority of the applied load is supported by the two outer load arms. The specimen is mounted on a special stage which accurately maintains the specimen surface parallel to the horizontal direction of travel of the specimen stage, which is translated manually to make the scratch. The apparatus is restricted to tests made outside the SEM. It simulates fixed-depth abrasion conditions as described earlier, because when the tool comes into contact with a large carbide particle in a Class II material it tries to deflect upward over it. This raises the rest of the loading head also, so that the normal load on the tool increases sharply. For example, if the ratio of carbide to matrix hardness is 10, simple indentations made with the same depths would differ in load by the same factor, and the same ratio of normal loads in carbide and matrix phases would be expected in the fixed-depth scratch test, to a first approximation.

In a study of abrasion of Class II alloys, Kosel et al [*13*] used the same Al_2O_3 abrasive particles in both the RWAT (a fixed-load test) and a test utilizing a low-speed grinding wheel that had loosely bonded abrasive so that fresh abrasive particles were always encountered by the specimen. They found that, in

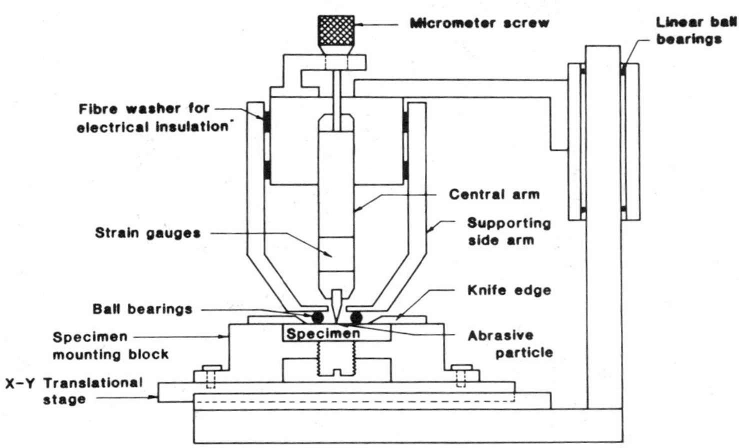

FIG. 7—*Schematic drawing of the fixed-depth scratch test apparatus developed by Prasad and Kosel* [20]. *See the text for details.*

the Stellite alloys studied, the fixed-load RWAT procedure did not produce appreciable carbide fracture but that in the fixed-depth procedure gross fracture and pitting of the carbides frequently occurred. Since the difference did not appear to be attributable to different loads per abrasive particle, it was suggested that the gross fracture of the carbides was due to the fixed-load test conditions, which can impose very large loads on a carbide.

Prasad and Kosel [*20*] performed both *in situ* and fixed-depth scratch tests on the same specimen using the same Al_2O_3 abrasive particle as a scratch tool, with loading conditions which produced similar scratch depths. The *in situ* test produces essentially fixed-load conditions, since the load is provided by the elastic bending of the load arm, and the increase in load due to the tool rising a few micrometres or less to pass over a carbide is negligible. The results show that extensive carbide fracture occurs under fixed-depth scratch conditions but not under the fixed-load conditions in the *in situ* test. An example of the carbide cracking induced by the fixed-depth tests is included in Fig. 8. Comparisons of fixed-depth scratches such as that in Fig. 8 with the *in situ* fixed-load scratches show that carbide cracking is evident only in the fixed-depth test unless rather high loads are used in the fixed-load test. The fixed-depth scratch test results therefore support the suggested explanation

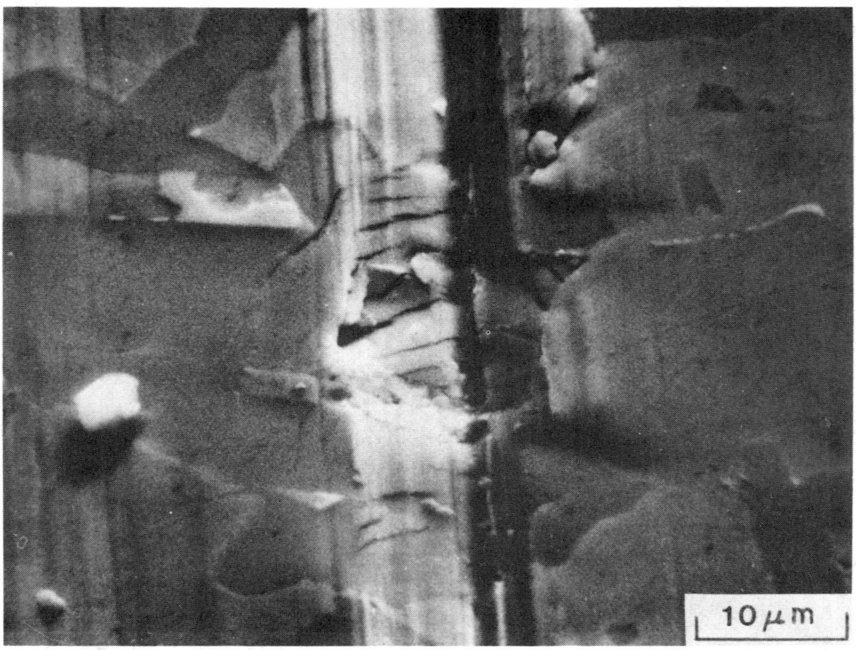

FIG. 8—*Fixed-depth scratch made on Stellite 19 pm. Note the extensive cracking of the carbide* [20].

of Kosel et al [*13*] for their observations of gross carbide cracking with a fixed-depth abrasion test.

The present apparatus also includes a strain-gage section on the central load arm, which allows measurement of the change in the tangential force on the tool when a carbide is encountered, and use of this has shown that load increases occur at intervals approximately corresponding to the carbide spacing. In order to measure the change in the normal force experienced by the tool when it encounters a carbide, it is planned to insert a piezoelectric force transducer into the load arm.

Conclusion

Studies of the material removal mechanisms during abrasion of Class I and Class II materials have been reviewed. Studies of Class I materials using single-pass scratch tests on metallographically prepared surfaces have produced interesting evidence of the existence of a critical rake angle, α_c, which divides the cutting and ploughing ranges.

The use of actual abrasive particles, of multiple-pass scratch tests, and especially of scratch tests on preworn surfaces provides a much more successful simulation of the material removal mechanisms in Class II alloys than the use of regularly shaped tools and metallographic surfaces.

In situ scratch tests in the SEM provide the opportunity to make repeated observations of the same point on the surface after successive scratches, thus making detailed observations of the development of surface damage much more efficient.

Scratch tests on Class II alloys with quartz particles have shown that direct cutting damage to carbides does not occur and that carbide fracture is therefore the controlling factor in quartz abrasion of Class II alloys.

The essential difference between fixed-load and fixed-depth abrasion conditions for Class II alloys has been demonstrated by the use of fixed-depth scratch tests with a specially designed new apparatus.

Acknowledgments

This work was supported by the Office of Basic Energy Sciences of the U.S. Department of Energy under Contract No. DE/DE/AC02-81ER10963. The author is grateful to the program monitor, Dr. Stanley M. Wolf, for this support.

References

[*1*] Sedriks, A. J. and Mulhearn, T. O., "The Effect of Work-Hardening on the Mechanics of Cutting in Simulated Abrasive Processes," *Wear*, Vol. 7, 1964, pp. 451–459.

[*2*] Murray, M. J., Mutton, P. J., and Watson, J. D., "Abrasive Wear Mechanisms in Steels,"

Proceedings of the International Conference on Wear of Materials, Dearborn, MI, 16–18 April 1979, American Society of Mechanical Engineers, New York, 1979, pp. 257–265.

[*3*] Aghan, R. L. and Samuels, L. E., "Mechanisms of Abrasive Polishing," *Wear*, Vol. 16, 1970, pp. 293–301.

[*4*] Mulhearn, T. O. and Samuels, L. E., "The Abrasion of Metals: A Model of the Process," *Wear*, Vol. 5, 1962, pp. 478–498.

[*5*] Sedriks, A. J. and Mulhearn, T. O., "Mechanics of Cutting and Rubbing in Simulated Abrasive Processes," *Wear*, Vol. 6, 1963, pp. 457–466.

[*6*] Shetty, H. R., Kosel, T. H., and Fiore, N. F., "A Study of Abrasive Wear Mechanisms in Cobalt-Base Alloys," *Wear*, Vol. 84, 1983, pp. 327–343.

[*7*] Shetty, H. R., Kosel, T. H., and Fiore, N. F., "A Study of Abrasive Wear Mechanisms Using Diamond and Alumina Scratch Tests," *Wear*, Vol. 80, 1982, pp. 347–376.

[*8*] Shetty, H. R., Kosel, T. H., and Fiore, N. F., "A Scratch Test Study of the Effects of Abrasive Hardness and Particle Shape," report to U.S. Department of Energy, Washington, DC, 16 March 1983.

[*9*] Prasad, S. V. and Kosel, T. H., "A Study of Carbide Removal Mechanisms During Quartz Abrasion, I: *In Situ* Scratch Test Studies," *Wear*, Vol. 92, 1983, pp. 253–268.

[*10*] Prasad, S. V. and Kosel, T. H., "A Study of Carbide Removal Mechanisms During Quartz Abrasion, II: Effect of Abrasive Particle Shape," *Wear*, Vol. 95, 1984, pp. 87–102.

[*11*] Rao, C. M., Kosel, T. H., and Fiore, N. F., "On a Deformation-Induced Phase Transformation During Abrasive Wear of Co-Base Alloys," Appendix 3, report to U.S. Department of Energy, Washington, DC, Contract No. DE-AC02-77ET 10460 Mod. A004, 24 February 1982.

[*12*] Desai, V. M., Rao, C. M., Kosel, T. H., and Fiore, N. F., "Effect of Carbide Size on the Abrasion of Co-Base Alloys," *Wear*, Vol. 94, 1984, pp. 89–101.

[*13*] Kosel, T. H., Fiore, N. F., Coyle, J. P., Udvardy, S. P., and Konkel, W. A., "Microstructure and Wear in High Alloys" in *Proceedings of the Symposium on Corrosion-Erosion Behavior of Materials*, K. Natesan, Ed., The Metallurgical Society of the American Institute of Mining, Metallurgical and Petroleum Engineers, Warrendale, PA, 1980, pp. 190–208.

[*14*] Fiore, N. F., Coyle, J. P., Udvardy, S. P., Kosel, T. H., and Konkel, W. A., "Abrasive Wear-Microstructure Interactions in a Ni-Cr White Iron, " *Wear*, Vol. 62, 1980, pp. 387–404.

[*15*] Fulcher, J. K., Kosel, T. H., and Fiore, N. F., "The Effect of Carbide Volume Fraction on the Low-Stress Abrasion Resistance of High Cr-Mo White Cast Irons," *Wear*, Vol. 84, 1983, pp. 313–325.

[*16*] Bates, T. R., Ludema, K. C., and Brainard, W. C., "A Rheological Mechanism of Penetrative Wear," *Wear*, Vol. 30, 1974, pp. 365–375.

[*17*] Bates, T. R., "The Influence of Eutectic Carbide Spacing on the Wear of White Iron," Ph.D. dissertation, University of Michigan, Ann Arbor, 1975.

[*18*] Prasad, S. V. and Kosel, T. H., "The Design and Some of Applications of an *In Situ* SEM Scratch Tester," *Journal of Materials Science*, Vol. 3, 1984, pp. 133–136.

[*19*] Avery, H. S., "An Analysis of the Rubber Wheel Abrasion Test," *Proceedings of the International Conference on Wear of Materials*, San Francisco, CA, 30 March–1 April 1981, American Society of Mechanical Engineers, New York, 1981, pp. 367–378.

[*20*] Prasad, S. V. and Kosel, T. H., "A Comparison of Carbide Removal Mechanisms During Fixed-Depth and Fixed-Load Scratch Tests," *Proceedings of the International Conference on Wear of Materials*, Vancouver, BC, 14–18 April 1985, American Society of Mechanical Engineers, New York, pp. 59–66.

T. Robert Shives[1] *and Leonard C. Smith*[1]

Microindentation Hardness Measurements on Metal Powder Particles

REFERENCE: Shives, T. R. and Smith, L. C., **"Microindentation Hardness Measurements on Metal Powder Particles,"** *Microindentation Techniques in Materials Science and Engineering, ASTM STP 889*, P. J. Blau and B. R. Lawn, Eds., American Society for Testing and Materials, Philadelphia, 1986, pp. 243–256.

ABSTRACT: The National Bureau of Standards has undertaken a rather extensive metal processing program whereby metal powders will be produced by the inert gas atomization process. Powder particle sizes are expected to range from about 10 to 100 μm in diameter. Part of the overall program is the determination of the mechanical properties of the individual powder particles in the as-produced condition. Conventional microindentation hardness is one of the techniques that is planned to be employed for the characterization of mechanical properties. In order to determine the applicability of this technique to very small particles, commercially available and experimental powders ranging in size from 3 to 165 μm have been mounted in various metallographic mounting media as well as in an electrodeposited metal mounting medium in order to evaluate the effects of the properties of the mounting materials on the measured hardness values. The effects of different applied loads and variations in particle size are also evaluated.

KEY WORDS: metal alloy powder, metal powder, microindentation hardness, rapid solidification, Vickers microindentation hardness, microindentation hardness testing

The properties of an alloy depend on its composition and microstructure which, in turn, depend on its processing history. Thus, the durability and performance of alloy components depend strongly on kinetic changes such as solidification processes, transformation, and diffusion which the alloys undergo during preparation and use. The National Bureau of Standards (NBS) has undertaken a rather extensive metal processing program that includes such rapid solidification techniques as dynamic atomization and inert gas atomization. Alloy powders produced by these techniques exhibit unusually

[1] Metallurgist and physical science technician, respectively, Fracture and Deformation Division, National Bureau of Standards, Gaithersburg, MD 20899.

good homogeneity and can also yield amorphous alloy structures (metallic glasses). It is anticipated that the particle size of the powders to be produced will range between 10 and 100 μm.

Part of the overall NBS program is an evaluation of the mechanical properties of the individual powder particles in the as-produced condition. Microindentation hardness is one of the tools planned to be used for this mechanical properties evaluation.

The purpose of this study is to demonstrate the feasibility of conventional microindentation hardness techniques for determining the hardness of individual metal powder particles produced by the rapid solidification processes.

Hardness Testers

Probably the two microindentation hardness techniques most widely used in the United States are Knoop and Vickers. For this work, Vickers was chosen in preference to Knoop primarily because of the indenter (and resulting indentation) geometry. The symmetrical configuration of the Vickers indenter makes it more suitable for use on small particles. The greater ratio of depth to diagonal length of the Vickers indentation also makes it more suitable for use at very low loads because elastic recovery is not likely to affect the hardness measurement as significantly as it would the Knoop indenter geometry. Vickers hardness numbers are, therefore, more load independent than Knoop hardness numbers [*1–3*].

Initially, Vickers hardness measurements were made on a certified test block using two different commercially available microindentation hardness testers. One of these testers employs a dash-pot system to apply the load whereas the other employs an electrically operated incremental loading system.

The hardness calibration measurements of the test block ranged from 715 to 743 HV certified for an applied load of 100 gf. Using the electrically operated tester, microindentation hardness measurements were made on the test block over a load range extending from 10 to 1000 gf. (Ten grams is the lightest load that can be used on this tester.) There was good agreement between the measurements and the calibrated values for loads of 100 gf and higher, but at lower loads, the measured values deviated significantly from the calibrated values. It was found that, at low loads, the electrically operated incremental loading tester applied an effective load greater than the nominal load. For instance, a 10-gf nominal load resulted in about a 12-gf effective load, an increase of 20%. For the tester with the dash-pot loading system, there was good agreement between the measured values and the calibrated value over a range of 2 to 1000 gf. With few exceptions, the data reported here were obtained from the tester with the dash-pot loading system.

Most of the microindentations were measured at a magnification of $\times$625 with a filar measuring eyepiece that is affixed to the hardness tester. The inaccuracies associated with the measurement of the indentation are essentially constant for a given operator [*4*], regardless of the size of the indentation. Therefore, the percentage error in indentation measurements increases as the size of the indentation decreases. Any frictional or vibrational effects inherent in the tester are accentuated at very light loads [*5*]. For improved accuracy, it is advantageous to use the largest load appropriate for the hardness of the powder particle and the particle size. It should be noted that no attempt was made to adhere to the 3:1 ratio of particle diameter to impression diagonal, as called for in the ASTM Test for Vickers Hardness of Metallic Materials (E 92-82); indeed, some of the impression diagonals are considerably larger than one third of the particle diameter.

Test Materials

Microindentation hardness measurements were made on powder particles of several different alloys. These included commercially produced nickel alloy spheres and two stainless steel powders and experimentally produced powders of aluminum alloys.

The nickel alloy spheres were obtained in two nominal size ranges: 3 to 17 μm and 60 to 75 μm. Actually, a number of the particles in the smaller size range measured more than 20 μm in diameter. Energy-dispersive X-ray analysis (EDAX) semiquantitative chemical analysis on specimens from both size groups indicated that both were similar in composition. By weight percent, they contained about 97.7% nickel, 0.7% aluminum, 1.3% silicon, and 0.3% iron. Scanning electron microscope (SEM) micrographs of the 3 to 17-μm and the 60 to 75-μm nickel spheres are shown in Figs. 1 and 2, respectively. The larger particles appeared to be much more uniform in size than the smaller ones. Some of the larger particles also exhibited varying amounts of porosity near their centers.

One of the stainless steels was austenitic (Type 310) and one was ferritic (Type 430). Both were produced by gas atomization. The particles of both materials tended to be oblong and were similar in shape and size distribution. An example of a 430 powder particle is shown in Fig. 3. The short dimension of the particles falls generally into the 35 to 75-μm range.

Hardness measurements were made on three aluminum alloys. The particles of all three alloys were generally spherical. One of the alloys, Al-4.4Cu-1.5Mg-1Fe-1Ni-0.2Zr (VA5), was produced by vacuum atomization. The particle size ranged from about 55 to 160 μm in diameter. Varying amounts of porosity were evident in these particles. The second aluminum alloy, Al-4Cu-1Mg-1.5Fe-0.75Ce (No. 3), was produced by gas atomization in flue gas. The particle size ranged from about 40 to 85 μm. The powder particles for the third alloy, Al-3Li-1.5Cu-1Mg-0.5Co-0.2Zr (USGA No. 4), were produced by ultrasonic gas atomization in helium. The particle size ranged from about 20 to 165 μm for this alloy.

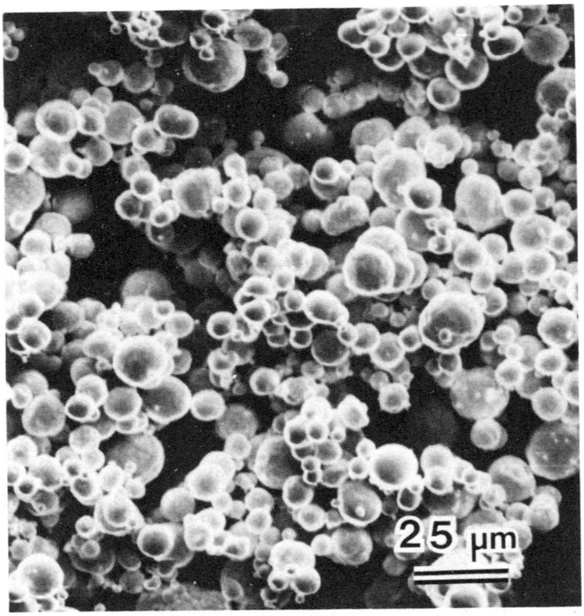

FIG. 1—*Scanning electron photomicrograph of nickel alloy spheres nominally 3 to 17 µm in diameter.*

Specimen Preparation

In all but one case, the metal powder particles on which the microindentation hardness measurements were made were embedded in a conventional metallurgical specimen mounting material which was then polished to produce a metallographic finish suitable for microindentation hardness measurements. Several commercially available mounting materials having different physical and mechanical properties were used to determine what effect the mounting material might have on measured hardness values and whether the mounting medium would adequately support the small metal powder particles under load during microindentation hardness testing. The mounting materials used included a thermoplastic resin [polymethyl methacrylate (PMMA)], several thermosetting resins [wood flour with a phenolic resin binder, short glass-fiber-filled diallylphthalate, copper-powder-filled epoxy, silicon carbide (SiC)-fiber-filled diallylphthalate], and an epoxy. The one exception to the use of conventional mounting materials was a mount prepared by trapping nickel spheres in electrodeposited copper.

Except for the electrodeposited copper mount, the specimens were prepared by mixing a small quantity of metal powder with a small amount of mounting material, placing this mixture in a mold, and then adding more

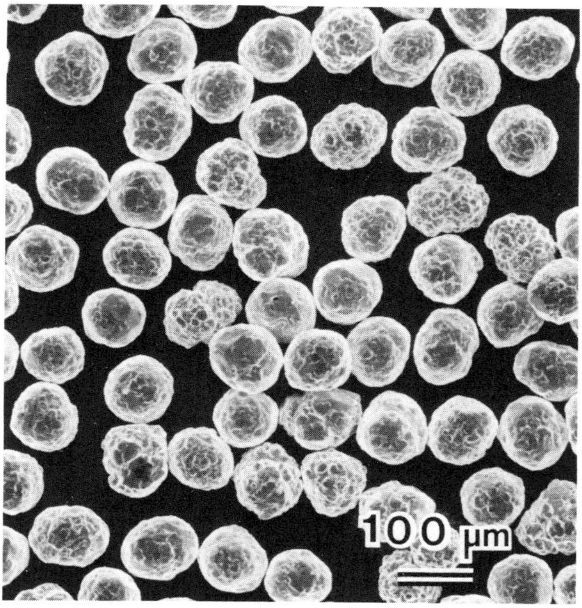

FIG. 2—*Scanning electron photomicrograph of nickel alloy spheres nominally 60 to 75 µm in diameter.*

mounting material on top of this mixture followed by the prescribed curing procedure for the material being used. This technique resulted in most of the powder particles being near the surface of the cured mount. The cured mounts were then polished to obtain a metallographic surface. The final polishing medium was 0.05-µm alumina.

The electrodeposited copper mount was prepared by placing 1 g of spherical nickel powder particles in 200 mL of copper pyrophosphate solution. The temperature of the plating bath was maintained at 50°C and the current density was 2.4 A/dm^2. Copper from this solution was electrodeposited onto a brass strip anode. The current passing through the continuously stirred plating bath induced the nickel spheres to the brass strip, which resulted in electrodeposited copper containing the nickel powder particles. A metallographic surface was obtained on this electrodeposited mount using magnesium oxide mixed with small amounts of ammonium hydroxide and 3% hydrogen peroxide as the final polishing medium. In addition to producing the desired surface finish, this procedure also chemically etched the material. The powder particles were thus optically "ringed," so that they could be easily distinguished from the copper matrix.

When the mounts are polished to prepare them for the hardness measurements, the surface layer is, of course, removed. The plane of the final pol-

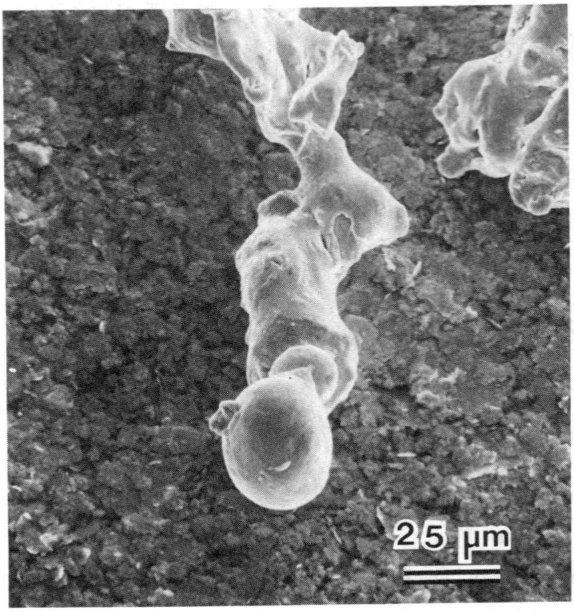

FIG. 3—*Scanning electron photomicrograph of Type 430 stainless steel powder particle.*

ished surface "slices" through various powder particles at different depths below the particle surface. When spherical particles are prepared in a given mount, for example, the apparent diameter of a particle as it appears in the mount is actually the diameter of the particle at the polished plane of the mount and not necessarily the diameter of the original particle. Therefore, it is possible that what appears to be a small-diameter particle is in reality a plane through a much larger particle well away from its center. If more than half of the particle has been removed during the polishing process, the remaining portion could be quite shallow and thereby inappropriate for microindentation hardness measurements. Care must therefore be exercised when taking measurements on the apparently smaller particles.

Results

Evaluation of Mounting Materials

Microindentation hardness measurements at loads ranging from 10 to 500 gf were made on each of the mounting materials previously described. Many of the mounting materials used are rather inhomogeneous. Some, such as the fiber-filled materials, are composites. For these composites, locations for light load measurements were selected to avoid the filler material, insofar as is possible. There was little variation in measured hardness for any given load in any given material, and the differences among the measured values at different loads were insignificant for any given material. The average microindentation hardness values at a load of 500 gf for the various mounting materials are given in Table 1.

Metallographic specimens containing spherical particles of commercially produced nickel alloy powder specified to range in size from 3 to 17 μm were prepared using each of the mounting materials discussed earlier. Microindentation hardness impressions were made on the particles in each mount using loads from 5 to 25 gf to determine how well the mounting materials would perform. Even for the lighter loads, it was found that the softer mounting materials such as PMMA and epoxy would not adequately support even the larger of the small nickel powder particles under load. At higher loads, of course, the lack of support was accentuated. An extreme example is shown in Fig. 4, where a reasonably well-defined microindentation in a nickel particle is overshadowed by a large indentation in the PMMA mounting material. This microindentation was made using a 10-gf load. Although the example shown in Fig. 4 is rather dramatic, the effects may be far more subtle and could escape detection. The particle may "sink" under load, but not far enough for the indenter to contact the mounting material. Assuming that the polished surface and back of the mount are parallel, an unsymmetrical indentation such as that shown in Fig. 5 is one indication that the particle has moved under load. The particle may become skewed as it "sinks," resulting in the unsymmetrical indentation.

The nickel alloy powder particles were inadequately supported in the wood flour/phenolic resin, glass-fiber-filled diallylphthalate, and copper-filled epoxy mounts at loads of 25 gf, and some measurements were questionable at a load of 10 gf. The particles appear to have been adequately supported by all three of these mounting materials for indentation loads of 5 gf. Both the SiC-filled diallylphthalate and the electrodeposited copper mount provided adequate support for microindentation hardness measurements to be made at loads up to 25 gf. The amount of support required for the microindentation hardness test depends on the size of the particles and the applied load and, to

TABLE 1—*Vickers microindentation hardness results for mounting materials at an applied load of 500 gf.*

Material	Average Hardness, HV
Polymethyl methacrylate (PMMA)	14
Epoxy	17
Diallylphthalate (glass fiber filled)	38
Phenolic resin (wood flour filled)	41
Epoxy (copper filled)	46
Diallylphthalate (SiC fiber filled)	61
Electrodeposited copper	185

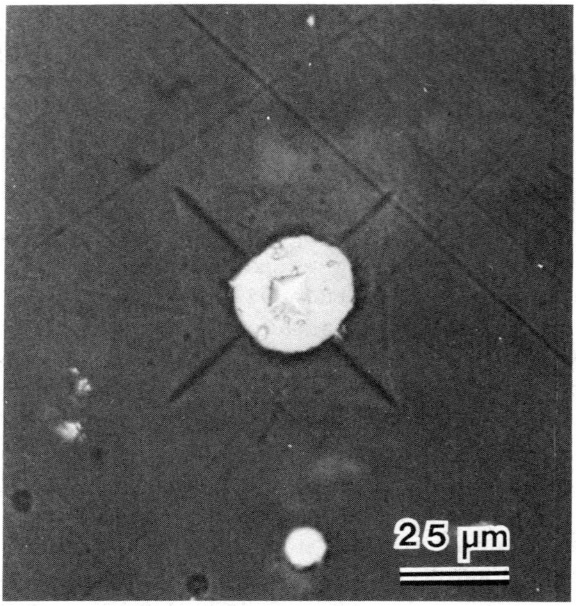

FIG. 4—*Vickers microindentation hardness impression in a nickel alloy spherical particle showing a superimposed microindentation in the PMMA mounting material.*

a lesser extent, on the shape of the particles. The SiC-filled diallylphthalate provided adequate support for the nickel alloy spherical particles that measured 23 μm or greater in the polished plane of the mount for loads up to 25 gf.

Hardness of Nickel Alloy Powder Particles

The results of the microindentation hardness measurements of the smaller spherical nickel alloy powder particles are given in Table 2 for the particles mounted in SiC-filled diallylphthalate and in electrodeposited copper. An example of a well-formed microindentation produced by a 25-gf load on a nickel sphere mounted in the electrodeposited copper is shown in Fig. 6. Even though the distances between the diagonal ends and the edge of the particle are less than one half the diagonal length, the measured hardness value agrees with values for larger particles. The range and average values at loads of 5, 10, and 25 gf are given.

The scatter within the results of measurements for each given load is very small for the particles mounted in the SiC-filled diallylphthalate; in fact, the difference between the high and low value in each case is never greater than the equivalent of one filar unit of the measuring eyepiece. The scatter of the

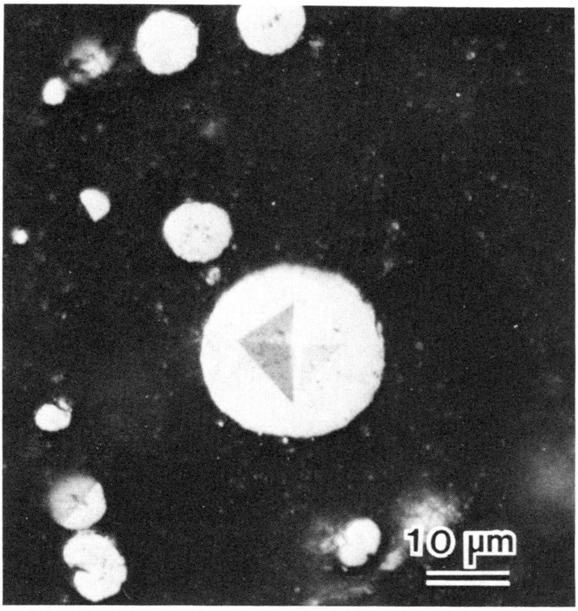

FIG. 5—*Improperly formed Vickers microindentation hardness impression in a nickel alloy spherical particle, indicating that the particle "sank" under the applied load.*

TABLE 2—*Vickers microindentation hardness results for nominal 3 to 17-μm-diameter spherical nickel alloy powder particles.*

Mounting Material	Applied Load, gf	Number of Measurements	Hardness, HV	
			Range	Average
SiC-fiber-filled diallylphthalate	25	3	329 to 340	333
	10	3	327 to 344	338
	5	3	289 to 309	296
Electrodeposited copper	25	5	291 to 388	337
	10	5	258 to 382	331
	5	5	212 to 385	286

results for any given applied load for particles mounted in electrodeposited copper was significantly greater.

The average hardness values for measurements in both mounts at loads of 10 and 25-gf agree quite well. For the SiC-filled diallylphthalate mounted particles, the measurements made with a 5-gf load resulted in hardness values significantly lower than those for loads of 10 and 25 gf. Although the average value for measurements at 5 gf on the electrodeposited copper mounted parti-

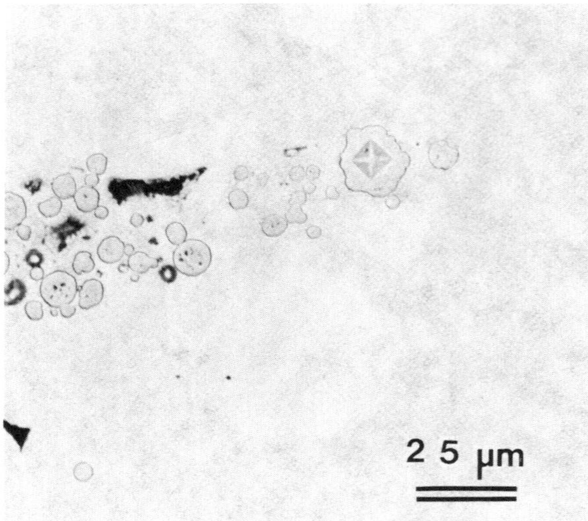

FIG. 6—*Well-formed microindentation hardness impression in a nickel alloy spherical particle mounted in electrodeposited copper.*

cles is lower than those for measurements at 10 and 25 gf, the large scatter precludes any conclusion regarding the significance of the difference.

The process by which this nickel alloy powder was produced is not known, but its hardness is much higher than expected for annealed material and is greater than normally expected for cold-worked material. The hardness is also somewhat greater than that found for a ferritic stainless steel, discussed later.

In the case of the 60 to 75-μm nickel alloy particles, the applied load during microindentation hardness measurements was distributed over a greater area, which permitted the use of mounting materials softer than the SiC-fiber-filled diallylphthalate. In this case, glass-fiber-filled diallylphthalate was found to be a satisfactory mounting material. The hardness measurements were made at loads ranging from 5 to 50 gf. The smallest measured diameter in the plane used for hardness measurement was 45 μm. The applied hardness load was 5 gf for this particle.

The results for the Vickers microindentation hardness measurements on the larger (60 to 75-μm) nickel alloy spheres are given in Table 3. There was a large variation in the hardness results at all loads except for the 25-gf load, and the differences among the average results for the various loads is significant. Some, and perhaps much, of this variability is due to the porosity of the material. It should be noted that there is much less scatter for the results at the higher loads than at the lighter loads. The larger indentation produced by

TABLE 3—*Vickers microindentation hardness results for nominal 60 to 75-μm-diameter spherical nickel alloy powder particles.*

Applied Load, gf	Number of Measurements	Hardness, HV	
		Range	Average
50	6	61 to 83	73
25	3	71 to 80	76
10	5	71 to 110	90
5	6	34 to 141	84

the heavier loads encompasses more material and reduces the effects of inhomogeneities and porosity.

The larger (60 to 75-μm) spheres have much lower hardness values than the smaller (3 to 17-μm) spheres, although they are similar in chemical composition. Microindentation hardness measurements were made on a specimen of annealed 99.4 weight percent nickel alloy. The average Vickers hardness was 96.7, which is greater than the hardness of the larger spheres. Based on this comparison, it would appear that the measured hardness of the nickel spheres is unrealistically low; however, porosity again may have contributed to these low values.

Hardness of Stainless Steel Particles

Vickers microindentation hardness measurements were made on particles of both 310 and 430 stainless steel powders mounted in copper-filled epoxy. The results of the microindentation hardness measurements made at loads of 5, 10, and 25 gf are given in Table 4. Either three or four measurements were made at each load. The variability among the results for each material is great.

TABLE 4—*Vickers microindentation hardness results for stainless steel powders.*

Applied Load, gf	Hardness, HV			
	Stainless Steel, Type 310		Stainless Steel, Type 430	
	Range	Average	Range	Average
25	136 to 154	147	226 to 319	286
10	124 to 186	161	296 to 344	312
5	148 to 180	163	289 to 289	289

TABLE 5—*Vickers microindentation hardness results for aluminum alloy powders.*

Alloy	Applied Load, gf	Number of Measurements	Hardness, HV	
			Range	Average
Al-3Li-1.5Cu-1Mg-0.5Co-0.2Zr (USGA No. 4)	100	3	96 to 103	99
	50	3	95 to 106	99
	25	3	97 to 102	99
	10	3	87 to 107	98
	5	5	93 to 118	105
	2	3	86 to 116	99
Al-4Cu-1Mg-1.5Fe-0.75Ce (No. 3)	50	4	71 to 93	82
	25	3	85 to 94	89
	10	4	77 to 98	87
	5	4	86 to 96	89
	2	4	66 to 102	81
Al-4.4Cu-1.5Mg-1Fe-1Ni-0.2Zr (VA5)	50	5	82 to 124	97
	25	4	76 to 104	94
	10	3	93 to 98	96
	5	4	86 to 96	93
	2	4	73 to 96	83

Hardness of Aluminum Alloy Powder Particles

Particles of the three aluminum alloys were mounted in SiC-fiber-filled diallylphthalate. For two of the alloys, the measurements were made at loads ranging from 2 to 50 gf. For the third alloy, the loads ranged from 2 to 100 gf. The results of the microindentation measurements on these aluminum alloy particles are given in Table 5.

The average values for both Alloys USGA No. 4 and No. 3 are rather consistent, although there is a significant spread for measurements at some loads. There is a larger variation in the average results of measurements on Alloy VA5. Metallographic examination of these particles indicated an apparent second phase in some and varying amounts of porosity near the centers of others. These inhomogeneities probably contributed to the scatter in the results for this material.

Discussion

The principal objective of this work was to evaluate the feasibility of using conventional microindentation hardness techniques for determining the hardness of individual small metal powder particles. It has been demonstrated that the Vickers microindentation hardness technique can be used, but there are the usual problems inherent in light-load, small-indentation measurements as well as problems peculiar to measurements of small particles. Inhomogeneities and porosity in the materials being evaluated can lead to significant variations in apparent hardness values, as demonstrated by the results for the large nickel alloy spheres and one of the aluminum alloys (VA5).

The error in the measurement of the indentation diagonal lengths increases as the size of the indentation decreases. The accuracy of these measurements could be improved somewhat by using a higher-magnification calibrated measuring microscope or by making the diagonal measurements in a scanning electron microscope.

Specimen preparation is critical for hardness measurements on very small particles. A mounting material must be selected that will adequately support the particle under load during hardness testing. For very small particles, the applied load during hardness measurements is distributed over a relatively small volume of material resulting in rather high stresses. For example, simplistically, a 20-μm-diameter particle with a 25-gf load would result in 779 MPa (113 ksi) on the mounting medium, based on the cross-sectional area of the particle. It was found, however, that for nickel alloy spheres with measured hardness numbers of about 300 HV, reliable microindentation hardness measurements could be made for particles as small as 14 to 16 μm in diameter (as they appeared on the polished surface of the mount) with an applied load of 10 gf in an electrodeposited copper mount and in a SiC-fiber-filled diallylphthalate mount. For powder of much lower hardness, either larger particles or lighter loads must be used.

Conclusions

1. It has been demonstrated that it is feasible to use conventional microindentation hardness techniques to determine the hardness of individual powder particles as small as about 14 μm in diameter, depending on the hardness.
2. For microindentation hardness measurements on mounted powder particles, the mounting medium must be hard enough to support the particles under the applied load.
3. Superimposed microindentation hardness impressions on the mounting material and improperly formed (skewed) impressions are indications of inadequate mount support during testing.
4. Inhomogeneities in the powder particles may produce large variability in the microindentation hardness results because of the small volume affected by the light loads.

Acknowledgments

The authors would like to thank Christian E. Johnson of the NBS Metallurgy Division for preparing the electrodeposited copper mount and Dr. Richard J. Fields of the NBS Fracture and Deformation Division for helpful discussions during the course of the work.

References

[1] Enberg, E. H., "A Guide to Selecting the Best Indentation Hardness Test," *American Machinist*, August 1964, p. 10.

[2] Taylor, D., *The Hardness of Metals*, Oxford University Press, London, 1951, p. 99.

[3] Lysaght, V. E., *Indentation Hardness Testing*, Reinhold Publishing, New York, 1949, p. 106.

[4] Ogburn, F., Shives, T. R., and Brady, C. H., "Calibration of a Filar Micrometer," *Journal of Materials*, Vol. 6, No. 1, 1971, p. 60.

[5] Mott, B. W., *Micro-Indentation Hardness Testing*, Butterworths, London, 1956, pp. 50–51.

Olof Vingsbo,[1] *Sture Hogmark,*[1] *Bo Jönsson,*[2] *and Anders Ingemarsson*[1]

Indentation Hardness of Surface-Coated Materials

REFERENCE: Vingsbo, O., Hogmark, S., Jönsson, B., and Ingemarsson, A., **"Indentation Hardness of Surface-Coated Materials,"** *Microindentation Techniques in Materials Science and Engineering, ASTM STP 889*, P. J. Blau and B. R. Lawn, Eds., American Society for Testing and Materials, Philadelphia, 1986, pp. 257–271.

ABSTRACT: A model for the mechanical interaction between a deposited surface film and its substrate during indentation is investigated for a number of material combinations. Linear relationships between the indentation microhardness and the inverse of the indentation diagonal are found for indentations larger than a limiting size, for both coated and uncoated materials. These relationships are used for measuring the hardening effect of a harder coating on a softer substrate and for finding separate hardness values characteristic of the film and the substrate materials.

KEY WORDS: indentation, microhardness, coating, microindentation hardness testing

Nomenclature

H	Hardness, kgf/mm^2
d	Indentation diagonal, mm
D	Indentation depth, mm
t	Film thickness, mm
α	Tip interior angle between opposite indenter facets, degrees
β	$(180° - \alpha)/2$
L	Applied load on indenter, kgf
H_c, H_s, H_f	Hardness of composite, substrate, and film material, respectively, kgf/mm^2
ΔH	$H_c - H_s$
A	Projected area of indentation $= d^2/2$, mm^2
A_s, A_f	Projected load-carrying area of substrate and film, respectively, mm^2

[1] Staff, Department of Technology, Uppsala University, S-751 21 Uppsala, Sweden.
[2] Staff member, Research and Development Department, Kanthal Electroheat AB, S-734 81 Hallstahammar, Sweden.

x Projected rim width of load-carrying film, mm
C Numerical constant defined by Eq 13 ($0.5 < C < 1.0$)
H_0 High load (conventional Vickers) hardness, kgf/mm^2
H_{s0}, H_{f0} High load (conventional Vickers) hardness of substrate and film material, respectively, kgf/mm^2
k Linear slope of $H(1/d)$ (Eq 17), kgf/mm
$k_{c,s,f}$ Linear slope of $H_c(1/d)$, $H_s(1/d)$, and $H_f(1/d)$, respectively, kgf/mm
Δk $k_c - k_s$
d_{lin} Lower limit of d for linear $H(1/d)$ relationship
t_{cr} Upper limit of t for influence on H_c by substrate material at a given indentation size
d_{cr}, D_{cr} Lower limit of d and D, respectively, for influence on H_c by substrate material at a given film thickness

The concept of "mechanical properties" of a material generally refers to well-defined bulk quantities, such as modulus of elasticity, yield stress, and fracture strain. Mechanical surface characteristics, however, are less easily defined and measured.

The resistance of a surface layer to deformation by a hard indenter, pressed into the material, is often described in terms of "hardness." The stress-strain situation during indentation is triaxial, and the relation between microstructure (grain morphology, dislocation distribution, and so on) and deformation mode (twinning, dislocation glide, and so on) is complex. Yet, indentation hardness measurement is a simple, reproducible, and cheap method, which is extensively used for the characterization and ranking of surfaces for mechanical applications.

The conceptual problem just indicated is even more obvious in relation to the combination of a coating on a substrate. When the indentation depth to film thickness ratio, D/t, is sufficiently low, the microhardness value, H, essentially corresponds to that of the material of the coating, but for D/t values exceeding a critical value, the measured H value is influenced by the substrate material as well. The critical D/t value, which generally varies between 0.07 and 0.2 [*1,2*], is not very distinct and, in fact, often varies with both D and t, as well as with the hardness contributions of both coating and substrate materials. Because the purpose of a surface coating often is to modify the mechanical properties of the substrate, it is also of interest to measure the composite hardness for indentation depths that give D/t ratios greater than the critical value in an attempt to quantify the film-substrate interaction.

The present paper considers the Jönsson and Hogmark model [*3*] for the interpretation of a composite microhardness, H_c, as a modification of the substrate hardness, H_s, expressed by

$$H_c = H_s + \Delta H \tag{1}$$

A new way of separating the film hardness contribution from the substrate hardness is suggested; this method generates microhardness values for the coating material characteristic of its function as a thin film.

Experimental Procedure

Materials

The chromium coating and four different substrate materials of Ref *3* were evaluated, together with two combinations of titanium nitride (TiN) on high-speed steel (see Table 1). The coatings were deposited by standard physical vapor-deposition methods [radio-frequency (chromium) and reactive magnetron (TiN) sputtering] to the thicknesses given in the table.

Hardness Measurements

Microhardness measurements were made with a Vickers microhardness tester (Leitz Durimet) at seven load levels ranging from 5 to 500 gf. To ensure that no hardness variations were introduced in the substrate during specimen preparation or film deposition, tapered sections were cut through the interface of the coated specimens, and initial microhardness depth profiles were recorded below the interface. Subsequently, all substrate hardness measurements were made on tapered sections at a given level immediately below the interface.

Some scatter between individual readings is inevitable, and the relative error will increase with decreasing indentation size. This is partly due to real variations in hardness, for instance, between grains of different orientation or composition, but the main contribution is most likely the finite accuracy of the diagonal reading. To ensure reasonably good accuracy, ten impressions were made, and all 20 diagonals were measured for each load level.

For comparison, conventional Vickers hardness numbers (HV) were obtained on the substrate materials with 10 or 100 kgf (see Table 1).

TABLE 1—*Substrate-coating combinations and characteristics.*

Material	Substrate HV, kgf/mm^2			Coating	
	L = 50 gf	L = 10 kgf	L = 100 kgf	Material	Thickness, μm
High-speed steel (HSS)	950	...	865	TiN	0.96, 1.17
Ball-bearing steel	650	...	600	chromium	1.0, 2.0
High-strength low-alloy (HSLA) steel	510	...	465	chromium	1.0
Austenitic stainless steel	190	165	...	chromium	1.0
Copper	90	85	...	chromium	1.0

A Model for the Interpretation of Composite Microhardness

In the Vickers method, which is adopted in the present study, the Vickers hardness number (HV) is defined as the ratio between the load, L, applied to a square-based diamond pyramid pressed into the specimen surface, and the area, A, of the resulting indentation.

The hardness, H, is expressed as a function of the indentation diagonal, d, by

$$H = 2 \cos \beta \cdot \frac{L}{d^2} \tag{2}$$

β is defined by

$$\beta = \frac{180° - \alpha}{2} \tag{3}$$

where α is the interior pyramid tip angle between opposite facets.

The indentation depth is similarly obtained as

$$D(d) = \frac{d}{2\sqrt{2}} \cdot \tan \beta \tag{4}$$

For a Vickers pyramid, the tip angle, α, is 136° and $D \approx d/7$.

Since the geometry of the indentation is independent of its size, the hardness is, in principle, independent of the applied load. In practice, however, there is a load dependence, particularly for small loads. This is generally attributed to the fact that dislocations and grain boundaries occur only at limited local densities in very small deformed volumes [*4*,*5*] and that the tip of any real indenter may not be perfectly formed to comply with geometric assumptions at very small scales.

In the case of a coating-substrate composite, an analytic expression of the individual hardness contributions would require a detailed knowledge of the elastic and plastic behavior of both coating and substrate materials, as well as of the interface characteristics and the friction between the indenter and the coating. In addition, the geometry of the pyramidal indenter implies a complex deformation. Consequently, solutions have been published only for some idealized materials and geometries [*6*,*7*].

Jönsson and Hogmark [*3*] have suggested a simple geometric approach, based on the idea that the indentation hardness represents a mean contact pressure, developed during the indentation process. They state that, for coated materials, the individual pressure contributions from film and substrate, respectively, match the composite hardness, H_c. The contributions are identified as the bulk hardness numbers for coating and substrate material, respectively, and the problem is how to distribute them over the indented area. This is simplified for thin films and high D/t ratios, for which observations in the scanning electron microscope (SEM) of the present coatings have revealed that cracks develop around the rim of the impression (see Fig. 1). This indicates that, during the final indentation process when the plastic deformation has penetrated into the substrate, the coating in the interior only transmits surface pressure to the substrate material, in which essentially all the plastic deformation takes place. Any increased ΔH in the measured hardness of the composite relative to the substrate thus originates from an area of the coating around the indentation rim outside the crack. With the notations of Fig. 2, the total projected area, A, of the indentation is equal to the sum of the projected areas of the rim, A_f, and the interior, A_s, respectively, so that

$$A = A_f + A_s \tag{5}$$

Accordingly, H_c can be expressed as a weighted mean of the film contribution, H_f, and the substrate contribution, H_s, related to the corresponding area fractions, A_f/A and A_s/A, by

$$H_c = \frac{H_f A_f}{A} + \frac{H_s A_s}{A} \tag{6}$$

Combinating Eqs 1 and 6 then gives the hardness difference

$$\Delta H = \frac{(H_f - H_s)A_f}{A} \tag{7}$$

and the composite hardness

$$H_c = H_s + \frac{(H_f - H_s)A_f}{A} \tag{8}$$

If the rim area is described as a band of width x, as shown in Fig. 2 (*bottom*), the A_f/A area fraction can be expressed as a function of x/d by

$$\frac{A_f}{A} = \frac{4\sqrt{2}x}{d} - \frac{8x^2}{d^2} \tag{9}$$

As suggested in Fig. 3, x can also be related to the film thickness, for a possible geometry of the deformed rim volume. If the whole volume, ABC, is strained without cracking [Fig. 3 (*left*)]

$$x = t \cos \beta \sin \beta \tag{10}$$

FIG. 1—*SEM micrographs of Vickers hardness indentations:* (top) *chromium film on copper substrate,* L = *10 gf,* d = *8 μm;* (middle) *chromium film on copper substrate,* L = *50 gf,* d = *27 μm;* (bottom) *TiN film on HSS substrate,* L = *300 gf,* d = *24 μm.*

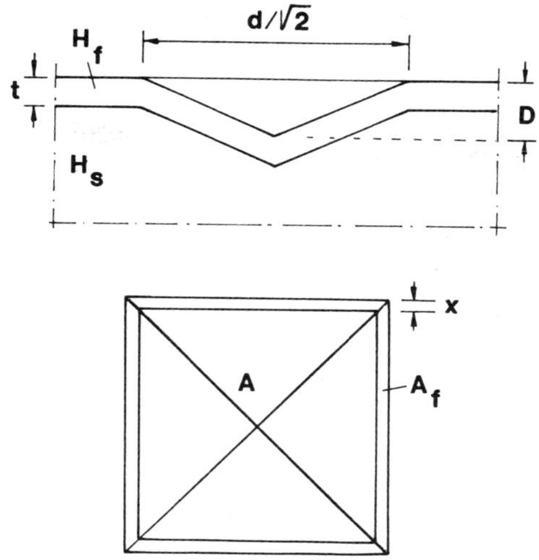

FIG. 2—*Horizontal* (top) *and vertical* (bottom) *projections of the hardness indentation.*

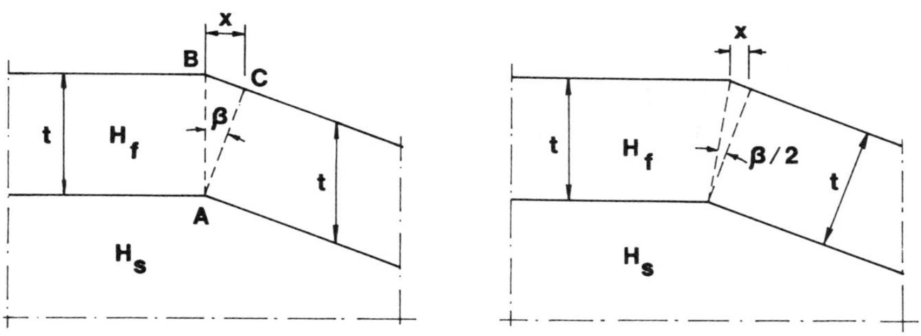

FIG. 3—*Suggested upper limit* (left) *and lower limit* (right) *for the width,* x, *of the load-carrying area of the film material.*

If, however, some of the stress is relaxed by crack formation, the "effective" rim width would be correspondingly reduced. The present SEM studies indicate a rather modest crack width, of the order of less than half of x. In the following estimates, the x interval

$$t \cos \beta \tan \frac{\beta}{2} \leq x \leq t \cos \beta \sin \beta \tag{11}$$

has been assumed. The lower limit of x corresponds to the geometry of Fig. 3 (*right*). (These limits correspond to the two models for rim deformation suggested by Jönsson and Hogmark [*3*].) By substituting Eqs 10 and 11 into 9, it is now possible to express the area fraction, A_f/A, as a function of t/d according to

$$\frac{A_f}{A} = \frac{2Ct}{d} - \frac{C^2t^2}{d^2} \tag{12}$$

where C is defined by

$$2\sqrt{2}\cos\beta\tan\frac{\beta}{2} \leq C \leq 2\sqrt{2}\cos\beta\sin\beta \tag{13}$$

For all practical purposes and with the Vickers pyramid ($\beta = 22°$), the value of C is restricted to

$$0.5 < C < 1.0 \tag{14}$$

In the present investigation, $t \approx 1$ μm, $d \geq 5$ μm, and $t/d \leq 0.2$. Neglecting the second-order term in Eq 12, then, would mean an overestimate of A_f/A. The relative error rapidly decreases toward $Ct/2d$ for decreasing t/d (increasing d) in the size interval of particular interest in the present study and is equal to 0.05 for $d = 10$ μm. Thus, for indentations not too small relative to the film thickness, the hardness difference and composite hardness (compare Eqs 7 and 8) can be expressed as

$$\Delta H = 2(H_f - H_s)\frac{Ct}{d} \tag{15}$$

and

$$H_c = H_s + \Delta H = H_s + 2(H_f - H_s)\frac{Ct}{d} \tag{16}$$

Results and Discussion

In view of the $H(d)$ relation, suggested by Eq 16, Figs. 4 through 6 show experimental plots of H_c and H_s versus $1/d$ for the investigated combinations. It can be seen that, within the limits of the measuring accuracy, there is an interval of linearity for $1/d$ values lower than a certain value—that is, for d values higher than a critical value d_{lin}—which probably is characteristic of the material combination. This observation reflects the well-known fact that the measured microhardness varies with load and suggests the following reinterpretation of the load dependence.

The established definition of hardness, according to Eq 2, implicitly contains a dependence on both load, L, and diagonal, d, but suffers from the

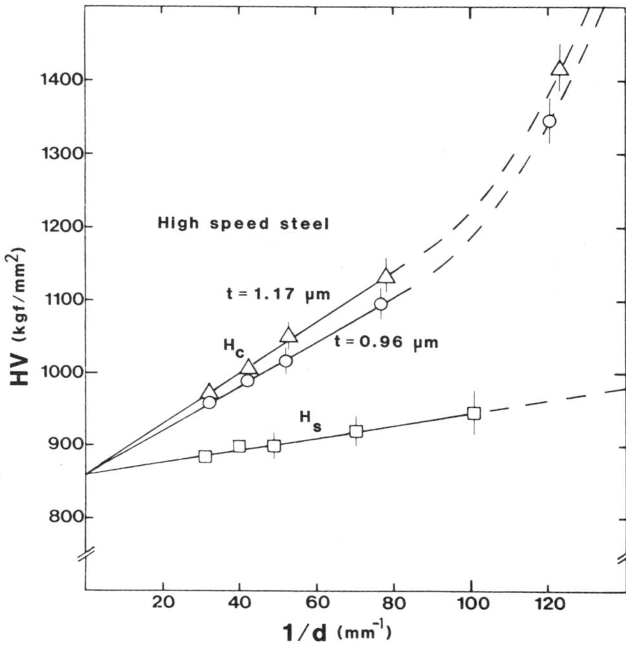

FIG. 4—*Vickers hardness (HV) versus the inverse indentation diagonal, 1/*d, *for two TiN films on the HSS substrate,* H_c, *and for the pure substrate material,* H_s. *The nonlinear part of the curves is dashed.*

weakness that the L–d relation is not easy to express analytically for the triaxial deformation process of the indentation. However, for high loads, that is, loads applied in conventional Vickers hardness measurements, the hardness value approaches a constant value H_0, which may be considered a material characteristic constant.

It is now possible to write the dependence of hardness on the indentation diagonal, suggested by the diagrams of Figs. 4 through 6, for coated as well as uncoated surfaces, as

$$H(d) = H_0 + \frac{k}{d} \tag{17}$$

where the intercept, H_0, as well as the slope, k, can be obtained from the linear parts of the $H(1/d)$ plots. For coated materials, the composite hardness according to Eq 8 could now be rewritten as

$$H_c(d) = H_s(d) + \frac{[H_f(d) - H_s(d)]A_f}{A} \tag{18}$$

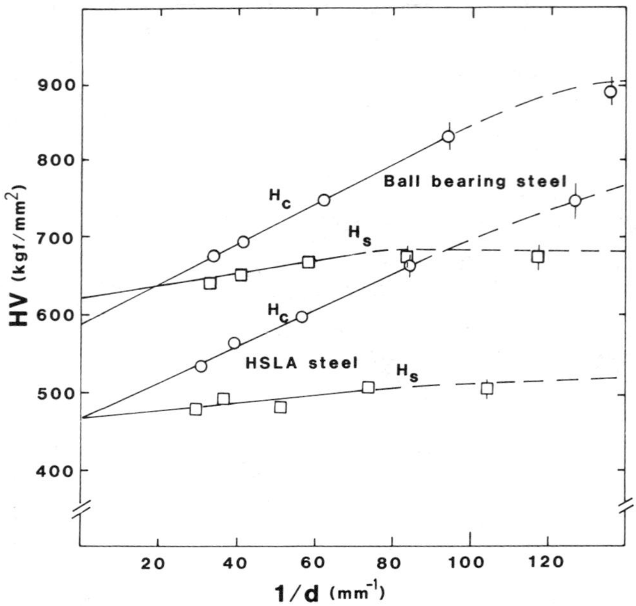

FIG. 5—*Vickers hardness (HV) versus the inverse indentation diagonal, 1/*d, *for chromium film on ball bearing and HSLA steel substrates,* H_c, *and for the pure substrate materials,* H_s. *The nonlinear part of the curves is dashed.*

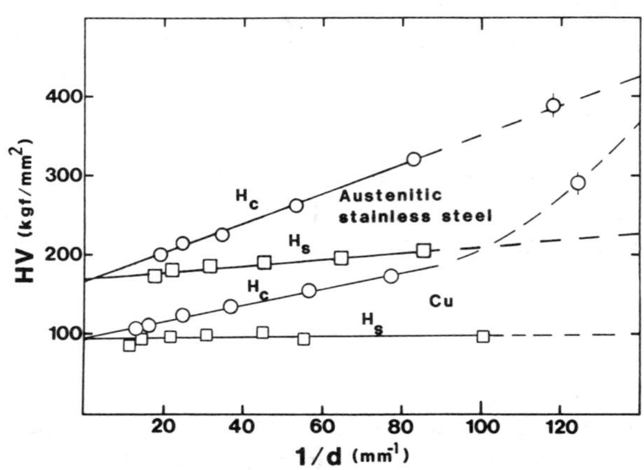

FIG. 6—*Vickers hardness (HV) versus the inverse indentation diagonal, 1/*d, *for chromium film on stainless steel and copper substrates,* H_c, *and for the pure substrate materials,* H_s. *The nonlinear part of the curves is dashed.*

Making use of Eq 17 and introducing the A_f/A expression (Eq 12) into Eq 18, neglecting second-order $1/d$ terms, then gives the composite hardness

$$H_c(d) = H_{s0} + \frac{k_s}{d} + \left(H_{f0} + \frac{k_f}{d} - H_{s0} - \frac{k_s}{d}\right)\frac{2Ct}{d} \tag{19}$$

Again, neglecting second-order $1/d$ terms gives the simplified expression of composite hardness

$$H_c(d) = H_{s0} + \frac{[k_s + 2Ct(H_f - H_{s0})]}{d} = H_{s0} + \frac{k_s + \Delta k}{d} \tag{20}$$

where

$$\Delta k = 2Ct(H_f - H_{s0}) \tag{21}$$

has been introduced. The subscript 0 of H_{f0} has been deliberately dropped in Eqs 20 and 21. Of course, an inherent d dependence of the film material for large d values cannot be experimentally determined from indentations in thin films because a variation in d changes only the length of the pressure-contributing rim, not its cross section or width (see Fig. 3). Yet, the value of H_f, implicitly given by Eqs 20 and 21, will be characteristic of the film material in its function as a coating, and probably also of the film thickness.

On the analogy of Eq 17, the composite hardness can now be expressed as

$$H_c(d) = H_{c0} + \frac{k_c}{d} = H_{s0} + \frac{k_c}{d} \tag{22}$$

with the coefficient, k_c, defined by

$$k_c = k_s + \Delta k \tag{23}$$

Similarly, the hardness difference according to Eq 15 is given by

$$\Delta H = \frac{\Delta k}{d} \tag{24}$$

From microhardness values in the linear ranges ($d > 10$ μm) of the diagrams of Figs. 4 through 6, H_{s0}, k_s, k_c, and Δk have been calculated for all the investigated coating–substrate combinations by linear regression (see Table 2). Hence, it is possible to solve H_f from Eq 21 as

$$H_f = H_{s0} + \frac{\Delta k}{2Ct} \tag{25}$$

corresponding to the film hardness (see Table 2).

In the present work, $t = 1$ μm, and Figs. 4 through 6 correspondingly represent variations with $1/d$ for constant t. However, the thickness of the TiN coating on high-speed steel (HSS) was varied slightly, and the corresponding H_c differences are clearly resolvable in Fig. 4. In a more general sense, it is possible to introduce a t dependence by plotting the hardness contributions versus the t/d ratio, since Δk is proportional to t according to Eq 21 and $\Delta H = \Delta k/d$.

As mentioned earlier, a critical D/t_{cr} value can be considered for the influence of the substrate material on the composite hardness. Below this value, the indentation process takes place entirely in the film material, and $H_c(d) \approx H_f(d)$. For $t = 1$ μm, this critical value should correspond to a critical magnitude, $d_{cr} \approx 1$ μm, of the indentation diagonal [1,2] ($d_{cr} \approx 7\, D_{cr}$ for Vickers indentations). Similarly, there is a linearity limit, $1/d_{lin} \approx 100$ mm^{-1}, as estimated from Figs. 4 through 6. The transition interval $1/d_{lin} < 1/d < 1/d_{cr}$ is generally characterized by a nonlinear $1/d$ dependence and, in terms of the applied model as expressed by Eq 12, an increasing domination of higher-order $1/d$ terms (the dashed curves in the figures).

Evaluation of the Applied Model

The most obvious feature of this application of the Jönsson and Hogmark model is that it gives an explicit analytical relation between hardness and indentation size for $d > d_{lin}$. The $1/d$ linearity allows extrapolation to the H_0

TABLE 2—*Numerical data obtained from the microhardness values within the linear part of the curves of Figs. 4 through 6.*

Composite	t, μm	H_{s0}, kgf/mm^2	k_c, kg/mm	k_s, kg/mm	Δk, kg/mm	H_f, kgf/mm^2 $C = 1.0$	$C = 0.5$
HSS/TiN	1.17	860	3.52	0.86	2.66	2000	3130
	0.96	860	3.08	0.86	2.22	2020	3170
Ball bearing steel/ chromium	1.0	620	2.60	0.60	2.00	1340	2090
HSLA/ chromium	1.0	470	2.34	0.51	1.83	1285	2300
Stainless steel/ chromium	1.0	170	1.85	0.41	1.44	890	1610
Copper/ chromium	1.0	90	1.04	0.09	0.95	565	1040

values, corresponding to "infinite" diagonals (high loads). H_0 represents a constant, characteristic of the indented material, or, in the case of coated specimens, of the substrate material (H_{s0}).

It is interesting to note that the H_0 values extrapolated from the microhardness values conform to the conventional high-load Vickers hardness numbers (see Tables 1 and 2). Note also that $H_{c0} \approx H_{s0}$, which was confirmed by high-load measurements.

The hardening effect of the coating is given by the Δk coefficient. The fact that the film material is harder than the substrate material implies a positive Δk, that is, a hardening, as expected. It is also obvious that the hardness contribution $\Delta H = \Delta k/d$ increases with decreasing d (load).

Comparing Δk for the chromium-coated materials, it can be seen from Table 2 that the hardening effect increases with substrate hardness. However, when the hardness of the substrate material approaches that of the film, a rapid drop of Δk has to be expected.

The effect of film thickness, described in Eq 21 as a direct proportionality between Δk and t, is accurately demonstrated by the two separate hardness lines of Fig. 4. Obviously, the composite hardness is sensitive to small variations in film thickness.

The suggested variation in relaxation of stresses in the indentation rim area by the formation of cracks is ultimately evaluated by the resulting limits of the calculated film hardness, H_f, for different C values. From a comparison of the chromium-coated specimens, it seems probable that C is close to 1.0 for the ball bearing and high-strength low-alloy (HSLA) steels, close to 0.5 for copper, and halfway between 1.0 and 0.5 for the stainless steel substrate. (The microhardness measurements with $d < d_{cr}$ on a 2-μm chromium film on the ball bearing steel indicated a "true" hardness of 1200 + 50 kgf/mm^2 [*3*].) These results simply conform to the SEM observation that the hard surface coating will crack more easily on a softer substrate. Consequently, for the TiN coating on hardened HSS steel, a literature value of $HV_{50gf} = 2100$ kgf/mm^2 [*8*] agrees well with a C limit of 1.0. The fact that the H_f hardness values of Table 2 vary with the film thickness as well as with the substrate material also reflects the idea that the mechanical properties of a thin surface film do not simply correspond to those of the film material in bulk dimensions.

It is of interest for the applicability of the present technique for the measurement of H_f that the order of magnitude of $d_{lin} \approx 10$ μm is considerably higher than the limit $d_{cr} \approx 1$ μm, below which the influence of the substrate would be vanishing.

The linear $H_s(1/d)$ relation for the bulk substrate materials as revealed by Figs. 4 through 6 is not yet understood. It can possibly be attributed to elastic relaxation on the unloading of the indenter. This may affect the optical diagonal reading because of size-dependent geometric alterations at the indentation rim.

Summary and Conclusions

For coated as well as uncoated materials, explicit hardness versus size functions are found to be valid for indentations larger than a limiting size, d_{lim}, according to

$$H(d) = H_{s0} + \frac{k}{d}$$

where the constants H_{s0} and k can be obtained from linear $H(d^{-1})$ plots. H_{s0} corresponds to the (uncoated) substrate hardness measured with a high-indentation load.

It is also possible to calculate the film hardness, H_f

$$H_f = H_{s0} + \frac{(k_c - k_s)}{2Ct}$$

which is, however, characteristic of the film material in its function as a coating, rather than in bulk dimensions.

In practice, estimation of H_f requires knowledge of the values of H_{s0}, the slopes k_c and k_s [from the linear $H(1/d)$ relations for coated and uncoated materials, respectively], the numerical constant C (the mode of film deformation), and the film thickness, t. This investigation indicates a value of $C \approx 0.5$ for high H_f/H_s ratios and $C \approx 1.0$ for lower ratios. In fact, for thin-film composites, H_{s0} is equivalent to the conventional high-load Vickers hardness number of either composite or substrate material. The k_c and k_s values are available from this number, together with one set of microhardness values from the linear range ($d > d_{\text{cr}}$) of the $H_c(1/d)$ and $H_s(1/d)$ curves, respectively.

Reversed, given the values of H_f, H_{s0}, k_s, C, and t, the composite hardness, $H_c(1/d)$, is obtained from

$$H_c = H_{s0} + \frac{[k_s + 2Ct(H_f - H_{s0})]}{d}$$

This relation can be utilized for selecting proper film–substrate combinations, for example, for tribological applications, in which the magnitude of surface damage can be estimated. The magnitude of surface damage can be related to a corresponding size of indentation diagonal, d, or depth, D.

The suggested relations have been verified for only a limited number of combinations of a harder coating on a softer substrate. If they are found to be of more general relevance, however, they will be very valuable because of their inherent simplicity, with easy-to-measure parameters. The applicability covers the d/t range of particular interest for studying the interaction between substrate and coating.

For coated materials, the recorded H versus $1/d$ relation is described by the suggested model as a hardening contribution of the coating localized to an area around the impression rim. For uncoated materials, however, the observed $H(1/d)$ linearity has not been satisfactorily explained. If found to be generally applicable, it would greatly contribute to the conventional technique for microhardness measurements.

References

[1] Il'inskii, A. I. and Lyakh, G. E., *Industrial Laboratory*, (U.S.S.R.), Vol. 44, No. 12, 1978, pp. 1719-1724.

[2] Palatnik, L. S., Il'inskii, A. I., Fedorov, G. V., and D'yachenko, V. S., *Izvestiya Vysshikh Uchebnykh Zavedenii, Fizika*, No. 1, 1966, pp. 122-128.

[3] Jönsson, B. and Hogmark, S., *Thin Solid Films*, Vol. 114, 1984, pp. 257-269.

[4] Mott, B. W., *Microindentation Hardness Testing*, Butterworths, London, 1957.

[5] Tabor, D., *Review of Physics in Technology*, Vol. 1, 1970, pp. 145-179.

[6] Kennedy, F. E. and Ling, F. F. *Transactions of the ASME, Journal of Engineering Materials and Technology*, Vol. 95, No. 2, 1974, pp. 97-103.

[7] Pauwelussen, J. P. and Alblas, J. B., *Acta Mechanica*, Vol. 35, 1980, pp. 109-126.

[8] Sundgren, J. E. "Formation and Characterization of Titanium Nitride and Titanium Carbide Films Prepared by Reactive Sputtering," Linköping Studies in Science and Technology, Dissertation No. 79, Technical University of Linköping, Linköping, Sweden, 1982.

Peter A. Engel[1] *and Mark D. Derwin*[1]

Indentation Test for Polymer-Film-Coated Computer Board Substrate

REFERENCE: Engel, P. A. and Derwin, M. D., **"Indentation Test for Polymer-Film-Coated Computer Board Substrate,"** *Microindentation Techniques in Materials Science and Engineering, ASTM STP 889*, P. J. Blau and B. R. Lawn, Eds., American Society for Testing and Materials, Philadelphia, 1986, pp. 272–285.

ABSTRACT: The phenomenon of indentation in coated substrates was checked by applying conical and spherical indenters of various geometries in a wide load range. Microscopic observation and cross sectioning of the indentations revealed three concentric regions: the inner contact zone, the indentation crater, and the debonding zone. The latter is formed if insufficient adhesion between the film and substrate gives way to the normal tensile stress on the interface, outside the contact area. In the contact zone the film is permanently compressed, and the indentation, accompanied by sinking in, is fundamentally a plastic deformation phenomenon. An analytical method is presented for estimating the "hardness" (defined as load per indentation crater area) of coated substrates from indentation and deflection measurements of the uncoated substrate plus a single deflection measurement of the coated substrate. Hardness curves (hardness versus force curves) for various indenters are plotted and compared.

KEY WORDS: adhesion, coating, conical indenter, spherical indenter, contact radius, debonding, displacement, film, glass epoxy, hardness, indentation radius, load, piling up, plastic deformation, prepreg, sinking in, substrate, microindentation hardness testing

Nomenclature

a	Indentation radius, mm
b	Debonding radius, mm
c	Contact radius, mm
H	Film thickness, mm
P	Indenting force, N
q_0	Hardness of substrate, MPa

[1] Senior engineer and engineer, respectively, IBM Systems Technology Division Laboratory, Endicott, NY 13760.

q_1 Hardness of coated substrate, MPa
q_c Contact hardness or flow pressure, MPa
r Radial coordinate, mm
R Ball radius (indenter), mm
w Film compression, mm
y_0 Contact displacement for uncoated substrate, mm
y_1 Contact displacement for coated substrate, mm

Relatively soft polymer coatings applied to substrates often play a crucial role in electronic packaging. They are used as protective, isolating elements, dielectrics, and so on, and as such they must have both sufficient strength and adhesion to the substrate in order to resist external mechanical actions without disintegrating or delaminating. Examples are photosensitive dielectric epoxy film adhered to copper or cloth in circuit board construction.

The indentation test described in this paper may be used to check the adhesive property of the film against debonding from the substrate; in the meantime, the indentation behavior itself is observed, revealing the mechanical interaction of the structure. Indentation is accomplished by a conical or spherical indenter, which is slowly pressed against the surface. The center of a contact is subjected to high compressive stress, while an annulus of coating just outside the contact area tends to carry tensile normal stress on its interface with the substrate; if adhesion in inadequate, then debonding of the film takes place in the latter region.

Previous papers [*1–3*] were devoted to a description, modeling, and industrial applications of the debonding phenomenon and the indentation debonding test. The debonding film acts as a thin plate outside the contact region, while the area under contact with the indenter is squeezed to the plastically deforming substrate. Outside the contact area the substrate usually exhibits sinking in [*4*], creating a depression or crater.

The present paper focuses on the indentation mechanics for a layered structure: the top film is an acrylated epoxy resin, and the substrate a sandwiched computer board or a copper plane. Computer board materials are typically of a hard, epoxy-glass sandwich structure [*5*]: the prepreg cloth consists of a weave of glass strands reinforcing the epoxy matrix. The adhesion bond between film and substrate was intentionally made inferior [*1*] to promote some debonding.

In this study, the indenters, which in earlier work were restricted to conical shapes, included spherical shapes, to facilitate generalization for both the practice and the theory of the test.

Experimental Procedure

Early indentation tests on polymer film coated circuit boards performed at the IBM Endicott Laboratory used needle indenters [*1,2*]: the purpose of that activity was to replace a scratch test used for quality control. Later, a Knoop hardness tester was adapted, in which the loading rate was also controllable. The time dependence of loading, however, has been found to be of very little consequence in most cases.

In recent laboratory experiments, the authors used an Instron tensile/compressive load tester, in which the indenter was mounted to the crosshead (Fig. 1). Microscopic observation of the contact was facilitated from the side during the loading process. The quasi-circular indentation and debonding regions could be measured by microscope after the loading was removed; this was possible because of permanent deformations of the film and substrate, discussed further on in this paper.

Four types of indenter were used in recent applications involving laminated epoxy-glass circuit boards as substrates and an acrylated epoxy resin, 0.058 mm in thickness, as an adhered film. These indenters were a hard brass conical indenter with a $2\,\phi = 90°$ cone angle and with a 0.143-mm head radius; a flatter cone ($2\,\phi = 120°$) made of hard steel and with a 0.20-mm tip radius (the Brale tip of Rockwell tests); a sharp ball indenter made of steel with a 1.588-mm (1/16-in.) diameter; and a flatter steel ball with a 3.175-mm diameter (1/8-in.). These indenters are referred to, respectively, as A, B, C, and D. The elastic or plastic deformation of the indenters was negligible compared with the deformations of the film and substrate.

In order to study these indentations, some tested specimens were potted and then cross sectioned. Figure 2 shows four cases under a 100-fold magnification. The photomicrographs Fig. 2*a* through *d* resulted from Indenters A, B, C, and D, respectively.

The film thicknesses in the contact area exhibited moderate variation from the original, undeformed thickness. The film compression, w, was proportionally larger for sharper indenters.

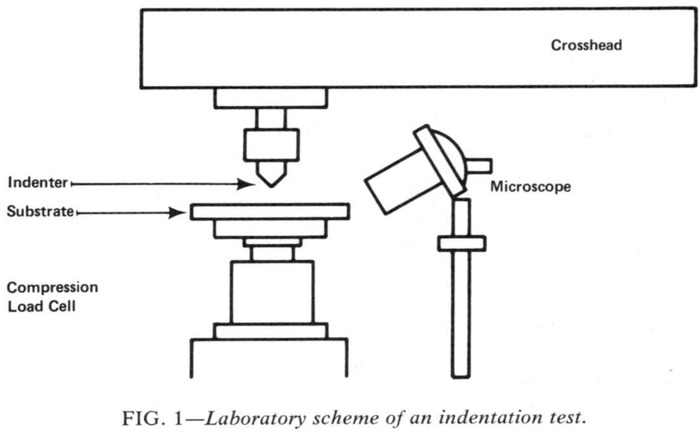

FIG. 1—*Laboratory scheme of an indentation test.*

Because the cross sections showed that the rigid indenters did not fill out the substrate craters, it was necessary to measure the displacement, y, of the indenter from the unloaded position. Dial gage measurements were far more accurate than the corresponding Instron deflection data, because the latter also include the machine constant; the contact stiffness is extremely high, therefore, the error may be sizable.

It has been verified by dial gage that contacts are plastically deformed during indentation, and no significant rebound would occur in time. This enables the experimental measurement of the quantities a and b (that is, the size of the indentation and of debonding) after the indenter has been retracted. The photograph in Fig. 3, the top view of an indentation made with a 120° cone (Case B) under a 134-N (30-lb) load, shows that the measurement of c, the contact radius, is not feasible by this means. Therefore, cross sectioning is probably needed to determine the latter dimension—still a difficult task. The method used for measuring c was laying a scale model of the respective indenter over the deformed film exhibited by the cross-sectional photomicrographs of Fig. 2.

Analysis

In the present study the variation of hardness with load was investigated. For our purposes hardness is defined as the ratio of load to the measurable indentation area

$$q = \frac{P}{\pi a^2} \tag{1}$$

Using various indenters, the hardness curves $q_0(P)$ and $q_1(P)$ for both uncoated and coated substrates, respectively, were plotted through the 600-N range. These plots generally showed a rising tendency, reminiscent of the effect of strain hardening, and a straight-line fit would be appropriate.

We shall set out to seek a relationship between $q_1(P)$ and $q_0(P)$. Assume that $q_0(P)$ has been experimentally obtained through the range of interest and that the deflections, $y_0(P)$, have also been measured for the uncoated substrate. A single deflection, $y_1(P^*)$, for the corresponding coated substrate is also considered available, at a load of $P = P^*$. The geometry of the latter, based on the experimental findings of Fig. 2 is shown in Fig. 4.

Writing

$$y_1(P) = y_0(P) + w(P) \tag{2}$$

we calculate the film compression, w, at load P^* as

$$w^* = w(P^*) = y_1(P^*) - y_0(P^*) \tag{3}$$

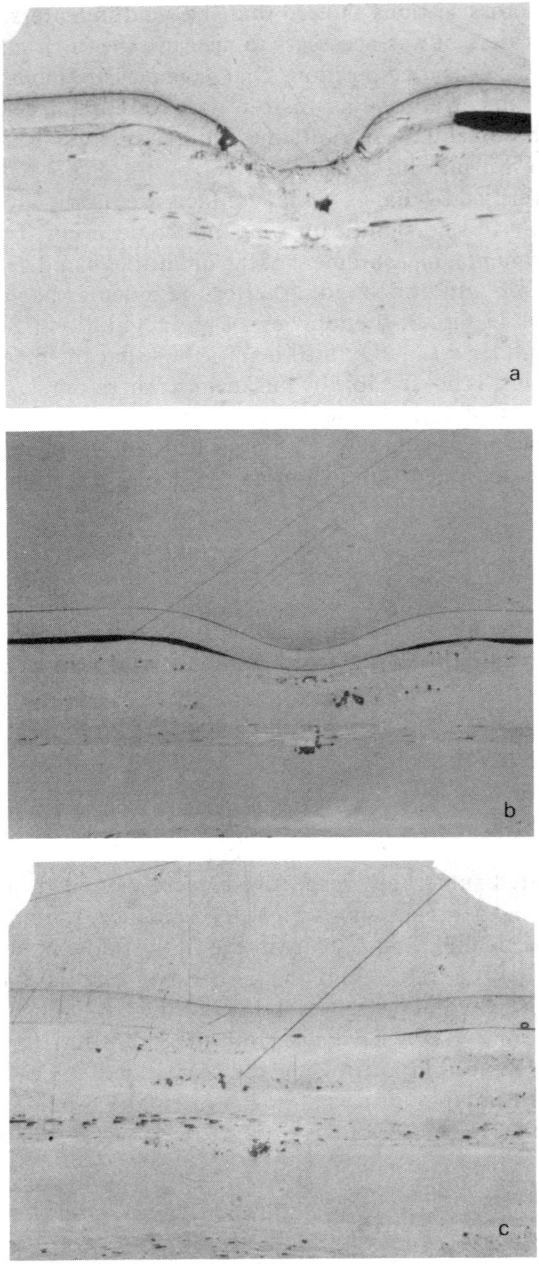

FIG. 2—*Photomicrographs of cross-sectioned indentations at* ×*100:* (a) *90° cone indenter, under 142 N (43 lb);* (b) *120° cone indenter, under 134 N (30 lb);* (c) *1.588-mm ϕ spherical indenter, under 156 N (36 lb).*

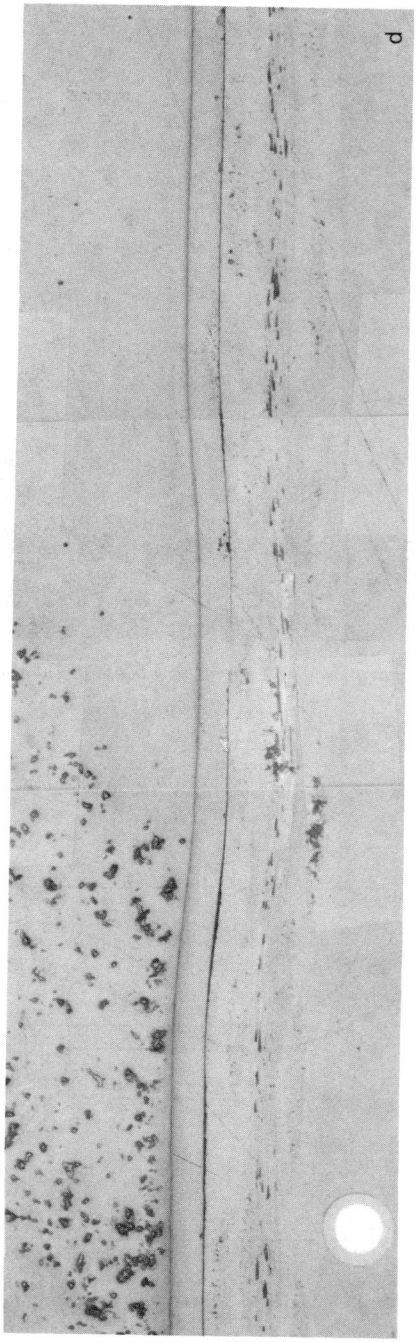

FIG. 2—*Continued:* (d) *3.175-mm* ϕ *spherical indenter, under 556 N (125 lb).*

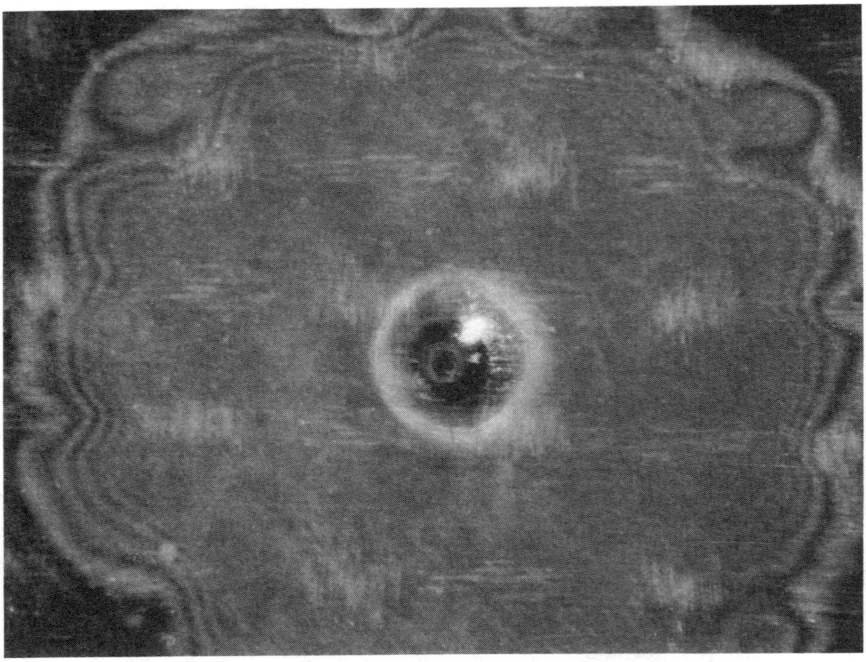

FIG. 3—*Photograph of an indentation at ×40, made with a 120° cone indenter under a 124-N (30-lb) load.*

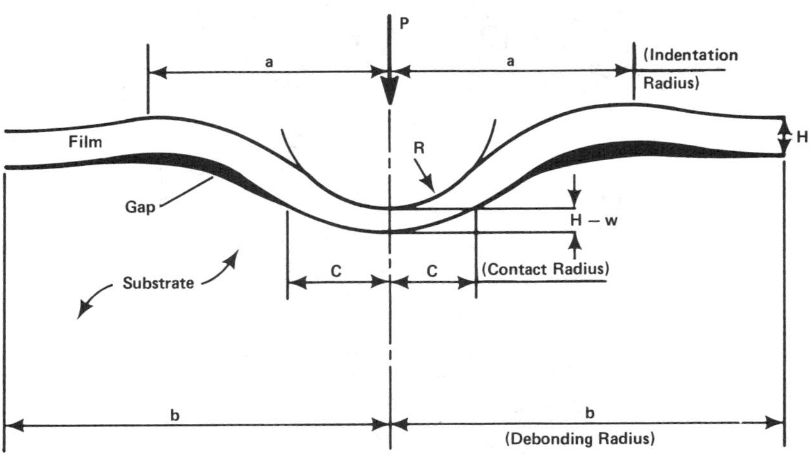

FIG. 4—*Schematic of a coated substrate loaded by a rigid indenter.*

It was found experimentally that, for the same indenter, w varies with the square root of P; thus, $w(P)$ can be estimated for any force

$$w(P) = w^*\left(\frac{P}{P^*}\right)^{1/2} \tag{4}$$

Plotting a_0 versus y_0, we get the functional relation

$$a_0 = f_0(y_0) \tag{5}$$

If a_1 were measured through the load range, another functional relationship would result

$$a_1 = f_1(y_1) \tag{6}$$

The hardness for the coated substrate at a force, P, by Eq 1, is

$$q_1(P) = \frac{P}{\pi a_1^2} \tag{7}$$

which is rewritten, by Eq 6, as

$$q_1(P) = \frac{P}{\pi f_1^2(y_1)} \tag{8}$$

On physical grounds the functions f_0 and f_1 of Eqs 5 and 6 may be assumed equal. This is because applying the same force, P, to a coated substrate, and comparing it with the corresponding uncoated substrate, both $a_1(P) > a_0(P)$ and $y_1(P) > y_0(P)$ can be found, so that

$$\frac{da_0}{dy_0} \approx \frac{da_1}{dy_1} \tag{9}$$

while both functions start out in the origin. Writing

$$f_0(y_0) = f_1(y_1) = f(y) \tag{10}$$

and applying Eq 8 to find a theoretical prediction, $Q_1(P)$, for the hardness in the coated substrate at a force, P, by using Eqs 2 and 4, we get

$$Q_1(P) = \frac{P}{\pi f^2[y_0(P) + w^*(P/P^*)^{1/2}]} \tag{11}$$

Last, a theoretical prediction of the deflections, Y_1, of the coated substrate is obtained by Eqs 2 and 4

$$Y_1(P) = y_0(P) + w^*\left(\frac{P}{P^*}\right)^{1/2} \tag{12}$$

Results and Discussion

Table 1 shows the results of a test matrix for the four indenters (Indenters A, B, C, and D mentioned in the Experimental Procedure section), and four values of load, 134, 223, 446, and 558 N (30, 50, 100, and 125 lb).

The assumption of Eq 10 is demonstrated in Fig. 5 for each indenter. It is also apparent that for conical indenters the deflection is larger for the same indentation area. The deflection per load is ranked A, C, B, D.

TABLE 1—*Matrix of measurements and analytical results for hardness tests using four types of indenters.*

	By Measurement								
	Uncoated			Coated			By Analysis		
Load, N	$y_0(P)$, mm	$a_0(P)$, mm	$q_0(P)$, MPa	$y_1(P)$, mm	$a_1(P)$, mm	$q_1(P)$, MPa	$w(P)$, mm	$Y_1(P)$, mm	$Q_1(P)$, MPa
				90° Cone Indenter					
134	0.0711	0.324	405	0.0762	0.3404	367	0.0051	0.0767	367
223	0.0940	0.401	440	0.0991	0.4318	312	0.0066	0.1006	380
446	0.1346	0.561	449	0.1422	0.5944	373	0.0094	0.1440	402
558	0.1524	0.638	436	0.1524	0.6707	391	0.0104	0.1628	394
				120° Cone Indenter					
134	0.0381	0.321	412	0.0381	0.3404	356	0.0036	0.0417	367
223	0.0521	0.413	416	0.0521	0.4420	347	0.0046	0.0567	363
446	0.0775	0.554	462	0.0737	0.6020	421	0.0065	0.0840	319
558	0.0838	0.603	486	0.0813	0.6426	436	0.0073	0.0911	429
				1.588-mm ϕ Ball Indenter					
134	0.0203	0.310	442	0.0279	0.3429	354	0.0046	0.0249	326
223	0.0330	0.401	440	0.0330	0.4293	355	0.0059	0.0389	385
446	0.0584	0.538	488	0.0787	0.5817	419	0.0084	0.0668	419
558	0.0686	0.597	497	0.0838	0.6452	446	0.0093	0.0779	446
				3.175-mm ϕ Ball Indenter					
134	0.0203	0.356	336	0.0203	0.3708	364	0.0031	0.0235	309
224	0.0300	0.429	384	0.0356	0.4597	339	0.0041	0.0340	335
446	0.0533	0.611	380	0.0559	0.6553	327	0.0057	0.0591	330
558	0.0610	0.667	398	0.0660	0.7112	374	0.0064	0.0674	350

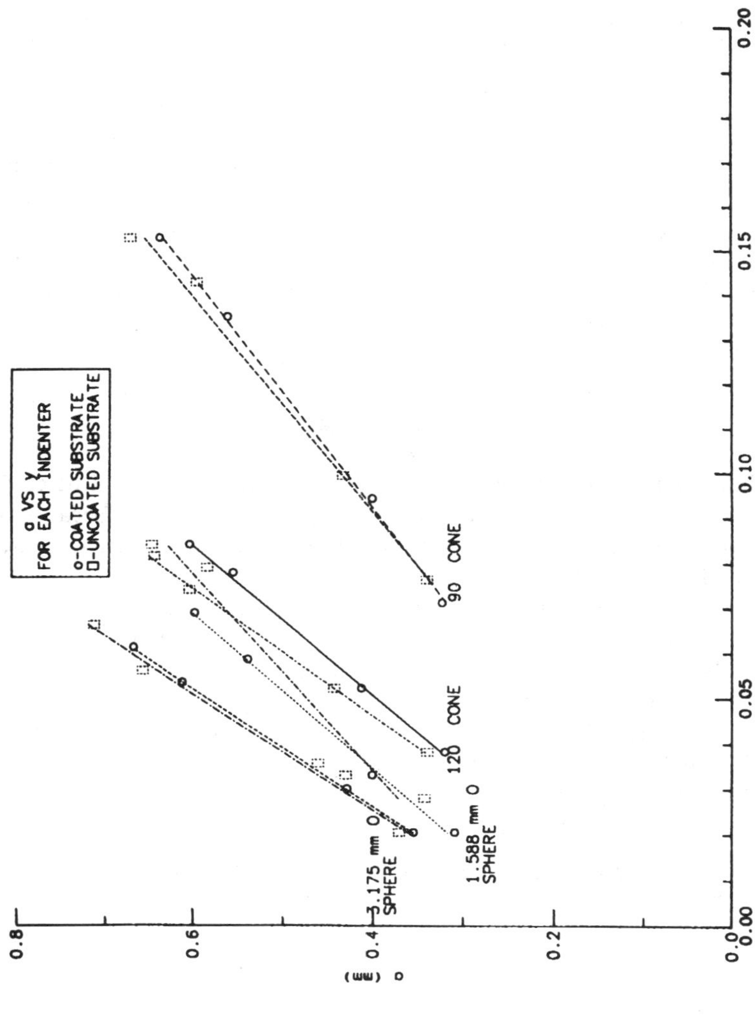

FIG. 5—*Indentation radius versus deflection measurements for both coated and uncoated substrates.*

The film compression $w(P)$ shows a good parabolic fit, justifying Eq 4 (see Fig. 6). The relative magnitude of w also shows the order A, C, B, D, as did the deflection. Because the only assumptions made in the Analysis section were those of Eqs 4 and 10, all the other deviations must be blamed on the measurability or the consistency of the indentation radii, a, or both, the indentation being usually a bit irregular rather than a perfect circle. The force for all but Indenter D (the flat sphere) was taken as 134 N. For the latter indenter, a small force yielded fuzzy $f(y)$ measurements, and therefore $P^* =$ 223 N was used in Eq 3 to get w^*.

The theoretical hardness curve, $Q_1(P)$, for the 120° cone indenter is shown in Fig. 7. It is also compared with an experimentally measured series of $q_1(P)$ hardnesses, showing reasonable agreement. A comparison of theoretically calculated displacements, Y_1, and the corresponding measured ones, y_1, also shows close agreement (Table 1).

The contact radii, c, were measured by the cross-sectioning technique, at a single force value for each indenter. Table 2 shows the contact hardness q_c, calculated from the formula

$$q_c = \frac{P}{\pi c^2} \tag{13}$$

A similar measurement without the presence of a film would not be easy, since c could not easily be discerned. It is thought, however, that because of the large pressures arising at the tip of the contact, c may not change greatly in the case of the coated substate.

The contact region measurements lead to q_c values quite near one another for all but the 90° conical indenter. The latter seemed to cause breaking up of the film in the tip region, as evidenced by the cross section (Fig. 2*a*), making it difficult to determine the c value.

Subsurface damage is characteristic of indentation processes [*6*]. Figure 2 shows that breakage of glass fibers occurs at a depth of approximately half the contact radius or greater.

TABLE 2—*Contact radii and contact pressure at a single load.*

	P		C		q_c	
Indenter	N	lb	mm	10^{-3} in.	MPa	10^6 psi
90° cone	142	32	0.155	6.10	1 890	0.27
120° cone	134	30	0.063	2.50	10 500	1.53
1.577-mm ϕ ball	156	35	0.065	2.56	11 700	1.70
3.175-mm ϕ ball	556	125	0.120	4.72	12 300	1.78

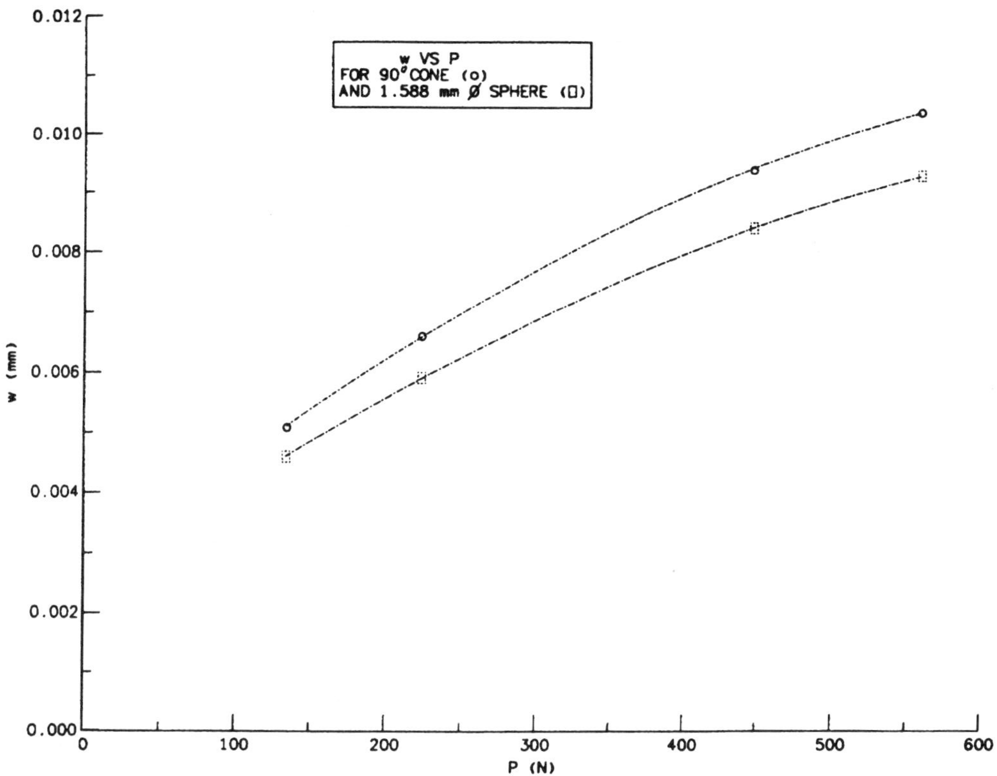

FIG. 6—*Analytically evaluated film compression values versus load, for a 90° cone and a 1.588-mm-diameter spherical indenter.*

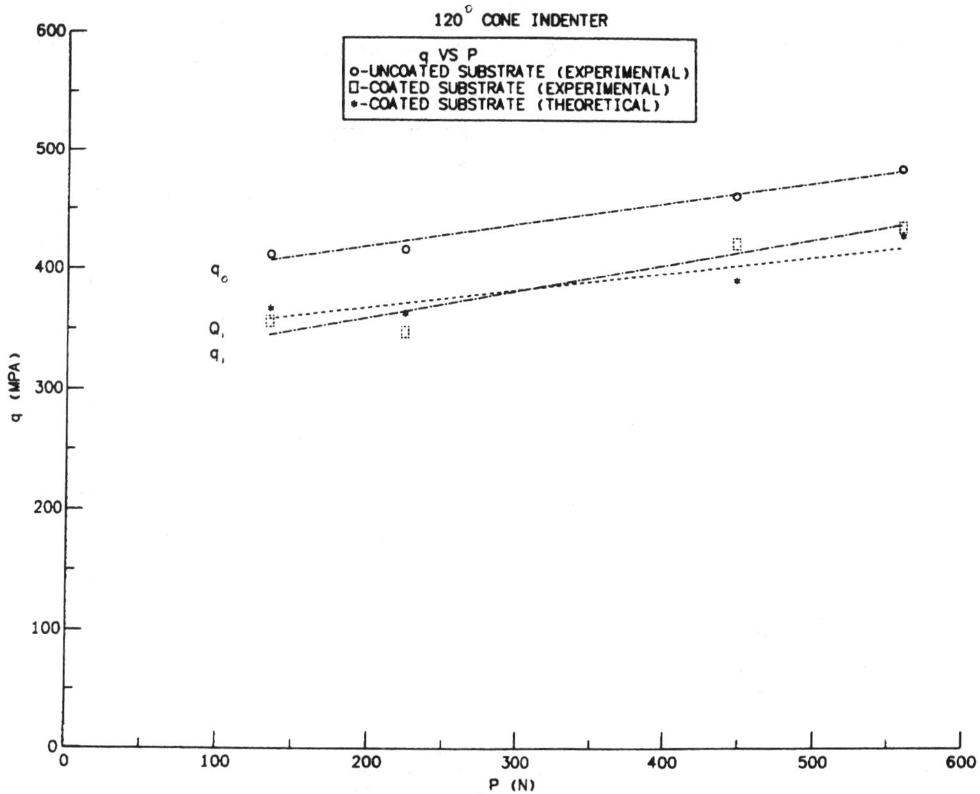

FIG. 7—*Hardness versus load curves for coated and uncoated substrates* (q_1 *and* q_0, *respectively*), *obtained using a 120° conical indenter. The analytically evaluated coated hardness curve,* Q_1, *is also shown.*

Conclusions

Hardness curves were obtained for a coated and corresponding uncoated computer board, using four types of indenters: two of a conical and two of a spherical outline. By using displacement measurements for the indenter, an analytical method could predict the hardness of the coated structure from the hardness of the uncoated one. The experimental measurements of both the indentation hardness and the contact hardness (that is, flow pressure) were shown.

Acknowledgments

The author is grateful to T. D. Jensen for his valuable contribution in the experimental and data processing work. J. J. Woods prepared the cross sections.

References

[*1*] Engel, P. A. and Roshon, D. D., *Journal of Adhesion*, Vol. 10, 1979, pp. 237-253.
[*2*] Engel, P. A. and Pedroza, G. C. in *Adhesion Aspects of Polymeric Coatings*, K. L. Mittal, Ed., Plenum, New York, 1983, pp. 583-595.
[*3*] Engel, P. A., "Indentation Debonding Test for Polymer Coatings Adhered to a Substrate," *Proceedings*, International Adhesion Conference, Nottingham, UK, 1984; also in *International Journal of Adhesion and Adhesives*, Vol. 5, No. 3, July 1985, pp. 129-132.
[*4*] Tabor, D., *The Hardness of Metals*, Clarendon Press, Oxford, 1951, p. 15.
[*5*] Coombs, C. F., Jr., Ed., *Printed Circuits Handbook*, 2nd ed., McGraw-Hill, New York, 1979, pp. 23-33.
[*6*] Engel, P. A., *Impact Wear of Materials*, Elsevier, New York, 1976, p. 44.

Walter W. Walker[1]

Knoop Microhardness Testing of Paint Films

REFERENCE: Walker, W. W., **"Knoop Microhardness Testing of Paint Films,"** *Microindentation Techniques in Materials Science and Engineering, ASTM STP 889*, P. J. Blau and B. R. Lawn, Eds., American Society for Testing and Materials, Philadelphia, 1986, pp. 286–289.

ABSTRACT: A standard method of determining the relative hardness of dried paint films is the ASTM Test for Film Hardness by Pencil Test [D 3363-74 (1980)]. Although the pencil test is useful in comparing the film hardness of paint systems, it does not yield specific numerical hardness data. Pencil hardnesses are alphanumerical and therefore pseudoquantitative. In order to quantify hardness results from the pencil test, a correlation was sought between pencil hardness and Knoop microindentation hardness. First, the Knoop microindentation hardness of a set of drawing leads meeting the requirements of ASTM Test D 3363-74 (1980) was determined. A linear correlation was found between pencil lead hardness numbers and Knoop microindentation hardness values. The Knoop microindentation hardness was next determined on a series of painted panels of known pencil hardness. A useful correlation was found between the film hardness as determined by the pencil test and that found by the Knoop microindentation hardness test.

KEY WORDS: paint films, hardness, pencil test, Knoop microindentation hardness, pencil lead hardness, microindentation hardness testing

A standard method of determining the relative hardness of dried paint films is the ASTM Test for Film Hardness by Pencil Test [D 3363-74 (1980)]. Another standard test method for determining the Knoop microindentation hardness of organic coatings is described in the ASTM Test for Indentation Hardness of Organic Coatings [D 1474-68 (1979)], Method A. As far as the author knows, no previous investigator has attempted to correlate paint film hardness by the pencil method with Knoop microindentation hardness. The author has made such a correlation study, and this paper presents the results of this investigation.

[1]Senior staff engineer, Hughes Aircraft Company, Tucson, AR 85734.

Equipment

Hardness Tester

The microindentation hardness tester used in this investigation was a Tukon microhardness tester Model LR (originally made by Wilson Mechanical Instrument Division, The American Chain & Cable Co., Inc., New York, NY). This tester was selected for its ready availability. Any standard microindentation hardness tester should be equally usable.

Pencil Leads

A set of calibrated drawing leads meeting the requirements of ASTM Test D 3363-74 (1980) was selected for this experiment.

Other Equipment

Standard metallographic laboratory equipment was used to mount and mechanically polish the leads for Knoop microindentation hardness testing.

Test Specimens

Three different coatings—(*a*) solvent-base polyurethane spray paint coating, (*b*) solvent-base epoxy spray paint coating, and (*c*) electrostatic epoxy powder spray coating—were applied to chemical conversion-coated aluminum alloy test panels. The panels were cut to a convenient size for Knoop microindentation hardness testing.

Specimens of pencil leads from the calibrated drawing lead set were cut to convenient sizes for mounting and metallographic polishing. Standard metallographic polishing procedures were used on the pencil leads.

Procedure

Pencil Testing

The three paint panel specimens were pencil tested for gouge hardness in accordance with ASTM Test D 3363-74 (1980).

Knoop Microindentation Hardness Testing Paint Panels

Knoop microindentation hardness testing of the paint panels was performed generally in accordance with ASTM Test D 1474-68 (1979). However, the machine was calibrated at a 100-g load using a metal Knoop hardness test block, which was the only test block available. All the hardness tests were carried out at a 200-g load since the diagonals were too difficult to measure with lower loads. The paint film thickness was 50.8 ± 5.08 μm.

Pencil Leads

The pencil leads were mounted in cast epoxy resin so that the plane of the polish was normal to the central axis of each lead. Five indentations were made on each lead at 200-g load, and the mean Knoop microindentation hardness was reported.

Results

The mean Knoop microindentation hardness numbers (HK) of a series of calibrated pencil leads increasing in hardness from B to 5H are given in Table 1.

The mean HK of the three test panels and the pencil gouge hardness and HK of the particular pencil lead are given in Table 2. The mean HK of the bare substrate was 195 ± 1 HK.

Discussion

The author recognizes that this paper is preliminary. The purpose of publishing at this time is to interest other investigators in this approach to paint film hardness testing.

No Knoop indentation hardness tests were performed with loads of less than 200-g because the indentation diagonal lengths were too difficult to mea-

TABLE 1—*Knoop indentation hardness of pencil leads per ASTM Test D 3363-74 (1980).*

Pencil Lead[a,b]	Mean Knoop Indentation Hardness (200-g load), HK	Estimated Standard Deviation, S, HK
5H	51.5	2.3
4H	47.0	1.6
3H	45.3	1.2
2H	38.7	0.8
H	31.7	1.0
F	27.0	1.2
HB	24.0	1.5
B	12.7	2.0

[a]No 6H lead available.

[b]Softer leads (6B to 2B) resulted in 200-g indentations, which were too large for the available cross section. Reducing the load introduced errors because of the low-load anomaly [*1*].

TABLE 2—*Comparison of pencil gouge hardness and Knoop indentation hardness for selected paint films.*

Paint Film	Pencil Gouge Hardness, Lead No.	Pencil Lead Hardness, HK[a]	Paint Knoop Hardness, HK[b]	Difference Between Pencil HK and Paint HK
Electrostatic powder epoxy	5H	51.5	30.2	21.3
Solvent base polyurethane	3H	45.3	22.7	22.6
Solvent base expoxy	H	31.7	8.9	22.8
Bare substrate	...	...	195 ± 1	...
	Mean difference = 22.2			

[a]From Table 1.
[b]200-g load.

sure on these rough, nonreflective surfaces. The effect of Mott's low-load anomaly [*1*] was, therefore, not determined in this investigation.

The thickness of the paint films used in this study was 50.8 ± 5.08 μm; thinner films may not respond similarly.

Simple abrasive wear theory postulates that for an indenter to scratch a surface, it must be harder than the surface. Table 2 shows an experimental average difference of 22.2 HK between the pencil lead and the paint. The physical basis for this difference is not yet understood.

Conclusion

This preliminary study has demonstrated that a useful correlation exists between 200-g Knoop indentation hardness and pencil gouge hardness on thick paint films. Further work needs to be done, particularly on the effect on this correlation of lower indenting loads and thinner paint films.

Reference

[*1*] Mott, B. W., *Micro-Indentation Hardness Testing*, Butterworths, London, 1956.

Summary

Summary

In the preparation of this book and the organization of the symposium from which it grew, there has been an attempt to look in some detail at various facets of a complex field. Some papers have dealt with mechanistic studies of interest to researchers in materials science. Other papers have addressed rather pragmatic questions involved in day-to-day engineering practice. Still others have provided examples of how microindentation testing may be used in creative, less conventional ways to probe the near-surface mechanical properties of materials.

All along the path from testing concepts to implementation and data analysis, there are fertile areas for study. Advances in the utility of microindentations can result from both improved machine designs and improved understanding of how properties and microstructures of materials affect their observed behavior.

Fundamentals of Indentation Testing

Samuels begins this section of the book with a review paper which questions some of the traditional models of the indentation process in metals. He examines indentation mechanisms involving cutting, elastic behavior, and compression. In this chapter, the strengths and limitations of modeling using the three mechanisms are considered, and the compression mechanism is favored. Implications from this model are discussed regarding matters of impression size, interconversion of hardness scales, impression shape, friction, and surface topography. The need to characterize better the deformation of the material adjacent to the impressions is also emphasized.

Marshall and Lawn review the use of microindentation methods in investigating the fracture and deformation behavior of brittle materials, such as ceramics and glasses. They argue against the shortcomings of traditional models of hardness phenomena, which are largely based on plasticity considerations which presume volume conservation and homogeneity. The authors present an interpretation based on intermittent "shear faulting" and also include, where appropriate, a contribution from structural compaction or expansion. Examples relating indentation behavior to elastic modulus and fracture toughness are provided, along with an illustration of surface residual stress determination in brittle materials.

The next three papers all involve some aspects of the load versus indenter displacement behavior of materials. *Pollock, Maugis, and Barquins* use a three-faced pyramidal indenter under loads of 0.01 to 30 mN to study a range of behavior in submicrometre surface layers. The authors also use a friction instrumented scratch hardness tester. Both instruments operate under computer control. Studies of small-scale microstructural features, such as grain boundaries and dislocation configurations, are summarized. The influences of elastic-plastic response, elastic recovery, and indentation creep are discussed.

Continuing on with the topic of indentation elastic-plastic relationships, *Loubet, Georges, and Meille* provide models for the interpretation of load versus microdisplacement curves. Young's modulus and Vickers indentation pressure are calculated for hardened 52100 steel and annealed aluminum from their analysis. The authors discuss the correlation between depth-based measurements and traditional light optical measurements for hardness determination. By examining the residual impression depth after off-loading, estimates of the ratio of work supplied to that retained in the material can be made.

Oliver, Hutchings, and Pethica complete the series of three papers dealing with load-displacement behavior by describing a set of experiments on hardness as a function of depth on gold, nickel, lithium fluoride, and silicon. Detailed examinations of impressions as shallow as 20 nm are made using both transmission and scanning electron microscopes. Modifications to Meyer's law are proposed, based on a consideration of the increasing influence of microstructural features (such as dislocation cells) when indentations become very small.

Dislocation aspects of indentations are the focus of the contribution from *Armstrong and Elban*, who model the crystallographic aspects of microindentations in two materials: magnesium oxide and cyclotrimethylenetrinitramine. These authors extend the earlier continuum mechanics approaches to account for the interactions of moving dislocations in accommodating the localized strains associated with the indentation process.

Techniques and Measurements

The paper by *Tabor*, which introduces the section on testing methodology, reviews many of the aspects of indentation processes and provides an overall perspective from which the limitations of microindentation testing can be viewed. It deals with the concepts of the expanding spherical cavity approach to elastic-plastic indentation behavior and with problems associated with load application and crack formation in brittle materials. The paper ends with some cautionary comments related to the interpretation of low-load penetration depth data.

Sargent carries forward some of the concerns of the previous authors on indentation size effects and introduces an empirical indentation size effect (ISE) index which is defined in terms of the applied load and indentation diameter (or some similar dimensional parameter). He applies the analysis of the ISE to a series of metals, ceramics, single crystals, and polycrystals at various temperatures to generate error ellipsoid plots. These plots are suggested as a compact way to display rigorous interpretations of the ISE in a range of materials.

The effects of external stresses imposed during microindentation testing procedures are investigated by *Vitovec*. This author uses a tensile stage on a bench-top microindentation machine to vary the externally applied load on the test specimen. He found that hardness decreased linearly with increasing imposed tensile stress. Yielding, surface films, and residual stresses all seemed to affect the hardness-load relationship and the trend of the stress effect.

Standardization of microindentation hardness testing procedures is extremely important if useful engineering data are to be obtained. *Kelley, Johnson, and Lashmore* discuss the fabrication and certification of copper and nickel electroformed standard reference materials at the National Bureau of Standards. Concerns relating to the choice of testing procedures for material certification are discussed. The need to prepare standard samples within well-known ranges of microindentation hardness numbers for both Knoop and Vickers testing presents particular problems in fabrication.

The testing of thin, hard coatings presents particular challenges to those who need to determine microindentation hardness numbers. Extremely small impressions tax traditional optical microscopy. *Westrich* describes an alternative approach to measuring very small impressions using a scanning electron microscope and a diffraction grating as an internal size standard. This greatly reduces the scatter in data frequently associated with hard coating testing. A comparison between light optical and electron optical measurements is made for titanium nitride and hafnium nitride coatings, and correction factors are provided. Computerized curve fitting is used to facilitate conversion of hardness scales.

Engineering Applications

The last section of the book begins with a paper by *Blau* on the application of microindentation techniques to tribology. Three aspects of microindentation methods are discussed: first, the use of indentations and scratch methods in measuring small amounts of wear loss from the surfaces of metals; second, the characterization of surfaces of test pieces prior to wear testing, in which it is shown that, depending on the type of wear, increasing hardness does not always lead to increased wear resistance; and third, several uses of microindentation tests in the study of the subsurface deformation of metals from wear.

The contribution by *Kosel* relates to the use of scratch tests of controlled depth in the study of the fracturing process of carbides during abrasive wear and machining. Instead of using conventional diamond indenters, particles similar in kind to the actual abrasives encountered in the field are used. This provides insight into the mechanisms of actual field abrasion conditions. A specially designed instrument for scratch testing in the scanning electron microscope permits observations of prescribed areas on carbide-laden specimens to be examined after successive scratches.

During powder metal processing, the earlier the properties of materials can be measured, the better one is able to monitor and adjust parameters for quality control of the final product. *Shives and Smith* discuss several of the practical problems of mounting and testing powder metal particles. A novel technique involving the embedment of particles into an electrodeposit prior to polishing and hardness testing is described. Commercial nickel alloy, stainless steels, and aluminum powders are used to illustrate various preparation techniques and sources of error in the data.

The final trio of papers in this book deals with various kinds of coating hardness and adhesion testing problems. *Vingsbo, Hogmark, Jönsson, and Ingemarsson* treat the problems of thin metal coatings by proposing a model in which a mixture-rule relationship is derived for coatings of various thickness. A linear relationship is seen to hold for indentations larger than a limiting value which depends on the materials being tested. Separate hardness values for the coating and substrate are estimated.

The adhesion of polymer films to glass-epoxy layered circuit-board substrates is quantified by *Engel* using conical or ball indenters. The extent of the annulus of de-adhered coating surrounding the impressions is the means by which relative film adhesion is obtained. The paper contains an analysis of the contribution of the coating to the hardness of the test piece surface. The methods described quantify debonding tendencies in laminar boards due to mechanical handling.

The final paper in this book describes an engineering approach to the quantification of what has been known as "pencil hardness." *Walker* describes a study in which Knoop microindentation hardness numbers for various grades of pencils are measured and correlated with data from painted surfaces. The linear correlation between Knoop and pencil hardness scales permits a useful method for assessing the hardness of painted surfaces.

Obviously, this book could not hope to cover the breadth and depth of a field so great as microindentation testing in all of materials science and engineering, but what it provides is a series of fundamental insights, testing guidelines, and creative applications of many techniques to engineering practice. The editors hope that similar texts will follow to stimulate greater scientific enlightenment and enable the development of effective solutions to the many challenges of surface science and engineering.

Peter J. Blau
Brian R. Lawn

National Bureau of Standards, Gaithersburg, MD, 20899; symposium cochairmen and editors.

Author Index

Subject Index

D

E

F

G

H

N

P

Q

R

S